Lecture Notes in Comp

129
Edited by G. Goos, J. Hartmani

Springer
Berlin
Heidelberg
New York
Barcelona
Hong Kong
London
Milan
Paris
Singapore
Tokyo

Michel Goemans Klaus Jansen
José D. P. Rolim Luca Trevisan (Eds.)

Approximation, Randomization, and Combinatorial Optimization

Algorithms and Techniques

4th International Workshop on Approximation Algorithms for Combinatorial Optimization Problems, APPROX 2001 and 5th International Workshop on Randomization and Approximation Techniques in Computer Science, RANDOM 2001
Berkeley, CA, USA, August 18-20, 2001
Proceedings

Volume Editors

Michel Goemans
Massachusetts Institute of Technology, MIT, Department of Mathematics
Cambridge, MA 02139, USA
E-mail: goemans@math.mit.edu

Klaus Jansen
University of Kiel, Institute for Computer Science and Applied Mathematics
Olshausenstr. 40, 24098 Kiel, Germany
E-mail: kj@informatik.uni-kiel.de

José D. P. Rolim
Université de Genève, Centre Universitaire d'Informatique
24, Rue General Dufour, 1211 Genève 4, Switzerland
E-mail: Jose.Rolim@cui.unige.ch

Luca Trevisan
University of California at Berkeley, Computer Science Division
615 Soda Hall, Berkeley, CA 94720-1776, USA
E-mail: luca@eecs.berkeley.edu

Cataloging-in-Publication Data applied for

Die Deutsche Bibliothek - CIP-Einheitsaufnahme

Approximation, randomization and combinatorial optimization : algorithms and techniques ; proceedings / 4th International Workshop on Approximation Algorithms for Combinatorial Optimization Problems, APPROX 2001 and 5th International Workshop on Randomization and Approximation Techniques in Computer Science, RANDOM 2001, Berkeley, CA, USA, August 18 - 20, 2001. Michel Goemans ... (ed.). - Berlin ; Heidelberg ; New York ; Barcelona ; Hong Kong ; London ; Milan ; Paris ; Singapore ; Tokyo : Springer, 2001
(Lecture notes in computer science ; Vol. 2129)
ISBN 3-540-42470-9

CR Subject Classification (1998): F.2, G.1, G.2

ISSN 0302-9743
ISBN 3-540-42470-9 Springer-Verlag Berlin Heidelberg New York

Springer-Verlag Berlin Heidelberg New York
a member of BertelsmannSpringer Science+Business Media GmbH

http://www.springer.de

Printed in Germany

Typesetting: Camera-ready by author
Printed on acid-free paper SPIN 10840004 06/3142 5 4 3 2 1 0

Foreword

This volume contains the papers presented at the *4th International Workshop on Approximation Algorithms for Combinatorial Optimization Problems* (APPROX'01) and the *5th International Workshop on Randomization and Approximation Techniques in Computer Science* (RANDOM'01), which took place concurrently at the University of California, Berkeley, from August 18–20, 2001. APPROX focuses on algorithmic and complexity issues surrounding the development of efficient approximate solutions to computationally hard problems, and is the fourth in the series after Aalborg (1998), Berkeley (1999) and Saarbrücken (2000). RANDOM is concerned with applications of randomness to computational and combinatorial problems, and is the fifth workshop in the series following Bologna (1997), Barcelona (1998), Berkeley (1999) and Geneva (2000).

Topics of interest for APPROX and RANDOM are: design and analysis of approximation algorithms, inapproximability results, on-line problems, randomization and de-randomization techniques, sources of randomness, average-case analysis, approximation classes, randomized complexity theory, scheduling problems, routing and flow problems, coloring and partitioning, cuts and connectivity, packing and covering, geometric problems, network design, and various applications.

The volume contains 14+11 contributed papers, selected by the two program committees from 34+20 submissions received in response to the call for papers, together with abstracts of invited lectures by Michel Goemans (MIT), Russell Impagliazzo (San Diego), Anna Karlin (Washington), Luca Trevisan (Berkeley), and Salil Vadhan (MIT-Harvard).

We would like to thank all of the authors who submitted papers, our invited speakers, the members of the program committees

APPROX'01
Michel Goemans, MIT, Chair
Moses Charikar, Google - Princeton
Uriel Feige , Weizmann
Naveen Garg, IIT, Dehli
Dorit Hochbaum, Berkeley
Howard Karloff, ATT
Claire Kenyon, LRI Paris
Seffi Naor, Technion
Ramamoorthi Ravi, Pittsburgh
Baruch Schieber, IBM
Santosh Vempala, MIT Cambridge

RANDOM'01
Luca Trevisan, Berkeley, Chair
Shafi Goldwasser, MIT-Weizmann
Jon Kleinberg, Cornell
Mike Luby, Digital Fountain
Peter Bro Miltersen, BRICS
Alessandro Panconesi, Bologna
Dana Randall, Georgia Tech
Omer Reingold , ATT
Ronitt Rubinfeld, NEC
Salil Vadhan, MIT-Harvard
Umesh Vazirani, Berkeley

and the external subreferees Rohit Khandekar, Tracy Kimbrel, Jeong Han Kim, Satish Rao, Andreas Schulz, David Shmoys, Aravind Srinivasan, Maxim Sviridenko, Nisheeth Vishnoi, David Wilson and Gerhard Woeginger.

We gratefully acknowledge support from the EU research training network ARACNE, the Computer Science Department of the University of California at Berkeley, the Institute of Computer Science of the Christian-Albrechts-Universität zu Kiel and the Department of Computer Science of the University of Geneva. We also thank Marian Margraf and Brigitte Preuss for their help.

June 2001

Michel Goemans and Luca Trevisan, Program Chairs
Klaus Jansen and José D. P. Rolim, Workshop Chairs

Table of Contents

Invited Talks

Contributed Talks of APPROX

Contributed Talks of RANDOM

Using Complex Semidefinite Programming for Approximating MAX E2-LIN3

Michel X. Goemans

MIT, Dept. of Mathematics, Room 2-351, Cambridge, MA 02139,
goemans@math.mit.edu

Abstract. A number of recent papers on approximation algorithms have used the square roots of unity, -1 and 1 to represent binary decision variables for problems in combinatorial optimization, and have relaxed these to unit vectors in real space using semidefinite programming in order to obtain near optimal solutions to these problems. In this talk, we consider using the cube roots of unity, 1, $e^{i2\pi/3}$, and $e^{i4\pi/3}$, to represent ternary decision variables for problems in combinatorial optimization. Here the natural relaxation is that of unit vectors in complex space. We use an extension of semidefinite programming to complex space to solve the natural relaxation, and use a natural extension of the random hyperplane technique to obtain near-optimal solutions to the problems. In particular, we consider the problem of maximizing the total weight of satisfied equations $x_u - x_v \equiv c \pmod 3$ and inequations $x_u - x_v \not\equiv c \pmod 3$, where $x_u \in \{0, 1, 2\}$ for all u. This problem can be used to model the MAX 3-CUT problem and a directed variant we call MAX 3-DICUT. For the general problem, we obtain a 0.79373-approximation algorithm. If the instance contains only inequations (as it does for MAX 3-CUT), we obtain a performance guarantee of $\frac{7}{12} + \frac{3}{4\pi^2} \arccos^2(-1/4) \approx 0.83601$. Although quite different at first glance, our relaxation and algorithm appear to be equivalent to those of Frieze and Jerrum (1997) and de Klerk, Pasechnik, and Warners (2000) for MAX 3-CUT, and the ones of Andersson, Engebretson, and Håstad (1999) for the general case.
This talk is based on a joint result with David Williamson, to appear in [1].

References

1. M.X. Goemans and D.P. Williamson, "Approximation Algorithms for MAX 3-CUT and Other Problems Via Complex Semidefinite Programming", in the Proceedings of the 33rd Symposium on the Theory of Computing, Crete, 2001, to appear.

M. Goemans et al. (Eds.): APPROX-RANDOM 2001, LNCS 2129, p. 1, 2001.

Hill-Climbing vs. Simulated Annealing for Planted Bisection Problems

Russell Impagliazzo*

Computer Science and Engineering,
UCSD 9500 Gilman Drive, La Jolla, CA 92093-0114
tcarson.russell@cs.ucsd.edu

While knowing a problem is NP-complete tells us something about a problem's worst-case complexity, it tells us little about how intractible specific distributions of instances really are, whether these distributions are mathematically defined or come from real-world applications. Frequently, NP-complete problems have been successfully attacked on "typical" instances using heuristic methods. Little is known about when or why some of these heuristics succeed.

An interesting class of heuristics are local search algorithms, a group that includes hill-climbing, Metropolis, simulated annealing, tabu-search, WalkSAT, etc. These methods are characterized by implicitly defining a search graph on possible solutions to an optimization problem and using some (often randomized) method for moving along the edges of this graph in search of good quality solutions. Of course, assuming $P \neq NP$, no such method will always succeed in quickly finding optimial solutions for NP-hard problems. However, many such methods have been successfully used in practice for different classes of NP-hard optimization problems.

While a large amount of effort has gone into both theoretical and experimental studies of such heuristics, large gaps in our knowledge remain. For example, it is not clear whether one of the methods is universally preferable to another, or whether all of the above methods are incomparable. Do some methods succeed where others fail? Or is one of the methods strictly better than the others, for all interesting problem domains?

These questions seem difficult to answer theoretically; there are very few successful analyses of local search heuristics for specific classes of problems, and even fewer comparisons of different methods. In fact, no natural examples of optimization problems where one method provably succeeds and another fails are known.

It is just as difficult to tackle these questions experimentally, because each general method has a large number of variations and parameters, and success seems quite sensitive to the details in implentation. While experiments showing a method succeeds are not uncommon, experimental studies showing that a

* Research Supported by NSF Award CCR-9734880, NFS Award CCR-9734911, grant #93025 of the joint US-Czechoslovak Science and Technology Program, and USA-Israel BSF Grant 97-001883

M. Goemans et al. (Eds.): APPROX-RANDOM 2001, LNCS 2129, pp. 2–5, 2001.

method fails or comparing two methods are both rare and hard to interpret. Does the method fail because it is by nature ill-suited to the problem domain, or because the implementation or parameters chosen were not optimized?

This talk will summarize joint work with Ted Carson addressing these questions, both theoretically and experimentally ([3,4,5]). In this talk, we concentrate on one NP-complete problem, the minimum graph bisection problem, and one class of distributions for instances: the planted bisection graph model ([2]). The planted bisection random graph model has been used as a benchmark for evaluating heuristics ([1,2,9,8,11,7,6]). In this model, a graph is constructed from two initially disjoint n node random graphs, drawn from $G_{n,p}$, by adding edges between cross graph vertex pairs with some lower probability $q < p$. If $p - q = \omega(\sqrt{logn}\sqrt{mn}/n^2)$, then with high probability the bisection separating the two is the unique only optimal solution.

A landmark paper of Jerrum and Sorken proves that the Metropolis algorithm succeeds with high probability for random instances of the graph bisection problem drawn from the planted bisection model $G_{n,p,q}$, when p is sufficiently greater than q. They proved their result for $p-q = n^{-1/6+\epsilon}$, significantly greater than needed for optimality. This is one of the few optimization problems for which Metropolis or simulated annealing are provably good. However, they left open the question of whether the full Metropolis algorithm was necessary, or whether degenerate cases of the algorithm such as random hill-climbing would also succeed.

We were expecting that these simpler heuristics would fail on this problem. However, our initial experimental work showed that hill-climbing methods succeeded at finding the planted bisection not just in the range of parameters above, but whenever the planted bisection was optimal.

Based on intuition gathered from these experiments, and some ideas from [6], we proved that a simple, polynomial-time, hill-climbing algorithm for this problem succeeds in finding the planted bisection with high probability if $p-q = \Omega(log^3 n^{-1/2})$, within polylog factors of the threshold ([4]). The above algorithm had one unnatural restriction. However, the same analysis shows a purely randomized hill-climbing algorithm succeeds in finding the planted bisection in polynomial time if $p - q = \Omega(n^{-1/4+\epsilon}\sqrt{mn})$, for any $\epsilon > 0$. This universal algorithm is a degenerate case of both Metropolis and go-with-the-winners, so this result implies, extends, and unifies those by Jerrum and Sorken, Dimitriou and Impagliazzo, and Juels [9,7,11].

Thus, there are no examples of planted graph bisections where sophisticated heuristic methods have been proven to work, but where simple hill-climbing algorithms fail. This result emphasises the need to find instance distributions for optimization problems that can be used to discriminate between local search heuristic techniques.

Returning to experimental results, we identified a candidate for such a discriminatory problem class. Namely, there exist parameters slightly below the threshold where the planted bisection is not optimal, but all "good" quality solutions are "near" the planted bisection. For these parameters, we were able to distinguish experimentally between various heuristic methods. We were able to find parameters where local optimization typically failed to find any good solution, but where an appropriate simulated annealing schedule finds solutions of near-optimal quality. We also showed experimentally that our simulated annealing performed better than Metropolis at any fixed temperature. ([5,3]).

To do these experiments, we first gathered statistics using a Go-with-the-winners sampling algorithm. This allowed us to characterize "random" bisections of different qualities, and to identify biases introduced by optimization methods. In particular, we showed that hill-climbing methods produce solutions that are overly locally-optimized, in the sense that they were smaller in cut than would be typical for their distance from the planted bisection. For sufficiently high temperatures, Metropolis avoids this type of bias, but for lower temperatures, it produces a similar bias. Thus, there is an optimal Metropolis temperature for these distributions; above this temperature, Metropolis seems to converge rapidly to its stationary distribution, but below this temperature it seems to reach locally optima and become stuck.

Even for the lowest temperature for which it has rapid convergence, Metropolis does not produce optimal solutions. On the other hand, the solutions it does produce at this temperature are significantly closer to the planted bisection. Starting a second, more greedy, phase of optimization from the results of the first allow further progress without introducing bias. Repeating this for a few steps leads to a cooling schedule for Simulated Annealing that significantly improves on Metropolis at any fixed temperature.

We hope that in future work this experimentally observed gap will be rigorously proven. This would give the first natural proven separation between Simulated Annealing, Metropolis and hill-climbing algorithms. More importantly, it would give insight into when and why some heuristic methods do better than others.

References

1. T. Bui, S. Chaudhuri, T. Leighton, and M. Sipser. Graph bisection algorithms with good average case behavior. In *Proceedings of the 25th IEEE Symposium on Foundations of Computer Science*, pages 181–192, 1984.
2. R. B. Boppana. Eigenvalues and graph bisection: An average case analysis. In *Proceedings of the 28th IEEE Symposium on Foundations of Computer Science*, pages 280–285, 1987.
3. T. Carson, *Empirical and Analytic Approaches to Understanding Local Search Heuristics*, Ph.D. thesis, University of California at San Diego, 2001.
4. T. Carson and R. Impagliazzo, Hill-climbing finds random planted bisections, SODA, 2001.
5. T. Carson and R. Impagliazzo. Determining regions of related solutions for graph bisection problems, *International Joint Conference on Artificial Intelligence, Workshop ML-1: Machine Learning for Large-Scale Optimization*, 1999.
6. A. Condon and R. M. Karp. Algorithms for Graph Partitioning on the Planted Bisection Model, RANDOM-APPROX'99, pages 221–32, 1999.

7. A. Dimitriou and R. Impagliazzo. Go-with-the-winnners algorithms for graph bisection, In *Proceedings of the 9th ACM-SIAM Symposium on Discrete Algorithms*, pages 510–520, 1998.
8. M. Dyer and A. Frieze. Fast solution of some random *NP*-hard problems. In *Proceedings of the 28th IEEE Symposium on Foundations of Computer Science*, pages 280–285, 1987
9. M. R. Jerrum and G. Sorkin. Simulated annealing for graph bisection. In *Proceedings 34th IEEE Symposium on Foundations of Computer Science (FOCS)*, pages 94–103, 1993.
10. D. S. Johnson, C. R. Aragon, L. A. McGeoch and C. Schevon. Optimization by Simulated Annealing: An Experimental Evaluation, Part I (Graph Partitioning). *Operations Research* 37:865–892, 1989.
11. A. Juels, *Topics in Black Box Optimization*, Ph.D. thesis, University of California at Berkeley, 1996.

Web Search via Hub Synthesis

Anna R. Karlin

Department of Computer Science & Engineering
University of Washington
114 Sieg Hall, Box 352350
Seattle WA 98195-2350,
karlin@cs.washington.edu

Abstract. We present a probabilistic generative model for web search which captures in a unified manner three critical components of web search: how the link structure of the web is generated, how the content of a web document is generated, and how a human searcher generates a query. The key to this unification lies in capturing the correlations between each of these components in terms of proximity in latent semantic space. Given such a combined model, the correct answer to a search query is well defined, and thus it becomes possible to evaluate web search algorithms rigorously. We present a new web search algorithm, based on spectral techniques, and prove that it is guaranteed to produce an approximately correct answer in our model. The algorithm assumes no knowledge of the model, and is well-defined regardless of the accuracy of the model.
Joint work with D. Achlioptas, A. Fiat and F. McSherry

M. Goemans et al. (Eds.): APPROX-RANDOM 2001, LNCS 2129, p. 6, 2001.

Error-Correcting Codes and Pseudorandom Projections

Luca Trevisan*

U.C. Berkeley, Computer Science Division, 615 Soda Hall, Berkeley, CA
luca@eecs.berkeley.edu

Abstract. In this talk we discuss constructions of hash functions, randomness extractors, pseudorandom generators and hitting set generators that are based on the same principle: encode the "input[1]" using an error-correcting code, select a random (or pseudorandom) subset of the bits of the encoding, and output the encoded codeword restricted to such bits. This general approach is common to constructions of very different combinatorial objects, and somewhat different strategies are used to analyse such different constructions.

An early application of the encode-and-project paradigm is in a paper of Miltersen [Mil98], applied to the construction of a family of hash functions with low collision probability.

Suppose we want to construct a family of hash functions $h : \{0,1\}^n \to \{0,1\}^m$, and suppose that we have an error-correcting code $C : \{0,1\}^n \to \{0,1\}^{\bar{n}}$ whose minimum distance is, say, $\bar{n}/3$. Let us introduce the following notation: if $y = (y_1, \ldots, y_k) \in \{0,1\}^k$ and $S = \{s_1, \ldots, s_l\} \subseteq [k]$, with $s_1 < s_2 < \cdots < s_l$ then $y_{|S} = (y_{s_1}, y_{s_2}, \ldots, y_{s_l}) \in \{0,1\}^l$. Then we can define a family of hash functions where each function of our family is indexed by a subset $S \subset [\bar{n}]$ of size m, and $h_S(x) = C(x)_{|S}$.

It is immediate to see that the collision probability is at most $(2/3)^m$. The advantage of this construction is that both the encoding and the projection can be evaluated in constant time on a unit-cost RAM (see [Mil98]), so that the hash functions in the family can be evaluated in constant time.

One can see a very similar construction at work in the pseudorandom generator construction in [STV01] and in the randomness extractor construction in [Tre99], with some fundamental difference. One difference is that the "projection" is not chosen uniformly at random, but it is rather generated from a seed of logarithmic length using the "combinatorial design construction" of Nisan and Wigderson [NW94]. Another difference is that, in the case of pseudorandom generators, it is not enough for the error-correcting code to have large minimum distance, but a certain type of sublinear-time list-decoding algorithm must also exist [STV01]; in the case of random extractors, howver, any code with

* Work supported by a Sloan Research Fellowship and an NSF Career Award.

[1] By "input" we mean the actual input for hash functions, the weakly random input for randomness extractors, and the description of a computationally hard problem for pseudorandom generators and for hitting set generators.

M. Goemans et al. (Eds.): APPROX-RANDOM 2001, LNCS 2129, pp. 7–9, 2001.

a relative minimum distance close to 1/2 can be used [Tre99]. Constructions of pseudorandom generators and/or randomness extractors in [RRV99,ISW00,TSUZ01] use error-correcting codes and the Nisan-Wigderson combinatorial designs, with improvements in the construction, in the analysis, and in the composition of the basic construction with other tools (and with itself).
The Nisan-Wigderson approach yields a randomness-efficient but somewhat counter-intuitive way of generating projections. When the input (or hard problem) is encoded as a multivariate polynomial (which is done in [STV01] and is a possible implementation of [Tre99]), a more natural approach to projection is to consider lines. Miltersen and Vinodchandran [MV99] show that by encoding a hard problem as a multivariate polynomial, and then restricting it to axis-parallel lines, one can get a *hitting set generator* construction, which in turn can be used to derandomize complexity classes. The approach of [MV99] does not replicate the result of [IW97] ([MV99] can prove $P = BPP$ only under a stronger assumption than the one postulated in [IW97]), however it can prove a result on AM that is stronger than the best known result based on Nisan-Wigderson [KvM99]. The analysis in [MV99] appears to be substantially different from the analysis in [STV01], although the hard function is encoded using the same error-correcting code, and the "only" difference is in the way the encoding is projected (lines versus the approach based on Nisan-Wigderson).
Ta-Shma, Zuckerman and Safra [TSZS01] showed how to construct randomness extractors by encoding the input using multivariate polynomials and then restricting it to a subset of a line (the line is selected using the seed of the extractor). While the construction of [TSZS01] is virtually identical[2] to the one in [MV99], the analysis is completely different.

References

ISW00. R. Impagliazzo, R. Shaltiel, and A. Wigderson. Extractors and pseudo-random generators with optimal seed length. In *Proceedings of the 32nd ACM Symposium on Theory of Computing*, pages 1–10, 2000.

IW97. R. Impagliazzo and A. Wigderson. $P = BPP$ unless E has sub-exponential circuits. In *Proceedings of the 29th ACM Symposium on Theory of Computing*, pages 220–229, 1997.

KvM99. A. Klivans and D. van Milkebeek. Graph non-isomorphism has subexponential size proofs unless the polynomial hierarchy collapses. In *Proceedings of the 31st ACM Symposium on Theory of Computing*, pages 659–667, 1999.

Mil98. P.B. Miltersen. Error-correcting codes, perfect hashing circuits, and deterministic dynamic dictionaries. In *Proceedings of the 9th ACM-SIAM Symposium on Discrete Algorithms*, 1998.

MV99. P.B. Miltersen and N.V. Vinodchandran. Derandomizing Arthur-Merlin games using hitting sets. In *Proceedings of the 40th IEEE Symposium on Foundations of Computer Science*, pages 71–80, 1999.

[2] However it should be noted that the analysis of [MV99] works for a large class of codes, of which multivariate polynomials are a special case, while it appears that the analysis of [TSZS01] requires the code to be a multivariate polynomial.

NW94. N. Nisan and A. Wigderson. Hardness vs randomness. *Journal of Computer and System Sciences*, 49:149–167, 1994. Preliminary version in *Proc. of FOCS'88*.

RRV99. R. Raz, O. Reingold, and S. Vadhan. Extracting all the randomness and reducing the error in Trevisan's extractors. In *Proceedings of the 31st ACM Symposium on Theory of Computing*, pages 149–158, 1999.

STV01. M. Sudan, L. Trevisan, and S. Vadhan. Pseudorandom generators without the XOR lemma. *Journal of Computer and System Sciences*, 62(2):236–266, 2001.

Tre99. L. Trevisan. Construction of extractors using pseudo-random generators. In *Proceedings of the 31st ACM Symposium on Theory of Computing*, pages 141–148, 1999.

TSUZ01. A. Ta-Shma, C. Umans, and D. Zuckerman. Loss-less condensers, unbalanced expanders, and extractors. In *Proceedings of the 33rd ACM Symposium on Theory of Computing*, 2001.

TSZS01. A. Ta-Shma, D. Zuckerman, and S. Safra. Extractors from Reed-Muller codes. Technical Report TR01-036, Electronic Colloquium on Computational Complexity, 2001.

Order in Pseudorandomness

Salil P. Vadhan

Division of Engineering and Applied Sciences, Harvard University
33 Oxford Street, Cambridge, MA 02138
salil@eecs.harvard.edu
http://eecs.harvard.edu/~salil

We survey recent works which unify four of the most important and widely studied objects in the theory of pseudorandomness, namely: *pseudorandom generators*, *expander graphs*, *error-correcting codes*, and *extractors*. Since all of these objects are "pseudorandom" in some sense, it is not surprising that constructions of one of these objects may be useful in constructing another. Indeed, there are many examples of this in the past, which we do not attempt to summarize here. Instead, we focus on how recent works show that some of these objects are almost *the same*, when interpreted appropriately. All of the connections we describe involve *extractors*, lending further credence to Nisan's assertion that they "exhibit some of the most 'random-like' properties of explicitly constructed combinatorial structures" [Nis96].

Extractors and Pseudorandom Generators. In 1999, Trevisan [Tre99] discovered a completely unexpected connection, showing that every "black-box" construction of pseudorandom generators from hard Boolean functions *is also* a construction of extractors. This unified two previously distinct lines of research, opened the door to many improved extractor constructions [Tre99, RRV99, ISW00, TUZ01], and provided new ways of thinking about both kinds of objects.

Extractors and Expander Graphs. It has been known since the defining paper of Nisan and Zuckerman [NZ96] that an extractor can be viewed as a certain kind of expander graph, and indeed this was exploited in [WZ99] to construct expanders that "beat the eigenvalue bound." Recently, this connection between extractors and expander graphs has been exploited more fully, yielding solutions to two classic problems about expander graphs: 1. a simple, iterative construction of constant-degree expanders [RVW00], and 2. graphs with expansion greater than 1/2 times the degree [TUZ01].

Extractors and Error-Correcting Codes. Several recent constructions of extractors (such as Trevisan's construction, its successors, and [RSW00]) make important use of error-correcting codes. This is no coincidence, as Ta-Shma and Zuckerman [TZ01] have shown that extractors can be viewed as a generalization of list-decodable error-correcting codes. This viewpoint has led to new extractor constructions, which more directly exploit the properties of algebraic codes [TZS01].

M. Goemans et al. (Eds.): APPROX-RANDOM 2001, LNCS 2129, pp. 10–11, 2001.

References

The monograph by Goldreich [Gol99] provides a comprehensive overview of the theory of pseudorandomness, and the survey by Nisan [Nis96] contains background on extractors and their applications.

Gol99. Oded Goldreich. *Modern Cryptography, Probabilistic Proofs, and Pseudo-randomness.* Number 17 in Algorithms and Combinatorics. Springer-Verlag, 1999.

ISW00. Russell Impagliazzo, Ronen Shaltiel, and Avi Wigderson. Extractors and pseudo-random generators with optimal seed length. In *Proceedings of the Thirty-Second Annual ACM Symposium on Theory of Computing*, pages 1–10, Portland,Oregan, 21–23 May 2000.

Nis96. Noam Nisan. Extracting randomness: How and why: A survey. In *Proceedings, Eleventh Annual IEEE Conference on Computational Complexity*, pages 44–58, Philadelphia, Pennsylvania, 24–27 May 1996. IEEE Computer Society Press.

NZ96. Noam Nisan and David Zuckerman. Randomness is linear in space. *Journal of Computer and System Sciences*, 52(1):43–52, February 1996.

RRV99. Ran Raz, Omer Reingold, and Salil Vadhan. Extracting all the randomness and reducing the error in Trevisan's extractors. In *Proceedings of the Thirty-First Annual ACM Symposium on Theory of Computing*, pages 149–158, Atlanta, Georgia, 1–4 May 1999.

RSW00. Omer Reingold, Ronen Shaltiel, and Avi Wigderson. Extracting randomness via repeated condensing. In *41st Annual Symposium on Foundations of Computer Science*, pages 22–31, Redondo Beach, CA, 17–19 October 2000. IEEE.

RVW00. Omer Reingold, Salil Vadhan, and Avi Wigderson. Entropy waves, the zig-zag graph product, and new constant-degree expanders and extractors. In *41st Annual Symposium on Foundations of Computer Science*, pages 3–13, Redondo Beach, CA, 17–19 October 2000. IEEE.

TUZ01. Amnon Ta-Shma, Christopher Umans, and David Zuckerman. Loss-less condensers, unbalanced expanders, and extractors. In *Proceedings of the Thirty-Third Annual ACM Symposium on Theory of Computing*, Crete, Greece, 6–8 July 2001.

TZ01. Amnon Ta-Shma and David Zuckerman. Extractor codes. In *Proceedings of the Thirty-Third Annual ACM Symposium on Theory of Computing*, Crete, Greece, 6–8 July 2001.

TZS01. Amnon Ta-Shma, David Zuckerman, and Shmuel Safra. Extractors from Reed–Muller codes. Technical Report TR01-036, Electronic Colloquium on Computational Complexity, May 2001.

Tre99. Luca Trevisan. Construction of extractors using pseudo-random generators. In *Proceedings of the Thirty-First Annual ACM Symposium on Theory of Computing*, pages 141–148, Atlanta, Georgia, 1–4 May 1999.

WZ99. Avi Wigderson and David Zuckerman. Expanders that beat the eigenvalue bound: explicit construction and applications. *Combinatorica*, 19(1):125–138, 1999.

Minimizing Stall Time in Single and Parallel Disk Systems Using Multicommodity Network Flows

Susanne Albers and Carsten Witt

Dept. of Computer Science, Dortmund University, 44221 Dortmund, Germany
albers@ls2.cs.uni-dortmund.de, carsten.witt@udo.edu

Abstract. We study integrated prefetching and caching in single and parallel disk systems. A recent approach used linear programming to solve the problem. We show that integrated prefetching and caching can also be formulated as a min-cost multicommodity flow problem and, exploiting special properties of our network, can be solved using combinatorial techniques. Moreover, for parallel disk systems, we develop improved approximation algorithms, trading performance guarantee for running time. If the number of disks is constant, we achieve a 2-approximation.

1 Introduction

In today's computer systems there is a large gap between processor speeds and memory access times, the latter usually being the limiting factor in the performance of the overall system. Therefore, computer designers devote a lot of attention to building improved memory systems, which typically consist of hard disks and associated caches. Caching and prefetching are two very well-known techniques for improving the performance of memory systems and, separately, have been the subject of extensive studies. Caching strategies try to keep actively referenced memory blocks in cache, ignoring the possibility of reducing processor stall times by prefetching blocks into cache before their actual reference. On the other hand, most of the previous work on prefetching tries to predict the memory blocks requested next, not taking into account that blocks must be evicted from cache in order to make room for the prefetched blocks. Only recently researchers have been working on an integration of both techniques [1,2,3,4,5,7].

Cao et al. [3] and Kimbrel and Karlin [7] introduced a theoretical model for studying "Integrated Prefetching and Caching" (IPC) that we will also use in this paper. We first consider single disk systems. A set S of memory blocks resides on one disk. At any time a cache can store k of these blocks. The system must serve a request sequence $\sigma = \sigma(1), \dots, \sigma(n)$, where each request $\sigma(i)$, $1 \leq i \leq n$, specifies a memory block. The service of a request takes one time unit and can only be accomplished if the requested block is in cache. If a requested block is not in cache, it must be fetched from disk, which takes F time units, where $F \in \mathbb{N}$. If a missing block is fetched immediately before its reference, then the processor has to stall for F time units. However, a fetch may also overlap with

M. Goemans et al. (Eds.): APPROX-RANDOM 2001, LNCS 2129, pp. 12–24, 2001.

the service of requests. If a fetch is started i time units before the next reference to the block, then the processor has to stall for only $\max\{0, F-i\}$ time units. In case $i \geq 1$, we have a real prefetch. Of course, at most one fetch operation may be executed at any time. Once a fetch is initiated, a block must be evicted from cache in order to make room for the incoming block. The goal is to minimize the processor stall time, or equivalently the *elapsed time*, which is the sum of the processor stall time and the length n of the request sequence.

In parallel disk systems with D disks we have D sets of memory blocks $S_1, \ldots, S_D$, where S_d is the set of blocks that reside on disk d, $1 \leq d \leq D$. We assume that each block in the system is located on only one of the disks. The main advantage of parallel disk systems is that blocks from different disks may be fetched in parallel. Thus if the processor has to stall at some point in time, then all the fetches currently being active advance towards completion. If a fetch is initiated, we may evict any block from cache, which corresponds to the model that blocks are read-only. Again the goal is to minimize the processor stall time.

Cao et al. [3,4] studied IPC in single disk systems. They presented simple combinatorial algorithms, called *conservative* and *aggressive*, that run in polynomial time and approximate the elapsed time. *Conservative* achieves an approximation factor of 2, whereas *aggressive* achieves a better factor of $\min\{2, 1+F/k\}$. Karlin and Kimbrel [7] investigated IPC in parallel disk systems and presented a polynomial-time algorithm whose approximation guarantee on the elapsed time is $(1 + DF/k)$. In [1] Albers, Garg and Leonardi developed a polynomial-time algorithm that computes an optimal prefetching/caching schedule for single disk systems. For parallel disk systems they developed a polynomial-time algorithm that approximates the stall time. The algorithm achieves an approximation factor of D, using at most $D-1$ extra memory locations in cache. All the results presented in [1] are based on a linear program formulation.

In this paper we show that IPC in single and parallel disk systems can be formulated as a min-cost multicommodity flow problem and, exploiting special properties of the network, can be solved using combinatorial methods. These results are presented in Sec. 2. We first investigate the single disk problem. We describe the construction of the network and establish relationships between min-cost multicommodity flows and prefetching/caching schedules. We prove that a combinatorial approximation algorithm by Kamath et al. [6] for computing min-cost multicommodity flows, when applied to our network, computes an optimal prefetching/caching schedule in polynomial time. We then generalize our multicommodity flow formulation to parallel disk systems. With minor modifications of the original network we are able to apply the algorithm by Kamath et al. [6] again. We derive a combinatorial algorithm that achieves a D-approximation on the stall time, using at most $D-1$ extra memory location in cache. Thus, the results presented in [1] can also be obtained using combinatorial techniques.

For parallel disk systems, D is the best approximation factor on the stall time currently known. This factor D is caused by the fact that the approach in [1] heavily overestimates the stall times in prefetching/caching schedules: Stall time is counted separately on each disk, i. e. no advantage is taken of the fact that prefetches executed in parallel simultaneously benefit from a processor stall time. In Sec. 3 we develop improved approximation guarantees that are bounded away from D. We are able to formulate a trade-off. For any $z \in \mathbb{N}$, we achieve an approximation factor of $2(D/z)$ at the expense of a running time that grows exponentially with z. If the number D of disks is constant, we obtain a 2-approximation. For the special case $D = 2$ we also give a better 1.5-approximation. Again, our solutions need $D-1$ extra memory locations in cache. The improved approximation algorithms can also be obtained using min-cost multicommodity flows. However, for the sake of clarity and due to space limitations we present an LP-formulation in this extended abstract.

2 Modeling IPC by Network Flows

We first consider single disk systems. We build up our combinatorial algorithm in several steps. Given a request sequence σ, we first construct a network $G = (V, E)$ with several commodities such that an integral min-cost flow corresponds to an optimal prefetching/caching schedule for σ, and vice versa. Of course, an algorithm for computing min-cost multicommodity flows does not necessarily return an integral flow when applied to our network. We show that a non-integral flow corresponds to a *fractional* prefetching/caching schedule in which we can identify an integral schedule using a technique from [1].

The main problem we are faced with is that we know of no combinatorial polynomial-time algorithm for computing a (non-integral) min-cost flow in our network. We solve this problem by applying a combinatorial approximation algorithm by Kamath et al. [6]. For any $\varepsilon \geq 0$, $\delta \geq 0$, the algorithm computes a flow such that a fraction of at least $1 - \varepsilon$ of each demand in the network is satisfied and the cost of the flow is at most $(1 + \delta)$ times the optimum. Unfortunately, the flow computed by the algorithm, when applied to our network, does not correspond to a feasible fractional prefetching/caching schedule: It is possible that (a) more than one block is fetched from disk at any time and (b) blocks are not completely in cache when requested. We first reduce the flow in the network to resolve (a). This reduces the extent to which blocks are in cache at the time of their request even further. We then show that, given such flow, we can still derive an optimal prefetching/caching schedule, provided that ε and δ are chosen properly.

2.1 The Network

Let σ be a request sequence consisting of n requests. We construct a network $G = (V, E)$ with $n + 1$ commodities. Associated with each request $\sigma(i)$ is a commodity i, $1 \leq i \leq n$. This commodity has a source s_i, a sink t_i and demand $d_i = 1$. Let a_i be the block requested by $\sigma(i)$. For each request $\sigma(i)$, we introduce vertices x_i and x'_i. These vertices are linearly linked, i. e. there are edges (x_i, x'_i), $1 \leq i \leq n$, and edges (x'_i, x_{i+1}), $1 \leq i \leq n-1$, each with capacity k and cost 0. Intuitively, this sequence of vertices and edges represents the cache. If

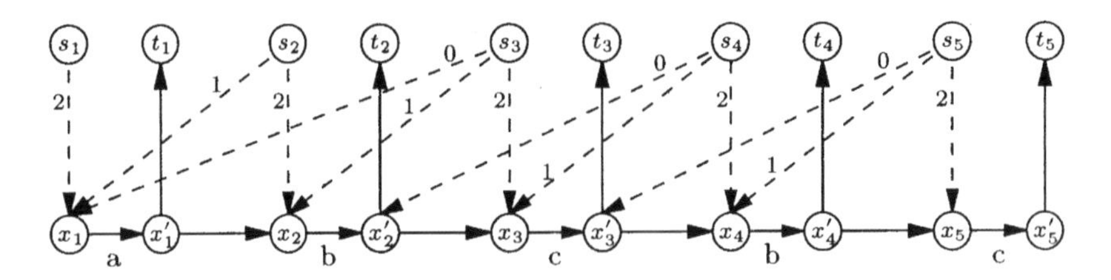

Fig. 1. Sketch of the network for request sequence *abcbc* and $F = 2$

commodity i flows through (x_j, x'_j), then block a_i is in cache when $\sigma(j)$ is served. To ensure that a_i is in cache when $\sigma(i)$ is served, we insert an edge (x'_i, t_i) with capacity 1 and cost 0, and there are no other edges into t_i or x'_i, i. e. commodity i must pass through (x_i, x'_i).

Let p_i be the time of the previous request to a_i, i. e. p_i is the largest j, $j < i$, such that a_i was requested by $\sigma(j)$. If a_i is requested for the first time in σ, then we set $p_i = 0$. To serve $\sigma(i)$, block a_i can (1) remain in cache after $\sigma(p_i)$ until request $\sigma(i)$, provided that $p_i > 0$, or can (2) be fetched into cache at some time before $\sigma(i)$. To model case (1) we introduce an edge (s_i, x'_{p_i}), if $p_i > 0$, with capacity 1 and cost 0. To model case (2) we essentially add edges (s_i, x_j), for $j = p_i+1, \ldots, i$, indicating that a fetch for a_i is initiated starting at the service of $\sigma(j)$. For the special case $j = i$ the edge represents a fetch executed immediately before $\sigma(i)$. If $i-j < F$, then the processor has to stall for $F-(i-j)$ time units and hence we assign a cost of $F - (i - j)$ to edge (s_i, x_j). Figure 1 illustrates this construction for the examplary request sequence $\sigma = abcbc$ and fetch time $F = 2$. Edges outgoing of a source s_i, $i \in \{1, \ldots, 5\}$, are labeled with their cost.

So far our construction allows a flow algorithm to saturate more than one of the edges that correspond to fetches executed simultaneously (consider, e. g. the edges (s_i, x_{i-1}) and (s_{i-1}, x_{i-2}) for some i such that $\sigma(i-2)$, $\sigma(i-1)$ and $\sigma(i)$ are pairwise distinct). However, we have to make sure that at most one fetch operation is executed at any time. Therefore, in our construction we split the "super edge" (s_i, x_j) into several parts. For any ℓ, $1 \leq \ell \leq n-1$, let $[\ell, \ell+1)$ be the time interval starting at the service of $\sigma(\ell)$ and ending immediately before the service of $\sigma(\ell+1)$. Interval $[0, 1)$ is the time before the service of $\sigma(1)$.

We have to consider all the fetches being active at some time in $[\ell, \ell+1)$, for any fixed ℓ. A fetch for a_i starting at $\sigma(j)$, $j < i$, is active during $[\ell, \ell+1)$ for $\ell = j, \ldots, \min\{j + F, i\} - 1$. For any fixed i and j with $1 \leq i \leq n$ and $p_i+1 \leq j < i$ we introduce vertices v^{ℓ}_{ij} and w^{ℓ}_{ij} where $\ell = j, \ldots, \min\{j+F, i\}-1$. These vertices are linked by edges of capacity 1 and cost 0. More specifically, we have edges $(v^{\ell}_{ij}, w^{\ell}_{ij})$, $\ell = j, \ldots, \min\{j + F, i\} - 1$, and edges $(w^{\ell}_{ij}, v^{\ell-1}_{ij})$, $\ell = j+1, \ldots, \min\{j+F, i\}-1$. The last vertex in this sequence, w^{j}_{ij}, is linked to x_j with an edge of cost 0 and capacity 1. Finally we add an edge (s_i, v^{ℓ}_{ij}), where $\ell = \min\{j + F, i\} - 1$, to the first vertex in this sequence with cost $F - (i - j)$ and capacity 1. In this construction we excluded the case $j = i$ because a fetch for a_i initiated at $\sigma(i)$ is somewhat special: The fetch is performed completely before $\sigma(i)$, i.e. it does not overlap with any request, and the processor stalls for F time units. The fetch is active at some time during $[i-1, i)$. We introduce vertices $v_{i,i}^{i-1}$ and w_{ii}^{i-1} linked by an edge of capacity 1 and cost 0. Vertex w_{ii}^{i-1} is linked to x_i with an edge of the same capacity and cost. Finally, we have an edge (s_i, v_{ii}^{i-1}) of capacity 1 and cost F.

Next we describe the role of the $(n+1)$-st commodity, which is used to ensure that no two prefetches are performed at the same time. More precisely, we ensure that at most one prefetch is executed in any fixed interval $[\ell, \ell+1)$, $1 \le \ell \le n-1$. For any fixed ℓ, let f_ℓ be the number of prefetches whose execution overlaps with $[\ell, \ell+1)$, i.e.

$$f_\ell = |\{v_{ij}^\ell \mid 1 \le i \le n,\ p_i + 1 \le j \le i\}| \ . \tag{1}$$

Commodity $n+1$ has a source s_{n+1}, a sink t_{n+1} and a demand of $d_{n+1} = \sum_{\ell=1}^{n-1}(f_\ell - 1)$. The flow from s_{n+1} to t_{n+1} is routed through the edges $(v_{ij}^\ell, w_{ij}^\ell)$ and newly introduced "subsinks" t_{n+1}^ℓ, $1 \le \ell \le n-1$. For any pair of vertices v_{ij}^ℓ and w_{ij}^ℓ we introduce edges (s_{n+1}, v_{ij}^ℓ) and $(w_{ij}^\ell, t_{n+1}^\ell)$ with capacity 1 and cost 0. Additionally, we insert edges (t_{n+1}^ℓ, t_{n+1}) with capacity $f_\ell - 1$ and cost 0. Now consider a fixed interval $[\ell, \ell+1)$, $1 \le \ell \le n-1$. Every prefetch for some a_i initiated at $\sigma(j)$ that is active at some time during $[\ell, \ell+1)$ is represented by a "super edge" (s_i, x_j) and contains an edge $(v_{ij}^\ell, w_{ij}^\ell)$. For fixed ℓ the network contains f_ℓ such edges. The capacities $f_\ell - 1$ of the edges (t_{n+1}^ℓ, t_{n+1}) ensure that only one of the edges $(v_{ij}^\ell, w_{ij}^\ell)$ can carry a flow of commodity i, $i \le n$. If two or more such edges were carrying flow of commodity $i \le n$, then the capacity constraint would be violated at some edge $(t_{n+1}^{\ell'}, t_{n+1})$, for some $\ell' \ne \ell$, or demand d_{n+1} would not be satisfied.

The following lemma states that our network correctly models IPC on a single disk. Its proof is omitted in this extended abstract.

Lemma 1. *Any feasible integral flow of cost C in G corresponds to a feasible prefetching/caching schedule with stall time C for σ, and vice versa.*

2.2 Properties of Optimal Flows

We show that a non-integral flow in our network corresponds to a fractional prefetching/caching schedule, defined in the following way.

Definition 2. *Given an instance of the problem IPC, we define the set of* fractional solutions *as a superset of the set of integral solutions to the instance. A fractional solution may deviate from an integral solution in the following way:*

- *The amount to which a block resides in cache may take a fractional value between 0 and 1. However, this amount must be 1 while the block is requested.*
- *Fractional parts of blocks in cache arise due to* partial evictions *or* partial fetches. *For each time interval, the net amount of blocks fetched must not be larger than the net amount of blocks evicted, and the net amount of blocks fetched must not exceed 1.*

- *Stall times are accounted as follows: If a fetch to $\delta \in [0, 1]$ units of block $\sigma(j)$ is initiated starting at the service of reference $\sigma(i)$ and $j - i < F$ holds, we incur a stall time of $\delta(F - (j - i))$ time units.*

Loosely speaking, the main difference between integral and fractional solutions lies in the possibility to interrupt fetches and to leave parts of a block in cache between consecutive requests to it. Regarding the second item in the above definition we may assume w. l. o. g. that between any two consecutive references to a specific block, the points of time where the block is evicted from cache precede the ones where the block is fetched back.

Lemma 3. *Let G be the network obtained by transforming a request sequence σ of the problem IPC according to the construction in Sec. 2.1. A valid multicommodity flow with cost C within the network G corresponds to a fractional prefetching/caching schedule with stall time C.*

The next lemma follows immeditately from [1]; it was shown that a fractional solution is a convex combination of polynomially many integral solutions.

Lemma 4. *Let L be a fractional solution to an input for IPC. There is a polynomial-time algorithm that computes an integral prefetching/caching schedule L^* from L where the stall time of L^* is less than or equal to the one of L.*

2.3 Applying the Approximation Algorithm

We show how to compute a flow in our network and how to derive an optimal prefetching/caching schedule. We apply the algorithm by Kamath et al. [6] by setting $\varepsilon := 1/(4F^2n^3)$ and $\delta := 1/(3nF)$. These settings have been derived from the easy-to-see upper bound $d_{n+1} \leq n^2F$ on the demand of commodity $n+1$.

As the approximation algorithm only satisfies a fraction of $1 - \varepsilon$ of each commodity, the flow out of each source s_i, $i \in \{1, \dots, n\}$, is lower bounded by $1 - \varepsilon$. Moreover, commodity $n+1$ might lack an amount of $\varepsilon d_{n+1} \leq \varepsilon F n^2$. We assume pessimistically that this leads to an additional "illegal" flow with value $\varepsilon F n^2$ during a time interval $[\ell, \ell+1)$, $\ell \in \{1, \dots, n-1\}$, in so far as edges representing fetches in that interval are not "congested" properly by commodity $n+1$.

Let $\varrho := 1 - \varepsilon d_{n+1} - \varepsilon$ be a crucial lower bound on the flow of commodities $1, \dots, n$. We can transform the flow ϕ output by the approximation algorithm into a uniform flow ϕ' which directs exactly ϱ units of flow from s_i to t_i for any commodity $i \in \{1, \dots, n\}$. The main idea is to reduce, for each edge, the flow of commodity i proportionally to the relative amount of flow of commodity i on the considered edge. Then we end up with a uniform flow ϕ' which does not "overflow" any interval $[\ell, \ell + 1)$ and delivers the same amount for each commodity.

In view of Definition 2 and the equivalence described in Lemma 3, the flow ϕ' corresponds to a fractional solution to IPC in which all blocks have size ϱ. Flow ϕ' corresponds to a fractional solution to IPC in which all blocks have size ϱ and the number of cache slots is upper bounded by k/ϱ—hereinafter we call such a solution a *ϱ-solution.* According to Lemma 4, we may interpret the fractional solution which corresponds to ϕ' as a convex combination of *integral* ϱ-solutions.

In order to analyze the quality of the above-described convex combination of integral ϱ-solutions, we have to establish a lower bound on ϱ. As $d_{n+1} \leq Fn^2$, we obtain $\varrho \geq 1 - \varepsilon d_{n+1} - \varepsilon \geq 1 - 2\varepsilon d_{n+1} \geq 1 - 1/2nF$. Next we estimate the cost C of the convex combination of ϱ-solutions. Since the approximation algorithm outputs flows with cost at most $(1+\delta)\,\mathrm{OPT}$, where OPT is the cost of an optimal schedule, and reducing flows to ϱ does never increase cost, the following upper bound on C holds:

$$C \leq (1+\delta)\,\mathrm{OPT} = \mathrm{OPT}/(3nF) + \mathrm{OPT} \leq \mathrm{OPT} + 1/3 \quad \text{since } \mathrm{OPT} \leq nF \ . \tag{2}$$

We underestimated the cost C of the convex combination of ϱ-solutions by an additive term of at most $n(1-\varrho)F$. This is due to the fact that each block corresponding to a specific commodity has size 1 in reality, but size ϱ in the convex combination. By increasing the block size (or, equivalently, the flow of the corresponding commodity), the cost can rise by at most $(1-\varrho)F$. Hence, the cost C' of the convex combination of integral solutions is at most

$$C' \leq C + n(1-\varrho)F \leq \mathrm{OPT} + 1/3 + nF/2nF < \mathrm{OPT} + 1 \ . \tag{3}$$

From $C' < \mathrm{OPT} + 1$ we conclude that the convex combination contains at least one integral solution with optimal costs. As the number of possible integral solutions is bounded by Fn^2 (see [1]), an optimal component, i. e. integral solution, within the convex composition can be computed in polynomial time. However, each integral solution originates from a ϱ-solution where a block has size ϱ. Since the cache is still k large, it remains to prove that no integral component of the convex composition does hold more than k blocks in cache concurrently.

Since, w. l. o. g., $k/(nF) < 1$ holds, the number of blocks of size ϱ held concurrently in cache is at most

$$\frac{k}{\varrho} \leq \frac{k}{1 - \frac{1}{2nF}} \leq k\left(1 + \frac{1}{nF}\right) < k + 1 \tag{4}$$

because $(1-\varepsilon')^{-1} \leq 1 + 2\varepsilon'$ for any $\varepsilon' \in [0, 1/2]$. Therefore, $(k+1)\varrho > k$ holds, and we would obtain a contradiction if an integral solution held more than k pages in cache concurrently. Finally, this implies that we have found a feasible and optimal prefetching/caching schedule. The overall running time of the approximation algorithm is $O^*(\varepsilon^{-3}\delta^{-3}c|E||V|^2)$, where c denotes the number of commodities and O^* means "up to logarithmic factors". As $|V| = O(n^2)$ and $|E| = O(n^2)$, we obtain the polynomial upper bound $O^*((nF)^3(n^3F^2)^3(n+1)n^2n^4) = O^*(n^{18})$. Now we state the main result of this section.

Theorem 5. *An optimal solution to an input for IPC can be computed by a combinatorial algorithm in polynomial time.*

2.4 Generalization to Multiple Disks

The solution developed for single disk systems can be generalized to multiple disks. Due to space limitations, we only state the main result here.

Theorem 6. *There is a combinatorial polynomial-time algorithm which computes a D-approximation to an input for IPC if the number of disks is D and there are $D-1$ slots of extra cache available.*

3 Improving the Approximation Factor

In this section we return to the linear program by Albers, Garg and Leonardi [1] for the multiple disk case and improve its approximation factor. If the number D of disks is constant, we achieve a 2-approximation. We know that our approach leads to a linear program which can also be stated as a min-cost multicommodity flow problem. We omit that representation as we consider the improved approximation guarantee to be the most important contribution.

3.1 Bundling Intervals

The drawback of the LP formulation by Albers, Garg and Leonardi [1] is that it overestimates the stall time of prefetching/caching schedules. We present an LP that counts stall time more accurately. As in [1] we represent time periods in which fetch operations are executed by open intervals $I = (i, j)$, with $i = 0, \ldots, n-1$ and $j = i+1, \ldots, n$, where $n = |\sigma|$ is the length of the given request sequence. Such an interval $I = (i, j)$ corresponds to the time period starting *after* the service of $\sigma(i)$ and ending before the service of $\sigma(j)$. Its length is $|I| = j - i - 1$. If $|I| < F$, then $F - |I|$ units of stall time must be scheduled in the fetch operation. Since fetches take F time units, we can restrict ourselves to intervals with $j \leq i + F + 1$. For each potential interval I we introduce a copy I^d for each disk $d \in \{1, \ldots, D\}$. Let $\mathcal{I}$ be the resulting set of all these intervals. The LP in [1] determines which intervals of $\mathcal{I}$ should execute prefetches. Stall times are counted separately for the intervals and disks, which causes the overestimate.

The main idea of our LP is to form *bundles* of intervals and treat each bundle as a unit: In any bundle either all the intervals or no interval will execute a fetch. We next introduce the notion of bundles and need one property of optimal prefetching/caching schedules. An interval $I = (i_1, i_2)$ *properly contains* interval $J = (j_1, j_2)$ (which is not necessarily associated with the same disk) if $i_1 < j_1$ and $j_2 < i_2$ hold. The proof of the next lemma is omitted.

Lemma 7. *An optimal (fractional or integral) prefetching/caching schedule for a system with D disks does not include fetch intervals properly containing each other.*

Definition 8. *A set of intervals B, $|B| \neq \emptyset$, is called a* bundle *if B contains at most one interval from each disk and is* overlapping. *A set of intervals B is* *called overlapping if it includes no intervals properly containing each other but has for all but one* $I = (i_1, i_2) \in B$ *some interval* $J = (j_1, j_2) \in B$, $J \neq I$, *such that* $j_1 \geq i_1$ *and* J *overlaps with* I. *Two intervals* $I = (i_1, i_2)$ *and* $J = (j_1, j_2)$, $i_1 \leq j_1$, *are called overlapping if either* $j_1 < i_2 - 1$ *is valid, or* $j_1 = i_2 - 1$ *and additionally* $i_2 - i_1 - 1 < F$ *hold.*

Fix a $z \in \mathbb{N}$ with $z \leq D$. We will bundle intervals from up to z disks. In this extended abstract we assume for simplicity that $D/z \in \mathbb{N}$. We partition the disk set into D/z sets $\{1, \ldots, z\}, \{z+1, \ldots, 2z\}, \ldots, \{D-z+1, \ldots, D\}$. Now let $\mathcal{B}_z$ be the set of all the bundles composed of intervals from $\mathcal{I}$, with the additional restriction that the intervals of a bundle must come from the same subset of the disk partition. One can show that $|\mathcal{B}_z| \leq n(F+1)^{2z}(D/z)z!$.

We are nearly ready to state the extended linear program for IPC and $D > 1$. For each bundle $B \in \mathcal{B}_z$, we introduce a variable $x(B)$ which is set to 1 if a prefetch is performed in all intervals in bundle B, and is set to 0 otherwise. In order to specify which blocks are fetched and evicted we use variables $f_{I^d,a}$ and $e_{I^d,a}$ for all $I^d \in \mathcal{I}$ and all blocks a. Variable $f_{I^d,a}$ (respectively $e_{I^d,a}$) is equal to 1 if a is fetched (respectively evicted) in I^d. Of course $e_{I^d,a} = f_{I^d,a} = 0$ if a does not reside on disk d. For a bundle $B \in \mathcal{B}_z$, let $s(B)$ be the minimum stall time needed to execute fetches in all the intervals of B assuming that no other fetch operations are performed in the schedule. The value $s(B)$ can be computed as follows. Let $(a_1, b_1), \ldots, (a_m, b_m)$ be the sequence of all intervals in B obtained by sorting them by increasing end index, where intervals with the same end index are sorted by increasing start index breaking ties arbitrarily. One can easily verify that in an optimal schedule for B, stall times occur at the end of intervals, the fetch in (a_1, b_1) is started at the latest point in time (i. e. immediately before request a_2 if $b_1 \neq a_1$ and after a_1 otherwise) whereas the fetches in (a_i, b_i), $i \geq 2$, are started at the earliest point in time. We determine the amounts of stall times needed at the end of intervals. Let $i_1, i_2, \ldots, i_{m'}$ with $m' \leq m$ be the sequence obtained from $b_1, \ldots, b_m$ by eliminating multiple occurrences of the same value and keeping only the indices i_j such that $b_{i_j+1} \neq b_{i_j}$. By definition, $i_0 := 0$ and $b_{i_0} := 0$. For $j = 1, \ldots, m'$, interval (a_{i_j}, b_{i_j}) is the shortest interval with end index b_{i_j} and determines the stall time to be inserted before that request. The function $h\colon \{b_{i_1}, \ldots, b_{i_{m'}}\} \to \mathbb{N}$ that indicates the actual stall time needed before request b_{i_j} is defined inductively, for $j = 1, \ldots, m'$, as follows:

$$h(b_{i_j}) := \max\Bigl\{0, F - (b_{i_j} - a_{i_j} - 1) - \sum_{r \in \{b_{i_1}, \ldots, b_{i_{j-1}}\}\colon r \in \{a_{i_j}+1, \ldots, b_{i_j}-1\}} h(r)\Bigr\} .$$

Using this definition we have $s(B) := \sum_{j=1}^{m'} h(b_{i_j})$.

In order to refer to individial disks, we need for $d \in \{1, \ldots, D\}$ the projections $\pi_d\colon \mathcal{B}_z \to \mathcal{I}$, where $\pi_d(B) = I$ if $I \in B$ and I resides on disk d, and $\pi_d(B) = \emptyset$ if B contains no interval associated with disk d. The value of π_d is well defined since at most one interval from each disk is part of a bundle. Now the **extended**

linear program reads as follows. Minimize the objective function

$$\sum_{B \in \mathcal{B}_z} x(B)s(B) \tag{5}$$

subject to

$$\forall i \in \{1,\ldots,n\}, \forall d \quad \sum_{B \in \mathcal{B}_z:\ \pi_d(B) \supseteq (i-1,i+1)} x(B) \leq 1 \tag{6}$$

$$\forall d, \forall I^d \quad \sum_a f_{I^d,a} = \sum_a e_{I^d,a} \leq \sum_{B \in \mathcal{B}_z:\ \pi_d(B) = I^d} x(B) \tag{7}$$

$$\forall a, \forall i \in \{1,\ldots,n_a\} \quad \sum_{I \in \mathcal{I}:\ I \subseteq (a_i,a_{i+1})} f_{I,a} = \sum_{I \in \mathcal{I}:\ I \subseteq (a_i,a_{i+1})} e_{I,a} \leq 1 \tag{8}$$

$$\forall a \quad \sum_{I \in \mathcal{I}:\ I \subseteq (0,a_1)} f_{I,a} = 1, \qquad \forall a \quad \sum_{I \in \mathcal{I}:\ I \subseteq (0,a_1)} e_{I,a} = 0 \tag{9}$$

$$\forall a, \forall i \in \{1,\ldots,n_a\} \quad \sum_{I \in \mathcal{I}:\ I \supseteq (a_i-1,a_i+1)} f_{I,a} = \sum_{I \in \mathcal{I}:\ I \supseteq (a_i-1,a_i+1)} e_{I,a} = 0 \tag{10}$$

$$\forall I \in \mathcal{I}, \forall a \quad f_{I,a}, e_{I,a} \in \{0,1\} \tag{11}$$

$$\forall B \in \mathcal{B}_z \quad x(B) \in \{0,1\}\ . \tag{12}$$

Here we have taken over some terminology from the original formulation in [1]. The first set of contraints ensures that for each disk and each point of time, the amount of fetch is at most 1. The second set of constraints guarantees for each interval on every disk that the amount of blocks fetched in the interval is at most the overall amount of blocks evicted in that interval. For a specific interval I^d, we allow a bundle variable $x(B)$, where B contains I^d, to take value 1; observe that B might consist of I^d as the only element. If a bundle variable $x(B)$ is 1, the second set of constraints allows fetches in all intervals belonging to the bundle. Please note that constraint (6) only ensures that at most one prefetch operation may be executed while serving a request. Especially, it allows prefetches to be started in the midst of stall times, such the exact point of time where a prefetch is started may be unspecified if there is stall time at the beginning of an interval. We will argue later that this freedom is justified. Constraints (8)–(11) have been adapted from the LP formulation in [1] and ensure that a block is in cache at the time of its reference. The objective function finally counts the s-values, which are related to parallel stall times, for bundles whose variables are 1. It remains to prove that a solution to the extended linear program induces a valid schedule whose stall time is counted at least once by the value of the objective function.

Consider an arbitrary integer solution to the extended LP, which specifies an assignment to the variables $f_{I,a}, e_{I,a}$ and $x(B)$. Using $f_{I,a}$ and $e_{I,a}$, we know between which requests a prefetch operation must be started, but may choose the exact point of time of the start if the related requests are intermitted by stall time. We inductively construct a schedule whose stall time is bounded from above by the value of the objective function. First, we sort the bundles B for which

$x(B) = 1$ holds by increasing maximum end index (of the intervals in the bundle) and, if equality holds, by increasing minimum start index. Let $B_1, \ldots, B_m$ be the resulting sequence. Suppose that we have already constructed a schedule for $B_1, \ldots, B_{r-1}$. For bundle B_r, we have to schedule fetches and evictions for those blocks whose variables $f_{I,a}$ and $e_{I,a}$ have been set to 1 and $I \in B_r$. We use the notation introduced for defining the stall times $s(B)$ on page 20. Let $(a_1, b_1), \ldots, (a_m, b_m)$ be the sequence of intervals in B_r. We first insert $h(b_{i_\ell})$ units of stall time before b_{i_ℓ}, $\ell \in \{1, \ldots, m'\}$, and then schedule the fetches in (a_i, b_i), $i \in \{1, \ldots, m\}$, as follows. The fetch in (a_1, b_1) is started at the latest possible point in time. More precisely, if $a_1 < b_1$, we start the fetch with the service of request $a_1 + 1$; otherwise the fetch is scheduled immediately before the service of b_1. The fetches in (a_i, b_i), $i \geq 2$, are started at the earliest point in time after a_i such that the required disk is available. The definition of the h-values ensures that we reserve at least F time units for each fetch irrespectively of stall times which are caused by fetches in intervals from bundles $B_1, \ldots, B_{r-1}$. In fact, a reserved time interval might even be longer. However, this is no problem. The fetch simply completes after F time units and the corresponding disk is then idle for the rest of the interval. Thus, the constructed schedule is feasible and we insert exactly $\sum_{j=1}^{m} s(B_j)$ units of stall time.

3.2 Achieving the $(2D/z)$-Approximation

Lemma 9. *The extended linear program for D disks has an integral solution of cost at most $(2D/z)\,\mathrm{OPT}$.*

Proof. Suppose we have been given an optimal integral prefetching/caching schedule of stall time OPT. We restrict ourselves to an arbitrary subset of the partition

$$\bigcup_{i=0}^{D/z-1} \{iz+1, iz+2, \ldots, iz+z\} \tag{13}$$

of the disk set $\{1, \ldots, D\}$. W.l.o.g., this subset is $\{1, \ldots, z\}$. We consider only stall times caused by fetches in intervals associated with disks $\{1, \ldots, z\}$. In the following, we specify an assignment to the variables associated with disks $\{1, \ldots, z\}$ such that the stall time that arises by executing only the fetches on disks $\{1, \ldots, z\}$ is counted at most twice in the objective funtion of the linear program. Repeating this process for all the subsets of the above partition, we obtain an assignment to all the variables $x(B)$ for $B \in \mathcal{B}_z$. As the objective function is separable with respect to the bundles in $\mathcal{B}_z$ and therefore with respect to the (D/z) subsets of the above partition, we count a specific stall time in the optimal prefetching/caching schedule at most $2(D/z)$ times.

By $\mathcal{I}' \subseteq \mathcal{I}$ we denote the set of all intervals associated with the disk set $\{1, \ldots, z\}$ in which prefetches are performed. According to Lemma 7, we have no intervals properly containing each other in the set $\mathcal{I}'$. Therefore, we can order the intervals in $\mathcal{I}'$ by increasing start points and (if these are equal) by increasing end points. Let $I_1, \ldots, I_m$ be the resulting sequence. We partition $\mathcal{I}'$ into bundles according to the following greedy algorithm.

$B := \emptyset$
for $j = 1, \ldots, m$ **do**
  if interval $I_j \cup B$ is a bundle **then** set $B := I_j \cup B$
  else output B as an element of the partition and set $B := \emptyset$.

Let $B_1 \cup \cdots \cup B_\ell$, $\ell \leq m$, be the partition of $\mathcal{I}'$ obtained by this process. Our solution to the linear program is constructed by setting $x(B_j)$ to 1 for $j \in \{1, \ldots, \ell\}$. The variables $f_{I,a}$ and $e_{I,a}$ are set according to which blocks are fetched in the intervals of the considered bundle. This process is repeated for each subset of the partition (13) of the disk set. All remaining variables are zero.

In the following, we revert to the subset $\{1, \ldots, z\}$ and denote by OPT* the stall time incurred if counting only fetch intervals associated with disks $\{1, \ldots, z\}$. For bundles B_j, $j \in \{1, \ldots, \ell\}$, let $s'(B_j)$ be the sum of the stall times in the optimal schedule between the start of the first fetch in B_j and the completion of the last fetch in B_j. Obviously, $s(B_j) \leq s'(B_j)$. One can show that $\sum_{j=1}^{\ell} s'(B_j) \leq 2 \cdot \mathrm{OPT}^*$, which implies that the value of the objective function is at most $2\,\mathrm{OPT}$. □

Lemma 9 implies that there is also a fractional solution to the extended LP with cost at most $(2D/z)\,\mathrm{OPT}$. Using techniques from [1], we can convert a fractional solution to the extended LP with cost C to an integral prefetching/caching schedule with stall time C if $D-1$ extra memory locations in cache are available. Thus we obtain an approximation algorithm of factor $2D/z$.

Theorem 10. *There is a polynomial $p(n)$ such that for each $z \in \{1, \ldots, D\}$ with $D/z \in \mathbb{N}$ there is a algorithm with running time $O(p(n) \cdot n \cdot (F+1)^z z!)$ that computes a $2D/z$-approximation for IPC provided that $D-1$ extra memory locations are available in cache.*

Finally, for $D = 2$ the extended linear program does not seem to give an improved approximation. However, in this special case, we can show an even better approximation factor of 1.5.

Lemma 11. *If $D = 2$, the extended linear program with $z = 2$ has an integral solution of cost at most $1.5 \cdot \mathrm{OPT}$.*

References

1. S. Albers, N. Garg, and S. Leonardi. Minimizing stall time in single and parallel disk systems. In *Proc. 30th Annual ACM Symp. on Theory of Computing*, pages 454–462, 1998.
2. B. Bershad, P. Cao, E. W. Felten, G. A. Gibson, A. R. Karlin, T. Kimbrel, K. Li, R. H. Patterson, and A. Tomkins. A trace-driven comparison of algorithms for parallel prefetching and caching. In *Proc. ACM SIGOPS/USENIX Assoc. Symp. on Operating System Design and Implementation (OSDI)*, 1996.
3. P. Cao, E. W. Felten, A. R. Karlin, and K. Li. A study of integrated prefetching and caching strategies. In *Proc. ACM Int. Conf. on Measurement and Modeling of Computer Systems (SIGMETRICS)*, pages 188–196, 1995.

4. P. Cao, E. W. Felten, A. R. Karlin, and K. Li. Implementation and performance of integrated application-controlled caching, prefetching and disk scheduling. *ACM Transaction on Computer Systems*, 14(4):311–343, 1996.
5. G. A. Gibson, E. Ginting, R. H. Patterson, D. Stodolsky, and J. Zelenka. Informed prefetching and caching. In *Proc. 17th Int. Conf. on Operating Systems Principles*, pages 79–95, 1995.
6. A. Kamath, O. Palmon, and S. Plotkin. Fast approximation algorithm for minimum cost multicommodity flow. In *Proc. 6th Annual ACM-SIAM Symp. on Discrete Algorithms*, pages 493–501, 1995.
7. R. Karlin and T. Kimbrel. Near-optimal parallel prefetching and caching. In *Proc. 37th Annual Symp. on Foundations of Computer Science*, pages 540–549. IEEE Society, 1996.

On the Equivalence between the Primal-Dual Schema and the Local-Ratio Technique

Reuven Bar-Yehuda and Dror Rawitz

Department of Computer Science, Technion, Haifa 32000, Israel
{reuven,rawitz}@cs.technion.ac.il

Abstract. We discuss two approximation approaches, the primal-dual schema and the local-ratio technique. We present two relatively simple frameworks, one for each approach, which extend known frameworks for covering problems. We show that the two are equivalent, and conclude that the integrality gap of an integer program serves as a bound to the approximation ratio when working with the local-ratio technique.

1 Introduction

The *primal-dual method* for solving linear programs was originally proposed by Dantzig, Ford, and Fulkerson [10] for solving linear programs. Over the years this method has become an important tool for solving combinatorial optimization problems. Many algorithms which solve combinatorial optimization problems such as Dijkstra's *shortest path* algorithm [11], Ford and Fulkerson's *network flow* (or *minimum* s,t*-cut*) algorithm [13], and Kuhn's assignment algorithm [20] either use the method or can be analyzed by it (see, e.g., [15] for more details).

Combinatorial optimization problems can often be formulated as integer programs. However, unlike the above mentioned problems, the LP-relaxation of many problems have non-integral optimal solutions (and an *integrality gap* which is greater than 1), therefore, the primal-dual method cannot be used to solve them. The *primal dual schema* for approximation is a modified version of the primal-dual method (see [15]). While both complementary slackness conditions are imposed in the classical setting, we impose the primal conditions but relax the dual conditions when working with the primal-dual schema. A primal-dual approximation algorithm constructs an approximate primal solution and a feasible dual solution simultaneously. The approximation ratio is derived from comparing the values of both solutions. Many LP-based algorithms can be analyzed by the primal-dual schema. However, the first algorithm using this schema is Bar-Yehuda and Even's [4] approximation algorithm for the *vertex cover* problem (and the *set cover* problem). Following the work of Goemans and Williamson [14], Bertsimas and Teo [8] proposed a primal-dual framework to design and analyze approximation algorithms for integer programming problems of the covering type. As in [14] this framework enforces the primal complementary slackness conditions while relaxing the dual conditions. A detailed survey on the primal dual schema was written by Goemans and Williamson [15].

M. Goemans et al. (Eds.): APPROX-RANDOM 2001, LNCS 2129, pp. 24–36, 2001.

Recently, Jain and Vazirani [19] presented a primal-dual approximation algorithm which uses a different approach from the one described here. Instead of the "usual" mechanism of relaxing the complementary slackness conditions, they relaxed the primal conditions. As far as we know, this is the only primal-dual algorithm which works with programs that have negative coefficients.

The local-ratio approach was developed by Bar-Yehuda and Even [5] in order to approximate the *weighted vertex cover* problem. Ten years later Bafna, et al. [1] extended the *Local-Ratio Theorem* from [5] by introducing the idea of minimal covers in order to construct a 2-approximation algorithm for the *feedback vertex set* problem. Later, Bar-Yehuda [3] presented a unified local-ratio approach for developing and analyzing approximation algorithms for covering problems. He further extended the Bafna et al. [1] extension of the Local-Ratio Theorem. His extension yields a generic r-approximation algorithm, which can explain most known optimization and approximation algorithms for covering problems. Lately, Bar-Noy et al. [2] have presented an approximation framework for solving resource allocation and scheduling problems. Their algorithms are based on the local-ratio technique, but can be interpreted within the primal-dual schema. This study was the first to present a primal-dual or a local-ratio approximation algorithm for a natural maximization problem.

We present a local-ratio framework which generalizes the one from [3], in which only non-negative weight subtractions were considered. We found this limitation to be redundant, and, therefore, our generic algorithm can be used to design and analyze a wider family of algorithms. We also present a primal-dual approximation framework which extends the generic algorithm from [8]. As done before, our framework relaxes the dual complementary slackness conditions. However, the use of valid inequalities with non-negative coefficients enables us to widen the variety of algorithms which fall with in the scope of our framework. Furthermore, due to [2], we are able to extend our frameworks to handle maximization problems. (This extension will appear in the full version.) By showing that primal-dual's valid inequalities and local-ratio's weight functions are equivalent notions we show that both frameworks are actually one and the same. A corollary to this equivalence is that the *integrality gap* of an integer program serves as a bound to the approximation ratio of a local-ratio algorithm (as in the case of a primal-dual algorithm). Also, note that in cases where a primal-dual algorithm serves as a subroutine (e.g., multyphase primal-dual [21], see also [8]) it can be replaced by the corresponding local-ratio algorithm. We demonstrate the combined approach on a variety of problems. First, we present a linear time approximation algorithm for the *generalized hitting set* problem. This algorithm achieves a ratio of 2 in the special case of the *generalized vertex cover* problem, and is better in time complexity than Hochbaum's [18] $O(nm \log \frac{n^2}{m})$ 2-approximation algorithm for this special case. We exhibit the first local-ratio algorithm which uses negative weights by presenting an analysis to Ford and Fulkerson's [13] algorithm for minimum s,t-cut. Furthermore, in order to explain this algorithm within the primal-dual schema, we use a linear program with inequalities which have some negative coefficients. We also analyze 2-approximation algorithms for the *minimum feedback vertex set* problem [7], and for a non-covering problem called the *minimum 2-satisfiability* problem [16,6]. Finally, we explain a $\frac{1}{2}$-approximation algorithm for a maximization problem called the *interval scheduling* problem [2]. This example provides some insight to the design and analysis of approximation algorithms for maximization problems, and to the equivalence between the two approaches in the maximization case.

We believe that this study contributes to the understanding of both approaches, and, especially, that it will help in the design of algorithms for maximization problems. Another point of interest is that while the use of a linear objective function and linear advancement steps are an integral part of both techniques, our frameworks are not limited to linear integer programs (though we did not find a natural non-linear application).

2 Definitions

We consider the following problem: Given a non-negative *penalty* (or *profit*) vector $p \in \mathbb{R}^n$, find a solution x that minimizes (or maximizes) the inner product $p \cdot x$ subject to some set $\mathcal{F}$ of feasibility constraints (not necessarily linear) on $x \in \{0,1\}^n$. Note that we will sometimes abuse the notation by treating a vector $x \in \{0,1\}^n$ as the set of its 1 coordinates, i.e., as $\{j : x_j = 1\}$. The correct interpretation should be clear from the context.

A minimization (maximization) problem $(\mathcal{F}, p)$ is called a *covering* (*packing*) problem if $\mathcal{F}$ is monotone, i.e., for all x' such that $x' \geq x$ ($x' \leq x$) if x satisfies $\mathcal{F}$ then x' satisfies $\mathcal{F}$. Note that a monotone set of linear constraints typically includes inequalities with non-negative coefficients.

Primal-dual approximation algorithms for covering problems traditionally reduce the size of the problem at hand in each iteration by adding an element whose corresponding dual constraint is tight to the primal solution (see [15,8]). Local-ratio algorithms for covering problems implicitly add all zero penalty elements to the solution, and, therefore, reduce the size of the problem in each step as well (see [3]). In order to implement this we alter the problem definition by adding a set (or vector), denoted by z, which includes elements which are considered to be taken into the solution. This makes it easier to present primal-dual algorithms recursively, and to present local-ratio algorithms in which the addition of zero penalty elements to the partial solution is explicit.

More formally, given a minimization (maximization) problem $(\mathcal{F}, p)$ and a vector z, we are interested in the following problem: Find a vector x such that (1) $z \cap x = \emptyset$; (2) $x \cup z$ satisfies $\mathcal{F}$ (x in the maximization case); And, (3) minimizes (maximizes) the inner product $p \cdot x$. (Note that when $z = \emptyset$ we get the original problem.) We define the following for a minimization problem $(\mathcal{F}, p, z)$. A vector x is called a *feasible solution* if $z \cap x = \emptyset$ and $z \cup x$ satisfies $\mathcal{F}$. We denote the set of feasible solutions with respect to $\mathcal{F}$ and z by $S(\mathcal{F}, z)$. A feasible solution x^* is called an *optimal solution* if every feasible solution x satisfies $p \cdot x^* \leq p \cdot x$. A feasible solution x is called an *r-approximation* if $p \cdot x \leq r \cdot p \cdot x^*$, where x^* is an optimal solution. The corresponding definitions for a maximization problem can be understood in a strait-forward manner. Given a minimization (maximization) problem $(\mathcal{F}, p, z)$, a feasible solution x is called *minimal* (*maximal*) if for all $j \in x$ $(j \notin x \cup z)$ the vector $x \setminus \{j\}$ $(x \cup \{j\})$ is not feasible.

3 Local-Ratio

The following is the Local-Ratio Theorem given in our terminology.

Theorem 1 ([2]). *Let $\mathcal{F}$ be a set of constraints, and let z be a boolean vector. Also, let p, p_1, and p_2 be penalty (or profit) vectors such that $p = p_1 + p_2$. Then, if x is an r-approximation with respect to $(\mathcal{F}, p_1, z)$ and $(\mathcal{F}, p_2, z)$, then x is an r-approximation with respect to $(\mathcal{F}, p, z)$.*

Note that $\mathcal{F}$ can include arbitrary feasibility constraints and not just linear, or linear integer, constraints. Nevertheless, all successful applications of the local ratio technique to date involve problems in which the constraints are linear.

Bar-Yehuda [3] defined the following for covering problems:

Definition 1. *Given a covering problem $(\mathcal{F}, p)$, a weight function δ is called* r-effective *with respect to $\mathcal{F}$, if every minimal solution x with respect to $(\mathcal{F}, \delta)$ satisfies $\delta x \leq r \cdot \delta x^*$, where x^* is an optimal solution.*

We prefer the following more practical definition:

Definition 2. *Given a minimization (maximization) problem $(\mathcal{F}, p, z)$, a weight function δ is called* r-effective with respect to $(\mathcal{F}, z)$, *if $\forall j \in z, \delta_j = 0$, and there exists β such that every minimal (maximal) solution x with respect to $(\mathcal{F}, z)$ satisfies: $\beta \leq \delta \cdot x \leq r\beta$ $(r\beta \leq \delta \cdot x \leq \beta)$. In this case we say that β is a* witness *to δ's r-effectiveness. If $z = \emptyset$ we say that δ is* r-effective with respect to $\mathcal{F}$.

Obviously, by assigning $\beta = \delta x^*$, where x^* is an optimal solution, we get that the first definition implies the latter. For the other direction, note that $\beta \leq \delta x^*$.

A local-ratio algorithm for a covering problem works as follows. First, construct an r-effective weight function δ, such that $p - \delta \geq 0$ and there exists some j for which $p_j = \delta_j$. Such a weight function is called *p-tight*. Subtract δ from the penalty vector p. Add all zero penalty elements to the partial solution z. Then, recursively solve the problem with respect to $(\mathcal{F}, p - \delta, z)$. When the empty set becomes feasible (or, when z becomes feasible with respect to $\mathcal{F}$) the recursion terminates. Finally, remove unnecessary elements from the temporary solution by performing a reverse deletion phase. Algorithm LRcov is a modified version of the generic local-ratio approximation algorithm for covering problems from [3]. (The initial call is LRcov$(p, \emptyset)$.) The main difference between the algorithm from [3] and the one given here is that in the latter the augmentation of the temporary solution is done one element at a time. By doing this we have the option not to include zero penalty elements which do not contribute to the feasibility of the partial solution z.

Algorithm LRcov(p, z)

1. If $\emptyset \in S(\mathcal{F}, z)$ return $\emptyset$
2. Construct a p-tight weight function δ which is r-effective w.r.t. $(\mathcal{F}, z)$
3. Let $j \notin z$ be an index for which $\delta_j = p_j$
4. $x \leftarrow \text{LRcov}(p - \delta, z \cup \{j\})$
5. If $x \notin S(\mathcal{F}, z)$ then $x \leftarrow x \cup \{j\}$
6. Return x

Theorem 2. *Algorithm LRcov outputs an r-approximate solution.*

Proof. Let p^k, δ^k and z^k be p, δ, and z, respectively, at the end of the k'th level, and let t be the maximal depth of the recursion. We prove by reverse induction on k that Algorithm LRcov returns an minimal r-approximation with respect to $(\mathcal{F}, p^k, z^k)$. For $k = t$, $\emptyset$ is a minimal optimal solution. Otherwise, examine x at the end of the k'th level. By the induction hypothesis $x \setminus \{j\}$ is an minimal r-approximation with respect to $(\mathcal{F}, p^{k+1}, z^{k+1})$. Therefore, x is a minimal solution with respect to $(\mathcal{F}, z^k)$. Moreover, due to the fact that $p_j^{k+1} = 0$ x is an r-approximation with respect to $(\mathcal{F}, p^{k+1}, z^k)$. By the r-effectiveness of δ^k and Theorem 1 x is an r-approximate solution with respect to $(\mathcal{F}, p^k, z^k)$.

Algorithm LRcov finds an r-effective weight function for the current problem, adds a zero penalty element to the solution, and then changes the problem. Such an algorithm heavily relies upon the fact that the constraint set $\mathcal{F}$ is monotone, and, therefore, the optimum of $(\mathcal{F}, p^{k+1}, z^{k+1})$ is not greater than the optimum of $(\mathcal{F}, p^{k+1}, z^k)$. Algorithm LRmin uses a slightly different approach. Instead of using r-effective weight functions with respect to $(\mathcal{F}, z)$, it uses weight functions which are r-effective with respect to $\mathcal{F}$, and updates p until a minimal zero-penalty solution is found. By using this "subtract first, ask questions later" approach we are able to approximate problems by using inequalities with some negative coefficients or weight functions with some negative weights. We illustrate this approach in the Sect. 6.

Algorithm LRmin(p)

1. Search for a feasible minimal solution $x \subseteq \{j : p_j = 0\}$
2. If such a solution was found return x
3. Construct a p-tight weight function δ which is r-effective w.r.t. $\mathcal{F}$
4. $x \leftarrow \text{LRmin}(p - \delta)$
5. Return x

Theorem 3. *Algorithm LRmin outputs an r-approximate solution.*

4 Primal-Dual

In [15] Goemans and Williamson presented a generic algorithm based on the *hitting set* problem which is defined as follows: Given subsets $T_1, \ldots, T_q \subseteq E$

and a non-negative cost c_e for every element $e \in E$, find a minimum-cost subset $x \subseteq E$ such that $x \cap T_i \neq \emptyset$ for every $i \in \{1, \ldots, q\}$. In turns out that many known problems are special cases of this problem. The LP-relaxation of its formulation as an integer program and the corresponding dual are:

$$\begin{array}{lll} \min \sum_{e \in E} c_e x_e & & \\ \text{s.t. } \sum_{e \in T_i} x_e \geq 1 & \forall i \in \{1, \ldots, q\} \\ \quad x_e \geq 0 & \forall e \in E \end{array} \qquad \begin{array}{lll} \max \sum_{i=1}^{q} y_i & & \\ \text{s.t. } \sum_{i: e \in T_i} y_i \leq c_e & \forall e \in E \\ \quad y_i \geq 0 & \forall i \in \{1, \ldots, q\} \end{array}$$

where $x_e = 1$ iff $e \in x$. The primal complementary slackness conditions are: $x_e > 0 \Longrightarrow \sum_{i: e \in T_i} y_i = c_e$, and the dual conditions are: $y_i > 0 \Longrightarrow |x \cap T_i| = 1$.

The algorithm from [15] starts with the dual solution $y = 0$ and the non-feasible primal solution $x = \emptyset$. Then it iteratively increases the both solutions until the primal solution becomes feasible. As in the primal-dual method, if we cannot find a feasible primal solution, then there is a way to increase the dual solution. In this case, if x is not feasible there exists a k such that $x \cap T_k = \emptyset$. Such a set is referred to as *violated*. In each iteration a *violation oracle* supplies a collection of violated subsets $\mathcal{V} \subseteq \{T_1, \ldots, T_q\}$[1]. Then we increase *simultaneously and at the same speed* the dual variables corresponding to subsets in $\mathcal{V}$. When x becomes feasible a *reverse delete step* is performed. This step discards as many elements as possible from the primal solution x without losing feasibility.

Let x^f and ϵ_j be the set output by the algorithm, and the increase of the dual variables corresponding to $\mathcal{V}_j$ in iteration j. Then, $y_i = \sum_{j: T_i \in \mathcal{V}_j} \epsilon_j$, $\sum_{i=1}^{p} y_i = \sum_{j=1}^{l} |\mathcal{V}_j| \epsilon_j$, and $c(x^f) = \sum_{e \in x_f} c_e = \sum_{i=1}^{q} |x^f \cap T_i| y_i = \sum_{i=1}^{q} |x^f \cap T_i| \sum_{j: T_i \in \mathcal{V}_j} \epsilon_j = \sum_{j=1}^{l} \epsilon_j \sum_{T_i \in \mathcal{V}_j} |x^f \cap T_i|$. From this, it is clear that the cost of x^f is at most the value of the dual solution times r if $\forall j \in \{1, \ldots, l\}$ $\sum_{T_i \in \mathcal{V}_j} |x^f \cap T_i| \leq r |\mathcal{V}_j|$. Examine iteration j of the reverse deletion step. We know that when e_j was considered for removal, no element $e_{j'}$ with $j' < j$ has been already removed. Thus, after e_j is considered for removal the temporary solution is $x^j = x^f \cup \{e_1, \ldots, e_{j-1}\}$. x^j is called a *minimal augmentation* of $\{e_1, \ldots, e_{j-1}\}$, i.e., x^j is feasible and $x^j \setminus \{e\}$ is not feasible for all $e \in x^j \setminus \{e_1, \ldots, e_{j-1}\}$. Moreover, $\sum_{T_i \in \mathcal{V}_j} |x^f \cap T_i| \leq \sum_{T_i \in \mathcal{V}_j} |x^j \cap T_i|$. Therefore, to achieve the above bound Goemans and Williamson have set the following requirement: $\sum_{T_i \in \mathcal{V}_j} |x' \cap T_i| \leq r |\mathcal{V}_j|$ for every $j \in \{1, \ldots, l\}$ and for every minimal augmentation x' of $\{e_1, \ldots, e_{j-1}\}$. We formalized this by the following.

Definition 3. *A set* $\mathcal{V} \subseteq \{T_1, \ldots, T_q\}$ *is called* r-effective with respect to $(\mathcal{F}, z)$, *if* $\sum_{T_i \in \mathcal{V}} |x \cap T_i| \leq r |\mathcal{V}|$ *for every minimal solution with respect to* $(\mathcal{F}, z)$, x.

As did Bertsimas and Teo [8] we prefer to speak in terms of inequalities.

Definition 4. *A set of valid inequalities* $\{\alpha^1 x \geq \beta^1, \ldots, \alpha^k x \geq \beta^k\}$ *(or* $\leq$ *in the maximization case) is called* r-effective with respect to $(\mathcal{F}, p, z)$, *if* $\alpha_j^k = 0$ *for all* k *and* $j \in z$, *and every integral minimal feasible solution with respect to*

[1] We could allow the oracle to return some sets which are not violated. See [15].

$(\mathcal{F}, z)$, x, satisfies: $\sum_{i=1}^{k} \alpha^i x \leq r \sum_{i=1}^{k} \beta^i$. ($\geq$ in the maximization case.) If the set contains a single inequality we will refer to this inequality as r-effective.

An r-effective collection $\mathcal{V}$ can be understood as the following r-effective set of valid inequalities: $\left\{\sum_{e \in T_i} x_e \geq 1 : T_i \in \mathcal{V}\right\}$. On the other hand, the latter definition allows the use of other kinds of inequalities (e.g., Sect. 6). Thus, our goal is to find a set of r-effective valid inequalities in order to increase the dual solution. However, it is enough to construct a single r-effective valid inequality for that purpose. This is because a set of valid inequalities $\left\{\alpha^i x \geq \beta^i\right\}_{i=1}^{k}$ satisfies $\sum_{i=1}^{k} \alpha^i x \leq r \sum_{i=1}^{k} \beta^i$ by definition. On the other hand, $\forall i, \alpha^i x \geq \beta^i$, and, therefore, $\sum_{i=1}^{k} \alpha^i x \geq \sum_{i=1}^{k} \beta^i$. Thus, we have found an r-effective inequality: $\sum_{j=1}^{n} (\sum_{i=1}^{k} \alpha^i)_j x_j \geq \sum_{i=1}^{k} \beta^i$.

Bertsimas and Teo [8] presented a primal-dual algorithm for linear covering problems which utilizes a single valid inequality in each iteration. This algorithm constructs a valid inequality in each iteration, and uses it to modify the current instance. After this modification at least one of the coordinates of the penalty vector p becomes zero, and this makes it possible to reduce the size of the problem in each iteration. Thus, the algorithm terminates after no more than n iterations. The performance of this algorithm depends on the choice of the inequalities. In fact, it corresponds to what Bertsimas and Teo call the *strength* of an inequality, which is the minimal r for which it is r-effective.

Algorithm PDcov is a recursive version of the algorithm from [8]. (The initial call is PDcov$(p, \emptyset, 1)$.) Informally, it can be viewed as follows: construct an r-effective inequality; update the corresponding dual variable and the vector p such that p remains non-negative; find an element j whose penalty is zero; add j to the temporary partial solution z; then recursively solve the problem with respect to $\mathcal{F}, z$ and the new penalty vector; the termination condition of the recursion is met when the empty set becomes a feasible solution; finally, a reverse deletion phase removes unnecessary elements from the temporary solution. There are two differences between the algorithm from [8] and our algorithm. First, we present our algorithm recursively, and, more importantly, our algorithm is not restricted to linear integer programs.

Algorithm PDcov(p, z, k)

1. If $\emptyset \in S(\mathcal{F}, z)$ return $\emptyset$
2. Construct a valid inequality $\alpha^k x \geq \beta^k$ which is r-effective w.r.t. $(\mathcal{F}, z)$
3. $y_k \leftarrow \max\left\{\epsilon : p - \epsilon\alpha^k \geq 0\right\}$
4. Let $j \notin z$ be an index for which $p_j = y_k \alpha_j^k$
5. $x \leftarrow$ PDcov$(p - y_k\alpha^k, z \cup \{j\}, k+1)$
6. If $x \notin S(\mathcal{F}, z)$ then $x \leftarrow x \cup \{j\}$
7. Return x

The following theorem is based on the corresponding theorem from [8].

Theorem 4. *Algorithm PDcov outputs an r-approximate solution.*

Proof. Denote the maximum depth of the recursion by t, and let z^k be the temporary partial solution of depth k. Examine the k'th inequality $\alpha^k x \geq \beta^k$ for some k. This is a valid inequality with respect to $(\mathcal{F}, z^k)$, and, therefore, it is valid with respect to $\mathcal{F}$. Thus, LP2 $= \min \{px : \forall k \in \{1, \ldots, t\}, \alpha^k x \geq \beta^k \text{ and } x \geq 0\}$ is a relaxation of $(\mathcal{F}, p)$. By the construction of x, it is a feasible solution for $\mathcal{F}$, and, therefore, for LP2. Also, y is a feasible solution for the dual of LP2.

Let x^k be the value returned by the k'th call to the recursion, let p^k be the penalty vector of depth k, and let $j(k)$ be the chosen element in the k'th call. We prove by induction that $p^k x^k \leq r \sum_{l \geq k} y_l b^l$. First, for $k = t$, we have $p^t x^t = 0$. For $k < t$ we have,

$$p^k x^k = (p^{k+1} + y_k \alpha^k) x^k \overset{(1)}{=} p^{k+1} x^{k+1} + y_k \alpha^k x^k \overset{(2)}{\leq} r \sum_{l \geq k+1} y_l \beta^l + y_k r \beta^k = r \sum_{l \geq k} y_l \beta^l$$

where (1) is because $p^{k+1}_{j(k)} = 0$, and (2) is implied by the induction hypothesis and the r-effectiveness of the inequality. Finally, $px = p^1 x^1 \leq r \sum_{l \geq 1} y_l \beta^l \leq r \cdot \text{Opt}(\text{LP2}) \leq r \cdot \text{Opt}(\mathcal{F}, p)$ and, therefore, x is an r-approximate solution.

Algorithm PDmin uses inequalities which are r-effective with respect to $\mathcal{F}$, instead of using inequalities which are r-effective with respect to $(\mathcal{F}, z)$ (as in the local-ratio case). It uses such inequalities, while updating p, and the dual solution, until a minimal zero-penalty solution is found.

Algorithm PDmin(p)

1. Search for a feasible minimal solution $x \subseteq \{j : p_j = 0\}$
2. If such a solution was found return x
3. Construct a valid inequality $a^l x \geq b_l$ which is r-effective w.r.t. to $\mathcal{F}$
4. $y_l \leftarrow \max \{\epsilon : p - \epsilon a^l \geq 0\}$
5. $x \leftarrow$ PDmin2$(p - y_l a^l)$
6. Return x

Theorem 5. *Algorithm PDmin outputs an r-approximate solution.*

5 Equivalence

Lemma 1. *$\alpha x \geq \beta$ is an r-effective inequality iff α is an r-effective weight function with the witness β.*

Proof. We prove the lemma for minimization problems. Similar arguments can be used in the maximization case. Let $\alpha x \geq \beta$ be an r-effective inequality. By definition every minimal feasible solution x satisfies: $\beta \leq \alpha x \leq r\beta$. Thus, α is an r-effective weight function. On the other hand, let α be an r-effective weight function with a witness β. Due to the r-effectiveness of α every minimal feasible solution x satisfies $\beta \leq \alpha x \leq r\beta$. Therefore, $\alpha x \geq \beta$ is an r-effective inequality.

By Lemma 1 the use of valid inequalities can be explained by utilizing the corresponding weight functions and vice versa. Thus, Algorithms PDcov and

LRcov are actually identical. The same goes for Algorithms PDmin and LRmin. From this we can also conclude that the integrality gap of an integer program serves as a bound to the approximation ratio of a local-ratio algorithm which can be explained by algorithms LRcov or LRmin.

6 Examples

Generalized Hitting Set. The *generalized hitting set* problem is defined as follows. Given a collection of subsets S of a ground set U, a non-negative penalty p_s for every set $s \in S$, and a non-negative penalty p_u for every element $u \in U$, find a minimum-penalty collection of objects $C \subseteq U \cup S$, such that for all $s \in S$, either there exists $u \in C$ such that $u \in s$, or $s \in C$. As in the *hitting set* problem our objective is to cover the sets in S by using elements from U. However, in this case, we are allowed not to cover a set s, provided we pay a tax p_s. The hitting set problem is the special case where the tax is infinite for all sets. The generalized hitting set problem can be formalized as follows:

$$\begin{array}{lll} \min & \sum_{u \in U} p_u x_u + \sum_{s \in S} p_s x_s & \\ \text{s.t.} & \sum_{u \in s} x_u + x_s \geq 1 & \forall s \in S \\ & x_t \in \{0,1\} & \forall t \in U \cup S \end{array}$$

Observe that paying the penalty p_s is required only when s is not covered. Thus, $\sum_{u \in s} x_u + x_s \geq 1$ is Δ_S-effective for any set $s \in S$, where $\Delta_S = \max\{|s| : s \in S\}$. The corresponding Δ_S-effective weight function would be: $\delta_s(t) = \epsilon$ if $t \in \{s\} \cup s$, and $\delta_s(t) = 0$ otherwise. The inequality remains Δ_S-effective if we use any value between 1 and Δ_S as x_s's coefficient. Analogously, $\delta_s(s)$ can take any value between ϵ and $\Delta_S \epsilon$. We can approximate this problem by using the "standard" approach for covering problems (when using the above inequalities or weight functions). However, we can also use a PDmin (or LRmin) to construct a linear time Δ_S-approximation algorithm. First, we use all the above inequalities in an arbitrary order; then a zero penalty minimal feasible solution can be found: all zero penalty elements and all the sets which are not covered by some zero penalty element. This would be a Δ_S-approximate solution.

Feedback Vertex Set. Let $G = (V, E)$ be an undirected graph, and let p be a penalty vector. The *feedback vertex set* problem is to find a minimum penalty set $F \subseteq V$ whose removal from G leaves an acyclic graph. Becker and Geiger [7] and Chudak et al. [9] proved that $\sum_{v \in V} \deg(v) \leq 2 \sum_{v \in F^*} \deg(v)$ for every minimal feedback vertex set F, where F^* is an optimal solution. Bar-Yehuda [3] indicated that this actually means that the weight function $\delta(v) = \deg(v)$ is 2-effective, and then used this weight function to simplify the presentation of Becker and Geiger's [7] 2-approximation algorithm. The corresponding 2-effective inequality is $\sum_{v \in V} \deg(v) x_v \geq \sum_{v \in F^*} \deg(v)$. Therefore, a primal-dual analysis to this algorithm can be given by using algorithm PDcov. It is important to note that you do not need to know the value $\sum_{v \in F^*} \deg(v)$ in order to execute the algorithm.

Minimum s,t-cut. Given an undirected graph $G = (V, E)$, a pair of vertices s, t, and a non-negative penalty for each edge $e \in E$, we wish to find a cut of minimum penalty. In order to solve this problem we solve the directed version: given a digraph $G = (V, E)$, a pair of vertices s and t, and a non-negative penalty p_e for each arc $e \in E$, find a directed cut of minimum penalty. Denote by Q the set of all (directed) paths from s to t. The minimum s,t-cut problem can be formalized by using the following integer program:

$$\begin{array}{ll} \min \sum_{e \in E} p_e x_e & \\ \text{s.t. } \sum_{e \in q} x_e \geq 1 & \forall q \in Q \\ \quad x_e \in \{0,1\} & \forall e \in E \end{array}$$

where $x_e = 1$ iff e is in the cut.

We analyze Ford and Fulkerson algorithm [13]. Examine the inequality: $\sum_{e \in q} x_e - \sum_{e \in \overline{q}} x_e \geq 1$ for some path $q \in Q$, where $\overline{q}$ is the corresponding path from t to s (i.e., the one going in the opposite direction). The corresponding weight function is: $\delta_q(e) = (|e \cap q| - |e \cap \overline{q}|) \cdot \epsilon$ for some ϵ. This means that arcs in the 'right' direction are given penalty ϵ, while arcs in the 'wrong' direction are given penalty $-\epsilon$. We show that δ_q and its corresponding inequality are 1-effective. Let F be some minimal feasible solution, i.e, F is a minimal set of arcs which induces an s,t-cut $(S, \overline{S})$. Any path going from s to t must go through F an odd number of times: $k \geq 1$ times from S to $\overline{S}$ and $k - 1$ times in the opposite direction. Therefore, $\delta_q(F) = \sum_{e \in F} \delta_e = k \cdot \epsilon + (k-1) \cdot (-\epsilon) = \epsilon$. The algorithm works as follows: iteratively find an unsaturated path q from s to t and subtract δ_q from the p, where ϵ is the value that saturates some arc $e \in q$; when no such path exists pick a minimal set of arcs from the set $\{e : p_e = 0\}$. Note that, in this case, we are allowed to change our minds about an arc, i.e., to increase the penalty of an arc during the execution of the algorithm.

Minimum 2SAT. Given a 2CNF formula φ with m clauses on the variables $x_1, \ldots, x_n$ and a penalty vector p, the *minimum weight 2-satisfiability* problem is to find a minimum penalty truth assignment which satisfies φ (or determine that no such assignment exists). Gusfield and Pitt [16] presented an $O(nm)$ time 2-approximation algorithm for 2SAT. We shall give an intuitive explanation of this algorithm in the context of the approximation frameworks. First, we can check whether φ is satisfiable by using the algorithm from [12]. Therefore, we may assume that φ is satisfiable. In order to construct a 2-approximation algorithm we need to find 2-effective inequalities or weight functions. Given a literal ℓ, let $T(\ell)$ be the set of variables which must be assigned TRUE whenever ℓ is assigned TRUE. Let x_i and x_j be variables such that $x_j \in T(\overline{x_i})$. For such variables the inequality $x_i + x_j \geq 1$ is valid. Moreover, it is easy to see that this inequality is 2-effective. Constructing $T(x_i)$, and a zero penalty feasible solution can be done efficiently by using constraint propagation (for a more detailed analysis see [6]).

Interval Scheduling. Bar-Noy et al. [2] have presented a framework for approximating resource allocation and scheduling problems. We analyze one of the algorithms from [2] which approximates a packing problem called the *interval scheduling* problem. This algorithm does not fall within the frameworks discussed earlier, but the analysis offers some intuition to the equivalence between the two approaches in the maximization case. An interesting point is that while a penalty vector is kept non-negative during the execution of an algorithm, a profit vector is expected to be non-positive when the subtraction phase is over. This means, in primal-dual terms, that in the maximization case the dual solution is initially not feasible, and that it becomes feasible at the end of the algorithm. The negative coordinates of the profit vector correspond to the non-tight constraints.

In the *interval scheduling* problem we are given a set of *activities*, each requiring the utilization of a given *resource*. The activities are specified as a collection of sets $\mathcal{A}_1, \ldots, \mathcal{A}_m$. Each set represents a single activity: it consists of all possible *instances* of that activity. An instance $I \in \mathcal{A}_i$ is defined by the following parameters: (1) A half-open time interval $[s(I), e(I))$ during which the activity will be executed. $s(I)$ and $e(I)$ are called the *start-time* and *end-time* of the instance; And, (2) the profit $p(I) \geq 0$ gained by scheduling this instance of the activity. A *schedule* is a collection of instances. It is feasible if it contains at most one instance of every activity, and at most one instance for all time instants t. In the interval scheduling problem our goal is to find a schedule that maximizes the total profit accrued by instances in the schedule. More formally, our goal is to find an optimal solution to the following integer program on the boolean variables $\{x_I : I \in \mathcal{A}_i, 1 \leq i \leq m\}$.

$$\begin{array}{lll} \max & \sum_I p_I x_I & \\ \text{s.t.} & \sum_{I: s(I) \leq t < e(I)} x_I \leq 1 & \forall t \\ & \sum_{I: I \in \mathcal{A}_i} x_I \leq 1 & \forall i \in \{1, \ldots, m\} \\ & x_I \in \{0, 1\} & \forall i \; \forall I \in A_i, \end{array}$$

The algorithm from [2] works as follows. First, delete all instances with non-positive profit. If no instances remain, return the empty schedule. Otherwise, let J be an instance with minimum end-time, and let $\mathcal{A}(J)$ and $\mathcal{I}(J)$ be the activity to which instance J belongs and the set of instances intersecting J (including J), respectively. Define $\delta(I) = p(J)$ if $I \in \mathcal{A}(J) \cup \mathcal{I}(J)$, and $\delta(I) = 0$ otherwise. Then solve the problem recursively with respect to a new profit vector, $p - \delta$. Let S' be the schedule returned. If $S' \cup \{J\}$ is a feasible solution return $S = S' \cup \{J\}$. Otherwise, return $S = S'$. δ is $\frac{1}{2}$-effective, and, therefore, this is a $\frac{1}{2}$-approximation algorithm. Bar-Noy et al. also presented a primal-dual interpretation to their algorithm. However, in order to do so they modified the original algorithm by using a different $\frac{1}{2}$-effective weight function: $\delta'(J) = p(J)$, $\delta'(I) = \frac{1}{2}p(J)$ if $I \in \mathcal{A}(J) \cup \mathcal{I}(J) \setminus \{J\}$, and $\delta'(I) = 0$ otherwise. We present the following alternative analysis. Consider the inequality $\sum_{I \in \mathcal{I}(J)} x_I + \sum_{I \in A(J)} x_I \leq 2$, which states that only one instance from $\mathcal{A}(J)$ and one from $\mathcal{I}(J)$ can belong to a feasible schedule. This inequality is $\frac{1}{2}$-effective because a maximal schedule contains at least one instance from $\mathcal{A}(J) \cup \mathcal{I}(J)$. The original algorithm can be explained by the $\frac{1}{2}$-effective inequality $\sum_{I \in \mathcal{I}(J) \cup \mathcal{A}(J)} x_I \leq 2$. The difference between δ and δ' (or between their corresponding inequalities) is the ratio between the weight of $\tilde{I}$ and the weights of the other instances. In fact, any value between 1 and 2 would have been acceptable.

Acknowledgments
We thank Eran Mann and especially Ari Freund for helpful discussions.

References

1. V. Bafna, P. Berman, and T. Fujito. A 2-approximation algorithm for the undirected feedback vertex set problem. *SIAM J. on Disc. Math.*, 12(3):289–297, 1999.
2. A. Bar-Noy, R. Bar-Yehuda, A. Freund, J. Naor, and B. Shieber. A unified approach to approximating resource allocation and scheduling. In *32nd ACM Symposium on the Theory of Computing*, 2000.
3. R. Bar-Yehuda. One for the price of two: A unified approach for approximating covering problems. *Algorithmica*, 27(2):131–144, 2000.
4. R. Bar-Yehuda and S. Even. A linear time approximation algorithm for the weighted vertex cover problem. *Journal of Algorithms*, 2:198–203, 1981.
5. R. Bar-Yehuda and S. Even. A local-ratio theorem for approximating the weighted vertex cover problem. *Annals of Discrete Mathematics*, 25:27–46, 1985.
6. R. Bar-Yehuda and D. Rawitz. Efficient algorithms for bounded integer programs with two variables per constraint. *Algorithmica*, 29(4):595–609, 2001.
7. A. Becker and D. Geiger. Approximation algorithms for the loop cutset problem. In *10th Conference on Uncertainty in Artificial Intelligence*, pages 60–68, 1994.
8. D. Bertsimas and C. Teo. From valid inequalities to heuristics: A unified view of primal-dual approximation algorithms in covering problems. *Operations Research*, 46(4):503–514, 1998.
9. F. A. Chudak, M. X. Goemans, D. S. Hochbaum, and D. P. Williamson. A primal-dual interpretation of recent 2-approximation algorithms for the feedback vertex set problem in undirected graphs. *Operations Research Letters*, 22:111–118, 1998.
10. G. B. Dantzig, L. R. Ford, and D. R. Fulkerson. A primal-dual algorithm for linear programs. In H. W. Kuhn and A. W. Tucker, editors, *Linear Inequalities and Related Systems*, pages 171–181. Princeton University Press, 1956.
11. E. W. Dijkstra. A note on two problems in connexion with graphs. *Numerische Mathematik*, 1:269–271, 1959.
12. S. Even, A. Itai, and A. Shamir. On the complexity of timetable and multicommodity flow problems. *SIAM J. on Comp.*, 5(4):691–703, 1976.
13. L. R. Ford and D. R. Fulkerson. Maximal flow through a network. *Canadian Journal of Mathematics*, 8:399–404, 1956.
14. M. X. Goemans and D. P. Williamson. A general approximation technique for constrained forest problems. *SIAM J. on Comp.*, 24(2):296–317, 1995.
15. M. X. Goemans and D. P. Williamson. The primal-dual method for approximation algorithms and its application to network design problems. In Hochbaum [17], chapter 4.
16. D. Gusfield and L. Pitt. A bounded approximation for the minimum cost 2-SAT problem. *Algorithmica*, 8:103–117, 1992.
17. D. S. Hochbaum, editor. *Approximation Algorithms for NP-Hard Problem*. PWS Publishing Company, 1997.
18. D. S. Hochbaum. A framework of half integrality and good approximations. Manuscript, UC berkeley, 1997.

19. K. Jain and V. V. Vazirani. Primal-dual approximation algorithms for metric facility location and k-median problems. In *40th IEEE Symposium on Foundations of Computer Science*, pages 2–13, 1999.
20. H. W. Kuhn. The Hungarian method of solving the assignment problem. *Naval Research Logistics Quarterly*, 2:83–97, 1955.
21. D. P. Williamson, M. X. Goemans, M. Mihail, and V. V. Vazirani. A primal-dual approximation algorithm for generalized Steiner network problems. *Combinatorica*, 15:435–454, 1995.

Online Weighted Flow Time and Deadline Scheduling*,**

Luca Becchetti[1], Stefano Leonardi[1], Alberto Marchetti-Spaccamela[1], and Kirk R. Pruhs[2]

[1] Dipartimento di Informatica e Sistemistica, University of Rome "La Sapienza"
{becchett,leon,marchetti}@dis.uniroma1.it
[2] Computer Science Department, University of Pittsburgh
kirk@cs.pitt.edu

Abstract. In this paper we study some aspects of weighted flow time on parallel machines. We first show that the online algorithm Highest Density First is an $O(1)$-speed $O(1)$-approximation algorithm for $P|r_i, pmtn|\sum w_i F_i$. We then consider a related Deadline Scheduling Problem that involves minimizing the weight of the jobs unfinished by some unknown deadline D on a uniprocessor. We show that any c-competitive online algorithm for weighted flow time must also be c-competitive for Deadline Scheduling. We finally give an $O(1)$-competitive algorithm for Deadline Scheduling.

1 Introduction

We consider several aspects of online scheduling to minimize the weighted flow time. In this problem a sequence of jobs has to be processed on a set of m identical machines. The ith job has a release time r_i, a processing time or length x_i and a non-negative weight w_i. At any time, only one processor may be running any particular job. Processors may preempt one job to run another. The ith job is completed after it has been run for x_i time units. The flow time F_i of the ith job is the difference between its completion time and its release date. The objective function to minimize is the weighted flow time $\sum w_i \cdot F_i$. Following the standard three fields notation for scheduling problems, we denote this problem by $P|r_i, pmtn|\sum w_i \cdot F_i$.

One motivation for studying weighted flow time is that by far the most commonly used measure of system performance is average flow time, or equivalently

* L. Becchetti, S. Leonardi and A. Marchetti-Spaccamela were partially supported by the IST Programme of the EU under contract number IST-1999-14186 (ALCOM-FT), IST-2000-14084 (APPOL), IST-1999-10440 (Brahms) and by the MURST Projects "Algorithms for Large Data Sets: Science and Engineering" and "Resource Allocation in Computer Networks".

** K. Pruhs was supported in part by NSF grant CCR-9734927 and by a grant from the US Airforce.

M. Goemans et al. (Eds.): APPROX-RANDOM 2001, LNCS 2129, pp. 36–47, 2001.

average user perceived latency, and weighted flow time is an obvious generalization. Another motivation for considering weighted flow time is that it is a special case of broadcast scheduling. A good overview of broadcast scheduling can be found in [1]. The setting is a client-server system where the server uses broadcast communication to download files to the clients. One notable commercial example of such a system is Hughes' DirecPC system; In this system the clients request web pages over phone lines, and the web pages are broadcasted via satellite to all clients (so it may be possible to satisfy several clients request with a single broadcast). The weighted flow time problem $1|r_i, pmtn|\sum w_i \cdot F_i$ is a special case of the broadcast problem where all requests for file i arrive at a single time r_i (however, there may be many requests for i at time r_i).

There have been some recent results on minimizing average flow time [7,8,9]; however, minimizing weighted flow time is not yet well understood. The uniprocessor offline problem $1|r_i, pmtn|\sum w_i \cdot F_i$ is known to be NP-hard [4]. But there is known no offline polynomial-time constant approximation algorithm. The one previous positive result for weighted flow time used resource augmentation analysis. Resource augmentation analysis was proposed [8] as a method for analyzing scheduling algorithms that are hard to approximate. Using the notation and terminology of [11], an *s-speed c-approximation* algorithm A has the property that the value of the objective function of the schedule that A produces with processors of speed $s \geq 1$ is at most c times the optimal value of the objective function for speed 1 processors. An algorithm is c-competitive if is an 1-speed c-approximation algorithm. It was shown in [11] that an LP-based online algorithm, Preemptively-Schedule-Halves-by-$\bar{M}_j$, is an $O(1)$-speed $O(1)$-approximation algorithm.

The first contribution of this paper, covered in Section 2, is that we simplify the resource augmentation analysis of $1|r_i, pmtn|\sum w_i F_i$ given in [11], while also extending the analysis to multiprocessors. Namely, we show that the polynomial time, online algorithm Highest Density First (HDF), which always runs the job that maximizes its weight divided by its length, is an online $O(1)$-speed $O(1)$-approximation algorithm. While this analysis of HDF is not stated in [11], upon reflection one can see that all the insights necessary to assert that HDF is an $O(1)$-speed $O(1)$-approximation algorithm on a single processor are inherent in their analysis of their proposed algorithm Preemptively-Schedule-Halves-by-$\bar{M}_j$. Besides making this result explicit, our proof has the advantages that (1) our analysis is from first principles and does not require understanding of a rather complicated LP lower bound, and (2) our analysis also easily extends to the multiprocessor problem $P|r_i, pmtn|\sum w_i F_i$, while the result in [11] apparently does not.

In section 3 we consider competitive analysis of $1|r_i, pmtn|\sum w_i F_i$. We first show that every c-competitive online algorithm A has to be *locally c-competitive*, that is, at every time t, the overall weight of the jobs that A has not finished by time t can be at most c times the minimum possible weight of the unfinished jobs at time t. The requirement of local competitiveness suggests that we consider what we call the Deadline Scheduling Problem (DSP).

The input to DSP consists of n jobs, with each job i having a length x_i and a non-negative weight w_i. Both the length of a job and its weight are revealed to the algorithm when the job is released. The goal is to construct a schedule (linear order) where the overall weight of the jobs unfinished by some initially unknown deadline D is minimized. In the standard three field scheduling notation, one might denote this problem as $1|d_i = D|\sum w_i \cdot U_i$. The requirement of local competitiveness means that any c-competitive online algorithm for $1|r_i, pmtn|\sum w_i F_i$ must be c-competitive for DSP. The DSP problem can be seen as a dual to the deadline scheduling problem considered in [6]. The setting considered in [6] was the same as DSP, except that the goal was to maximize the jobs completed before the deadline D on multiple machines. In [6] it is shown that no constant competitive algorithm for the maximization problem exixts. The second main contribution of this paper is an online $O(1)$-competitive algorithm for DSP on one machine.

2 Resource Augmentation Analysis of $P|r_i, pmtn|\sum w_i F_i$

In this section, we adopt the following notation. We use $A(s)$ to denote the schedule produced by algorithm A with a speed s processors, and abuse notation slightly by using A to denote $A(1)$. Let $U^A(t)$ be the jobs that have been released before time t, but not finished by algorithm A by time t. We define the *density* of a job j as the ratio $\mu_j = w_j/p_j$. We use $y_j^A(t)$ to denote the remaining unprocessed length of job j at time t according to the schedule produced by algorithm A. Similarly, we use $p_j^A(t) = p_j - y_j^A(t)$ to denote the processed length of job j before time t. So from time t, algorithm A, with a speed s processor, could finish j in $y_j^A(t)/s$ time units. We let $w_j^A(t) = y_j^A(t)\frac{w_j}{p_j}$ be the fractional remaining weight of job j at time t in A's schedule. We denote by $W^A(t) = \sum_{j \in U^A(t)} w_j$ and by $F^A(t) = \sum_{j \in U^A(t)} w_j^A(t)$ respectively the overall weight and the overall fractional weight of jobs that A has not completed by time t.

It is well known that the weighted flow time of a schedule A is equal to $\int W^A(t)dt$. Hence to prove that an algorithm A is a c-approximation algorithm for weighted flow time it is sufficient to show that A is locally c-competitive, this meaning that, at any time t, $W^A(t) \leq cW^{\mathrm{Opt}}(t)$. Observe that, in principle, A might be a c-approximation algorithm without being locally c-competitive. We show further in this section that this in fact is not the case for the problem we are considering.

Recall that, given m processors, the algorithm Highest Density First (HDF) always runs the up to m densest available jobs. We now prove, via a simple exchange argument, that giving HDF a faster processor doesn't decrease the time that HDF runs any unfinished job by any time.

Lemma 1. *For any number of processors, for any job instance, for any time t, and for any $j \in U^{\mathrm{HDF}(1+\epsilon)}(t)$, it is the case that $p_j^{\mathrm{HDF}(1+\epsilon)}(t) \geq (1+\epsilon)p_j^{\mathrm{HDF}(1)}(t)$, or equivalalently, before time t it is the case that* HDF*$(1+\epsilon)$ has run job j for more time than* HDF*(1).*

Proof. Assume to reach a contradiction that there is a time t and a job j where this does not hold. Further, assume that t is the first such time where this fails to hold. Note that obviously $t > 0$. HDF(1) must be running j at time t by the definition of t. Also by our assumptions, HDF$(1+\epsilon)$ can not have finished j before time t, and does not run j at time t. Hence, HDF$(1+\epsilon)$ must be running some job h at time t, that HDF(1) is not running at time t. HDF$(1+\epsilon)$ could not be idling any processor at time t since j was available to be run, but wasn't selected. Since HDF$(1+\epsilon)$ is running job h instead of job j, job h must be denser than job j. Then the only reason that HDF(1) wouldn't be running h is if it finished h before time t. But this contradicts our assumption of the minimality of t.

We now prove that $F^{\mathrm{HDF}}(t)$ is a lower bound to $W^{\mathrm{Opt}}(t)$ in the uniprocessor setting. (As for this issue we also refer to [3].) Define $W^{\mathrm{FHDF}}(t) = \sum_{j \in U^{\mathrm{HDF}(1)}(t)} w_j^{\mathrm{HDF}(1)}(t)$ to be the remaining fractional weight at time t for HDF(1). So if HDF(1) has run 1/3 of a job j by time t then j only contributes $\frac{2}{3}w_j$ to $W^{\mathrm{FHDF}}(t)$.

Lemma 2. *For any instance of the problem* $1|r_i, pmtn|\sum w_i F_i$, *and any time* t, *it is the case that* $F^{\mathrm{HDF}}(t) \leq F^{\mathrm{Opt}}(t) \leq W^{\mathrm{Opt}}(t)$.

Proof. The second inequality is clearly true. To prove $F^{\mathrm{HDF}}(t) \leq F^{\mathrm{Opt}}(t)$ we use a simple exchange argument. To reach a contradiction, let Opt be an optimal schedule that agrees with the schedule HDF(1) the longest. That is, assume that Opt and HDF(1) schedule the same job at every time strictly before some time t, and that no other optimal schedule agrees with HDF(1) for a longer initial period. Let i be the job that HDF(1) is running at time t and let j be the job that Opt is running at time t. The portion of job i that HDF(1) is running at time t must be run later by Opt at some time, say s. Now create a new schedule Opt$'$ from Opt by swapping the jobs run at time s and time t, that is Opt$'$ runs job i at time t, and job j at time s. Note that job i has been released by time t since it is run at that time in the schedule HDF(1). By the definition of HDF, it must be the case that $\mu_i \geq \mu_j$. Hence, since we are considering the fractional remaining weight of the jobs, it follows that for all times u, $F^{\mathrm{Opt}'}(u) \leq F^{\mathrm{Opt}}(u)$, and Opt$'$ is also an optimal solution. This is a contradiction to the assumption that Opt was the optimal solution that agreed with HDF(1) the longest.

Observe that the above Lemma is not true for parallel machines since HDF may underutilize some of the machines with respect to the optimum.

We now establish that HDF$(1+\epsilon)$ is locally $(1+\frac{1}{\epsilon})$-competitive against HDFin the uniprocessor setting.

Lemma 3. *For any* $\epsilon > 0$, *for any job instance, for any time* t, *and for any job* j, *it is the case that* $W^{\mathrm{HDF}(1+\epsilon)}(t) \leq (1+\frac{1}{\epsilon})F^{\mathrm{HDF}}(t)$

Proof. It follows from lemma 1 that the only way that $W^{\mathrm{HDF}(1+\epsilon)}(t)$ can be larger than $F^{\mathrm{HDF}}(t)$ is due to the contributions of jobs that were run but not completed by both HDF$(1+\epsilon)$ and HDF. Let j be such a job. Since HDFcan not have processed j for more than $p_j^{\mathrm{HDF}(1+\epsilon)}(t)/(1+\epsilon)$ time units before time t,

$$w_j^{\mathrm{HDF}(1)}(t) \geq \frac{1}{p_j}\left(p_j - \frac{p_j^{\mathrm{HDF}(1+\epsilon)}(t)}{1+\epsilon}\right) w_j.$$

The ratio $w_j/w_j^{\mathrm{HDF}(1)}(t)$ is then at most

$$\frac{p_j}{\left(p_j - \frac{p_j^{\mathrm{HDF}(1+\epsilon)}(t)}{1+\epsilon}\right)}$$

When $p_j^{\mathrm{HDF}(1+\epsilon)}(t) = p_j$, which is where this ratio is maximized, this ratio is $1 + 1/\epsilon$. Thus the result follows.

We are then able to conclude the following.

Theorem 1. *Highest Density First (*HDF*) is a $(1+\epsilon)$-speed $(1+\frac{1}{\epsilon})$-approximation algorithm for the problem $1|r_i, pmtn|\sum w_i F_i$.*

Proof. If follows from lemma 2 that $F^{\mathrm{HDF}}(t)$ is a lower bound to $W^{\mathrm{Opt}}(t)$. From Lemma 3 the claim follows.

Theorem 2. *Highest Density First (*HDF*) is a $(2+2\epsilon)$-speed $(1+\epsilon)$-approximation online algorithm for $P|r_i, pmtn|\sum w_i F_i$.*

Proof. Omitted.

3 Deadline Scheduling Problem

In this secton we first show that a c-competitive algorithm for minimizing weighted flow time must be also locally *c-competitive.* We know that local c-competitiveness is sufficient to establish c-competitiveness. We now show that for online algorithms, with speed 1 processors, it is the case that local c-competitiveness is also necessary.

Theorem 3. *Every c-competitive deterministic online algorithm A for $1|r_i, pmtn|\sum w_i F_i$ must be locally c-competitive.*

Proof. Omitted.

The negative result shown above motivates the DSP problem: the goal is to design an algorithm that, at each time instant t, completes a set of jobs whose overall weight is a constant factor away from the set of jobs completed by an optimal algorithm *that knows* t. We stress that this is equivalent to consider a single deadline t *unknown* to the algorithm but known to the off-line optimal adversary. The simplification with respect to minimizing weighted flow time is that in DSP all jobs are released at time 0 and that we consider a simplified objective function.

Number of Jobs	weight	length	Density = weight/length
1	k	k^2	$1/k$
k^3	1	1	1

Fig. 1. A non-trivial input for DSP. $k \gg 1$.

We now give an example to show that DSP is more subtle than one might at first think. Consider the following set of jobs, released at time 0:

Two natural schedules are: (1) First run the low density job followed by the k^3 high density jobs, and (2) First run the k^3 high density jobs followed by the low density job. It is easy to see that the first algorithm is not constant competitive at time $t = k^3$, and that the second algorithm is not constant competitive at time $t = k^3 + k^2 - 1$. In fact, what the online algorithm should do in this instance is to first run $k^3 - k$ of the high density jobs, then run the low density job, and then finish with the remaining k high density jobs. A little reflection reveals that this schedule is $O(1)$-competitive for this instance of DSP. This instance demonstrates that the scheduler has to balance between delaying low density jobs, and delaying low weight jobs. Further in this section we propose an $O(1)$-competitive algorithm for DSP. This algorithm has to recursively deal with difficulties similar to those that arise in the instance in table 3. It is possible to construct instances where one can show that essentially the full complexity of our algorithm description is necessary to achieve $O(1)$-competitiveness.

An alternative way to look at DSP is by reversing the time axis. This makes DSP equivalent to the problem minimizing the overall weight of jobs that have been started by some unknown deadline. This problem is more formally stated below.

NDSP Problem Statement: The input consists of n jobs released at time 0, with known processing times and weights, plus an a priori unknown deadline t. The goal is to find a schedule that minimizes the total weight of jobs that have been *started* before time t including the job running at time t. Intuitively, time t in this schedule corresponds to time $\sum p_i - t$ in the original schedule.

In this section, we use the notation $A(t)$ to refer to the schedule produced by algorithm A before time t, and $W^A(t)$ to refer to the weight of the jobs started by algorithm A before time t. (This matches the use of $W^A(t)$ done in Section 2 to denote the weight not completed by time t in the original time direction.) Similarly, $W^{\mathrm{Opt}}(t)$ refers to the minimum weight collection of jobs with length at least t, and the schedule that obtains this minimum. Note that in general there is not a single schedule that optimizes all t's simultaneously, that is, the schedules $\mathrm{Opt}(t)$ and $\mathrm{Opt}(t')$ may be very different. For a fixed instance of NDSP, the competitive ratio of an algorithm A is the maximum over all $t \in [0, \sum p_j]$ of $W^A(t)/W^{\mathrm{Opt}}(t)$.

Observe that NDSP is simply the minimum knapsack problem, for which an FPTAS exists [13].

We first describe a polynomial time offline algorithm Off for NDSP, and then show that Off is 3-approximate with respect to $\mathrm{Opt}(t)$. We then propose a polynomial time online algorithm called R for NDSP, and show that R is 8-competitive with respect to Off. Hence, we can conclude that R is a 24-competitive algorithm. Both R and Off at various points in their execution must select the lowest density job from some collection; Ties may be broken arbitrarily, but in order to simplify our analysis, we assume that Off and R break ties similarly (say by the original numbering of the jobs).

Description of Algorithm Off: Algorithm Off computes a set of busy schedules $\text{Off}_1(t), \ldots, \text{Off}_u(t)$. The schedule produced by Off is then the schedule among $\text{Off}_1(t), \ldots, \text{Off}_u(t)$ with minimum weight. When $\text{Off}(t)$ is started, $\mathcal{J}$ is the collection of all jobs, and in general $\mathcal{J}$ contains the jobs that maight be scheduled in the schedules $\text{Off}_h(t), \text{Off}_{h+1}(t), \ldots$. the algorithm will produce by the end of its execution. Variable s is initialized to $s = 0$, and in general indicates is the time that the last job in the current schedule ends. Variable h is initially set to 1, and in general is the index for the current schedule $\text{Off}_h(t)$ that $\text{Off}(t)$ is constructing.

1. Let j be the least dense job in $\mathcal{J}$
2. Remove j from $\mathcal{J}$
3. If $s + p_j < t$ then
   (a) Append j to the end of the schedule $\text{Off}_h(t)$
   (b) Let $s = s + p_j$
   (c) Go to step 1
4. Else if $s + p_j \geq t$ then
   (a) Let $\text{Off}_{h+1}(t) = \text{Off}_h(t)$. **Comment:** Initialize $\text{Off}_{h+1}(t)$.
   (b) Append j to the end of the schedule $\text{Off}_h(t)$. **Comment**: Schedule $\text{Off}_h(t)$ is now completed.
   (c) Let $h = h + 1$
   (d) Remove from $\mathcal{J}$ any job with weight greater than $w_j/2$
   (e) If the total length of the jobs in $\mathcal{J}$ is at least $t - s$ then go to step 1.

Algorithm Off intuitively tries to schedule the least dense jobs until time t. The algorithm Off must then be concerned with the possibility that the job that it has scheduled at time t, say j, is of too high a weight. So Off recursively tries to construct a lower aggregate weight schedule to replace j, from those unscheduled jobs with weight at most $w_j/2$. Observe that algorithm Off runs in polynomial time.

If a job j is scheduled in step 3a then we say that j is a *non-crossing job*. If a job j is scheduled in step 4b then we say that j is a *crossing job*, and that afterwards a Schedule $\text{Off}_h(t)$ *was closed on* j.

Theorem 4. $W^{\text{Off}}(t) \leq 3W^{\text{Opt}}(t)$.

Proof. The jobs in the schedule Off are ordered in increasing order of density. Assume that the jobs in $\text{Opt}(t)$ are ordered by increasing density as well. Let us introduce some notation. Let $f_1, \ldots, f_u$ be the sequence of jobs that close

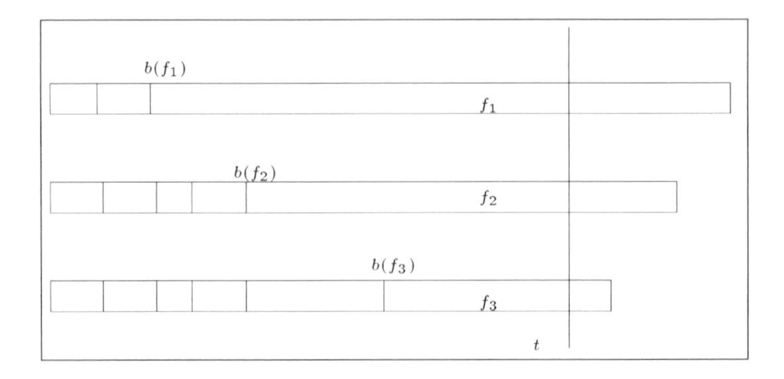

Fig. 2. A depiction of our notation with $k = 3$.

schedules $\mathrm{Off}_1(t), ..., \mathrm{Off}_u(t)$. Let $b(f_h)$ be the time at which j_h is started in $\mathrm{Off}_h(t)$. This notation is illustrated in figure 2.

We break the proof into two cases. In the first case we assume there is some schedule $\mathrm{Off}_h(t)$ that runs at some time instant a job of density higher than the job run at the same time by $\mathrm{Opt}(t)$. Denote by x the earliest such time instant. Since jobs scheduled by Opt are ordered by increasing density, we assume w.l.o.g. that $\mathrm{Off}_h(t)$ starts such job at time x. Denote by j the job started by $\mathrm{Off}_h(t)$ at time x and by i the job run by Opt at time x. Clearly, $\mu_j > \mu_i$. It is easy to see that $\mathrm{Off}_1(t)$ is always running a job with density at most the density of the job that $\mathrm{Opt}(t)$ is running at that time. Hence, it must be the case that $h > 1$. Since $i \neq j$, there must be some job g run by $\mathrm{Opt}(t)$ before i, that $\mathrm{Off}_h(t)$ does not run before j. Note that it may be the case that $g = i$. Then the only possible reason that $\mathrm{Off}(t)$ didn't select g instead of j, is that g was eliminated in step 4d, before $\mathrm{Off}(t)$ closed a schedule on some f_d, $d \leq h-1$. Hence, $w_g \geq w_{f_{h-1}}/2$, and $w_{f_{h-1}} \leq 2\mathrm{Opt}(t)$ since w_g is scheduled in $\mathrm{Opt}(t)$. By the minimality of x, it must be the case that, at all times before time $b(f_{h-1})$, the density of the job that $\mathrm{Off}_{h-1}(t)$ is running is at most the density of the job that $\mathrm{Opt}(t)$ is running. Hence, the aggregate weight of the jobs in $\mathrm{Off}_{h-1}(t)$, exclusive of f_{h-1}, is at most $\mathrm{Opt}(t)$. We can then conclude that $\mathrm{Off}_{h-1}(t) \leq 3\mathrm{Opt}(t)$, and hence $\mathrm{Off}(t) \leq 3\mathrm{Opt}(t)$ by the definition of Off.

In the second case assume that, at all times in all the schedules $\mathrm{Off}_h(t)$, it is the case that the job that $\mathrm{Off}_h(t)$ is running at this time is at most the density of the job that $\mathrm{Opt}(t)$ is running at this time. Hence, the weight of the jobs in the last schedule $\mathrm{Off}_u(t)$, exclusive of f_u, is at most $\mathrm{Opt}(t)$. Consider the time that $\mathrm{Off}(t)$ adds job f_u to schedule $\mathrm{Off}_u(t)$. The fact that $\mathrm{Off}_u(t)$ is the last schedule produced by $\mathrm{Off}(t)$ means that the total lengths of jobs in $\mathrm{Off}_u(t)$, union the jobs that are $\mathcal{J}$ after the subsequent step 4d is strictly less than t. The jobs in $\mathcal{J}$ at this time, are those jobs with weight at most $w_{f_u}/2$ that were not previously scheduled in $\mathrm{Off}_u(t)$. Hence, the total length of the jobs in the original input with weight at most $w_{f_u}/2$ is strictly less than t. Therefore at least one job in $\mathrm{Opt}(t)$ has weight at least $w_{f_u}/2$. Hence, $w_{f_u} \leq 2\mathrm{Opt}(t)$. Therefore, once again we can conclude that $\mathrm{Off}_u(t) \leq 3\mathrm{Opt}(t)$, and that $\mathrm{Off}(t) \leq 3\mathrm{Opt}(t)$.

We now turn to describing the algorithm R. R is an on-line algorithm that produces a complete schedule of $\mathcal{J}$ without knowing the deadline t. The cost incurred by R will be the total weight of jobs started by time t. Intuitively, the algorithm R first considers the lowest density job, say j. However, R can not immediately schedule j because it may have high weight, which might mean that R wouldn't be competitive if the deadline occurred soon after R began executing j. To protect against this possibility, R tries to recursively schedule jobs with weight at most $w_j/2$, before it schedules j, until the aggregate weight of those jobs scheduled before j exceeds w_j. We now give a more formal description of the algorithm R.

Description of Algorithm R**:** The algorithm R takes as input a collection $\mathcal{J}$ of jobs. Initially, $\mathcal{J}$ is the collection of all jobs, and in general, $\mathcal{J}$ will be those jobs not yet scheduled by previous calls to R. The algorithm R is described below:
While $\mathcal{J} \neq \emptyset$

1. Select a $j \in \mathcal{J}$ of minimum density. Remove j from $\mathcal{J}$.
2. (a) Initialize a collection $\mathcal{I}_j$ to the empty set. $\mathcal{I}_j$ will be the jobs scheduled during step 2.
   (b) Let $\mathcal{J}_j$ be those jobs in $\mathcal{J}$ with weight at most $w_j/2$.
   (c) If $\mathcal{J}_j$ is empty then go to step 3.
   (d) Recursively apply R to $\mathcal{J}_j$. Add the jobs scheduled in this recursive call to $\mathcal{I}_j$. Remove the jobs scheduled in this recursive call from $\mathcal{J}$.
   (e) If the aggregate weight of the jobs in $\mathcal{I}_j$ exceeds w_j then go to step 3, else go to step 2c.
3. Schedule j after all jobs of $\mathcal{I}_j$.

End While

We say a job j was *picked* if it was selected in step 1. If, after j is picked, a recursive call to R is made with input $\mathcal{J}_j$, we say that R *recurses* on j. Notice that like the algorithm Off, the algorithm R also runs in polynomial time.

We now explain how to think of the schedule R as a forest, in the graph theoretic sense. The forest has a vertex for every job in $\mathcal{J}$. For this reason we will henceforth use the terms job and vertex interchangeably. The jobs in the subtree $T(j)$ rooted at job j are j and those jobs scheduled in the loop on step 2 after j was selected in step 1. The ordering of the children is by non-decreasing density that is the order in which R selects and schedules these jobs. The roots of the forest are those vertices selected in step 1 at the top level of the execution of the algorithm. Similarly, the ordering of the roots of the forest is also by non-decreasing density, and in the order in which R selects and schedules these jobs. We adopt the convention that the orderings of trees in the forest, and children of a vertex, are from left to right.

Before proceeding we need to introduce some notation. Given a descendent y of a vertex x in some tree, we denote by $P(x, y)$ the path from x to y in T, x and y inclusive. If y is not a descendent of x, then $P(x, y)$ is the empty path. We

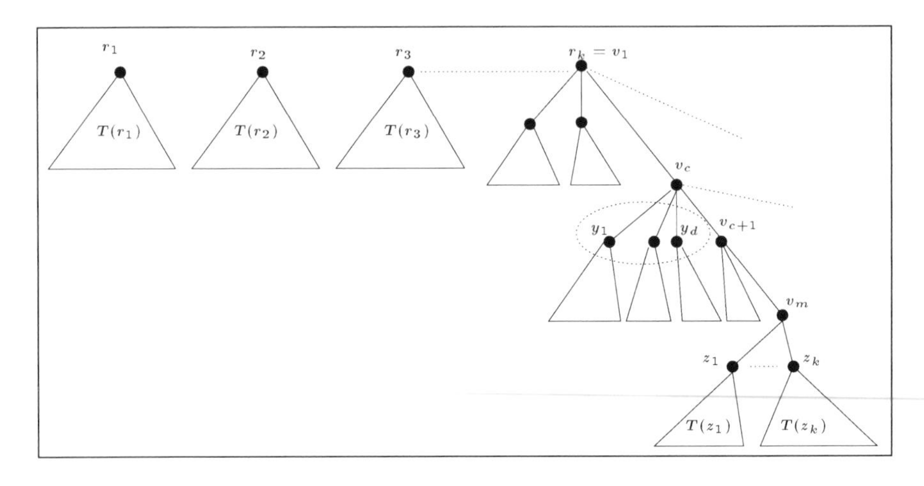

Fig. 3. Forest structure of R's schedule.

denote by $r_1, \dots, r_k$ the roots of the trees in R that contain at least one job that R started before time t. Let $v_1, \dots, v_m$ be the vertices in the path $P(r_k, v_m)$, where $v_1 = r_k$, and where v_m is the job that R is running at time t. Finally, denote by $W(S)$ the total weight of the jobs in a set S. See Figure 3 for an illustration of the forest structure of R and some of our definitions.

The following facts can easily be seen to follow from the definition of R. The post-order of the jobs within each tree is also the order that these jobs are scheduled by R. All jobs in a tree $T(r_h)$, $h < k$, are completed by R before time t. All jobs in a tree $T(r_h)$, $h > k$, are not started by R before time t. The jobs of $T(r_k)$ that are started by R before time t are: (i) The jobs in $T(v_m)$, and (ii) the jobs in $T(\ell)$, for each left sibling ℓ of a job on $P(v_2, v_m)$. All the jobs in the path $P(r_k, v_{m-1})$ are not started by R before time t. All the jobs in a $T(\ell)$, ℓ being a right sibling of a job on the path $P(v_2, v_m)$, are not started by R before time t.

The following lemma shows that the aggregate weight of the vertices in a subtree is not too large.

Lemma 4. *For any job x in R, $W(T(x)) \leq 4w_x$.*

Proof. The proof is by induction on the height of $T(x)$. The base case (i.e. x is a leaf) is straightforward. Let y be x's rightmost child. The aggregate weight of the subtrees rooted at the left siblings of y is at most w_x, or R would not have picked y in tree $T(x)$. Since R removes jobs of weight greater than $w_x/2$ in step 2b, we know that $w_y \leq w_x/2$. By induction $W(T(y)) \leq 4w_y$, and therefore $W(T(y)) \leq 2w_x$. Hence, by adding the weight of the subtrees rooted at x's children to the weight of x, we get that $W(T(x)) \leq w_x + w_x + 2w_x = 4w_x$.

Theorem 5. $R(t) \leq 8\text{Off}(t)$.

Proof. We will charge all the weight of the jobs started by R before time t to the jobs scheduled by Off in a way that no job in Off is charged for more than 8 times its weight.

First notice that Off completes the roots $r_1, \dots, r_{k-1}$ by time t. This is because t occurs in tree $T(r_k)$, all jobs in $T(r_h), h \geq k$ have density not smaller than μ_{r_k}, and therefore the total length of jobs of density less than μ_{r_k} is less than t. We then charge $W(T(r_i))$ to each r_i scheduled by Off, $1 \leq i \leq k-1$, and therefore by Lemma 4 we charge each r_i at most four times its weight.

We are left to charge the weight of the jobs in $T(r_k)$ started by R before time t. We consider three cases:

(i) Off schedules job $r_k = v_1$;
(ii) Off schedules a job on path $P(v_1, v_m)$;
(iii) Off schedules no job on path $P(v_1, v_m)$.

In case (i), we charge $W(T(r_k))$ to r_k. This way r_k is charged at most four times its weight. So now assume that r_k is picked but not scheduled by Off. To finish our analysis of case (ii) and case (iii) we need the following claim, which illuminates the similarities between the algorithms R and Off.

We slightly abuse notation by denoting $r_1, ..., r_{k-1}$ as the left siblings of r_k.

Claim. Assume Off schedules no job on $P(v_1, v_b)$, $b \in \{1, .., m\}$. Then, for every $c = 1, ..., b$, v_c is the crossing job that closes $\text{Off}_c(t)$, and $\text{Off}_c(t)$ contains all left siblings of v_l, $l = 1, .., c$, that are scheduled by R before time t. Moreover, if $b = m$ then Off schedules all the children of v_m.

Proof. We prove a stronger statement: in particular, we prove that for every $c = 1, ..., b$, v_c is the crossing job that closes $\text{Off}_c(t)$, $\text{Off}_c(t)$ contains, for every $l = 1, .., c$, all left siblings of v_l and possibly other jobs, but *only from* the trees rooted at the left siblings of v_l.

The proof is by induction. For the basis of the induction observe that the overall length of jobs of density less than μ_{v_1} is less than t while Off does not schedule v_1, therefore v_1 is the crossing job that closes $\text{Off}_1(t)$, that in turn also contains all the root vertices $r_1, ..., r_{k-1}$ and possibly other vertices, but only from $T(r_1), ..., T(r_{k-1})$. For the inductive hypothesis, we assume that the claim is true up to vertex v_c that closes solution $\text{Off}_c(t)$. We then prove the claim for vertex v_{c+1}. $\text{Off}_c(t)$ was closed on vertex v_c. Solution Off_{c+1} will contain all the jobs of density higher than μ_{v_c} but less than $\mu_{v_{c+1}}$, and weight at most $w_{v_c}/2$. Also these jobs are contained in trees that are rooted for some $l \in \{1, .., c\}$ at some left siblings of v_l. Therefore Off_{c+1} will contain all these jobs including all left siblings of v_{c+1}, since the limit t won't be reached. Therefore v_{c+1} will eventually be selected to be included in $\text{Off}_{c+1}(t)$ but since it is not scheduled by Off it will be the crossing job that closes $\text{Off}_{c+1}(t)$.

Finally, if $b = m$ then by the same argument as above it follows that Off_{m+1} contains all the children of v_m and none of these jobs is a crossing job, therefore they are all included in the final solution provided by Off.

The following corollary immediately follows from Theorems 4 and 5 by reversing the schedule given by R.

Corollary 1. *There exists a 24-competitive algorithm for DSP.*

4 Conclusions

In this paper we considered several aspects of the problem of minimizing the weighted flow time of a set of jobs on identical machines.

We interpret the results in this paper as suggesting that a reasonably simple 1-speed $O(1)$-competitive algorithm for the problem of minimizing the average broadcast time of a set of requests is unlikely, since minimizing the weighted flow time is a particular case of broadcast scheduling to minimize the average flow time.

References

1. S. Aacharya, and S. Muthukrishnan, "Scheduling on-demand broadcasts: new metrics and algorithms", 4th ACM/IEEE International Conference on Mobile Computing and Networking, 43 – 54, 1998.
2. Y. Bartal, and S. Muthukrishnan, "Minimizing Maximum Response Time in Scheduling Broadcasts", 558-559, Proceedings of the Eleventh Annual ACM/SIAM Symposium on Discrete Algorithms, pages 558-559, 2000.
3. M. Goemans, "A Supermodular Relaxation for Scheduling with Release Dates". In *Proceedings of the 5th International Conference on Integer Programming and Combinatorial Optimization (IPCO 96)*, LNCS 1084, Springer, pp. 288-300, 1996.
4. J. Labetoulle, E. Lawler, J.K. Lenstra, A. Rinnooy Kan, "Preemptive scheduling of uniform machines subject to release dates", in W.R. Pulleyblank (eds.), *Progress in Combinatorial Optimization*, 245–261, Academic Press, 1984.
5. E. Lawler, J.K. Lenstra, A. Rinnooy Kan, and D. Shmoys, "Sequencing and Scheduling: Algorithms and Complexity", *Logistics of Production and Inventory*, Handbooks in OR & MS 4, Elsevier Science, Chapter 9, 445–522, 1993.
6. Vincenzo Liberatore, "Scheduling jobs before shut-down", SWAT, 2000.
7. S. Leonardi and D. Raz. Approximating total flow time on parallel machines. In *Proceedings of the Twenty-Ninth Annual ACM Symposium on Theory of Computing*, pages 110–119, El Paso, Texas, 1997.
8. B. Kalyanasundaram, and K. Pruhs, "Speed is as powerful as clairvoyance", *IEEE Symposium on Foundations of Computation*, 214 – 221, 1995.
9. B. Kalyanasundaram, and K. Pruhs. Minimizing Flow Time Nonclairvoyantly. *In Proc. of IEEE Symposium on Foundations of Computer Science (FOCS '97)*, 1997, pages 345-352.
10. B. Kalyanasundaram, K. Pruhs, and M. Velauthapillai, "Scheduling broadcasts in wireless networks", *European Symposium on Algorithms* (ESA), 2000.
11. C. Phillips, C. Stein, E. Torng, and J. Wein "Optimal time-critical scheduling via resource augmentation", *ACM Symposium on Theory of Computing*, 140 – 149, 1997.
12. J. Shanmugasundaram, A. Nithrakashyap, R. Sivasankaran, and K. Ramamritham, "Efficient concurrency control for broadcast environments", Proceedings of the 1999 ACM SIGMOD International Conference on Management of Data (SIGMOD 99), pages 85–96, 1999.
13. G. Ausiello, P. Crescenzi, G. Gambosi, V. Kann and A. Marchetti-Spaccamela. Complexity and Approximation. Springer eds., 1999.

An Online Algorithm for the Postman Problem with a Small Penalty

Piotr Berman and Junichiro Fukuyama

Department of Computer Science and Engineering
The Pennsylvania State University
University Park, PA 16802
{berman,fukuyama}@cse.psu.edu

Abstract. *The Postman Problem* is defined in the following manner: Given a connected graph G and a start point $s \in V(G)$, we call a path J in G a *postman tour* if J starts and ends at s and contains every edge in $E(G)$. We want to find a shortest postman tour in G. Thus, we minimize the number $|J| - |E(G)|$, called the *penalty of J for G*. In this paper, we consider two online versions of The Postman Problem, in which the postman is initially placed at a start point and allowed to move on edges. The postman is given information on the incident edges and/or the adjacent vertices of the current vertex v. The first model is called The Labyrinth Model, where the postman only knows the existence of the incident edges from v. Thus, he/she does not know the other endpoint w of an edge from v, even if it is already visited. The second one is The Corridor Model, where the postman can 'see' the other endpoint w from v. This means that if w is already visited, he/she knows the existence of w before traversing the edge (v, w). We devised an algorithm for The Corridor Model whose penalty is bounded by $2|V(G)| - 2$. It performs better than the algorithm described in [8], in the case when the other visited endpoint of an edge from a current vertex is known to the postman. In addition, we obtained lower bounds of a penalty for The Corridor and Labyrinth Models. They are $(4|V(G)|-5)/3$ and $(5|V(G)|-3)/4$, respectively. Our online algorithm uses a linear time algorithm for the offline Postman Problem with a penalty at most $|V|-1$. Thus, the minimum penalty does not exceed $|V| - 1$ for every connected graph G. This is the first known upper bound of a penalty of an offline postman tour, and matches the obvious lower bound exactly.

1 Introduction

The Postman Problem is defined in the following manner: Given a connected graph G and a start point $s \in V(G)$, we call a path J in G a *postman tour* if J starts and ends at s and contains every edge in $E(G)$. We want to find a shortest postman tour in G. Thus, we minimize the number $|J| - |E(G)|$, called the *penalty of J for G*.

In this paper, we consider two online versions of The Postman Problem. In an online Postman Problem, the postman is initially placed at a start point and

M. Goemans et al. (Eds.): APPROX-RANDOM 2001, LNCS 2129, pp. 48–54, 2001.

allowed to move on edges. The postman is given information on the incident edges and/or the adjacent vertices of the vertex v, at which he/she currently locates.

There are several models worth considering that differ in the information given to the postman at v. We consider the following two of them.
1. The Labyrinth Model
The postman only knows the existence of the incident edges from v. Thus, he/she does not know the other endpoint w of an edge from v, even if it is already visited.
2. The Corridor Model
The postman can 'see' w from v. It means that if w is already visited, he/she knows the existence of w before traversing the edge (v, w).

The offline Postman Problem is known to be polynomially computable [10]. In [8], an algorithm for The Labyrinth Model based on the depth first search scheme (dfs) is given, so that a generated penalty is bounded by $3|V(G)|$. Note that an online algorithm with an $O(|V(G)|)$ penalty is not obviously obtained, while $2|E(G)|$ penalty can be achieved by the dfs algorithm. [8] also mentions that any simple heuristics using a dfs or greedy scheme cannot guarantee an $O(|V(G)|)$ penalty.

Typical applications of the online Postman Problem are the exploration and navigation problems for a robot in an unknown environment. Most of the known results [1,2,3,4,5,6,7,9] make use of geometric restrictions of the environments, not viewing the problems graph-theoretically on arbitrary undirected graphs.

We will present an algorithm for the Corridor Model with a penalty at most $2|V(G)| - 2$. Thus, we will achieve a better performance than [8], in the case when the other visited endpoint of an edge is known to the postman. In addition, we will show the lower bounds $(4|V(G)| - 5)/3$ and $(5|V(G)| - 3)/4$ of a penalty generated by an algorithm for The Corridor and The Labyrinth Models, respectively.

Our algorithm for The Corridor Model uses a linear time algorithm for the offline Postman Problem with a penalty at most $|V(G)| - 1$. Although we already have an exact polynomial time algorithm for the offline problem, nothing is known about a penalty upper bound for an arbitrary connected graph. Our algorithm is faster than the optimal algorithm that runs in super linear time using maximum matching. It should be noted that the upper bound $|V(G)| - 1$ equals the obvious lower bound, since any postman tour on a tree has size at least $2|E(G)|$.

The rest of this paper consists of four sections; Section 2 defines general terminology. Section 3 describes our algorithm for the Corridor Model. The lower bounds and conclusions are provided in Sections 4 and 5, respectively.

2 General Definitions

When we consider a graph $G = (V, E)$ that the postman is exploring, let H denote the subgraph of G consisting of the visited vertices and the traversed edges. A vertex $v \in V$ is called *saturated* if every edge incident to v is in H.

If the postman traverses an edge for the second or more time, it is a *penalty traversal.*

3 An Algorithm for the Corridor Model with a Penalty at Most $2|V| - 2$

In this section, we show an algorithm for the Corridor Model with a penalty not exceeding $2|V| - 2$. First, we need the following theorem.

Theorem 3.1 *The minimum penalty of a postman tour for a graph G is at most* $|V(G)| - 1$.

Proof. We claim that the following algorithm using Divide and Conquer computes a postman tour with a penalty at most $|V| - 1$ for an undirected graph.

Algorithm *OfflinePostman*(G, s)
Input: A connected graph $G = (V, E)$ and a start point $s \in V(S)$.
Output: A postman tour J of G with a penalty $|V| - 1$ or less.

Find a cycle $C = \{c_1, c_2, \ldots, c_l\}$ in G. If G is a tree, the dfs algorithm generates J. Remove the edges in C from G, which leaves connected components $S_1, S_2, \ldots, S_k$. Without loss of generality, assume $c_i \in V(S_i)$ for each $i \in \{1, 2, \ldots, k\}$ where $k \leq j$, and $s \in V(S_1)$.

In the divide phase, we recursively compute $J_1 =$*OfflinePostman*(S_1, s) and $J_i =$*OfflinePostman*(S_i, c_i) for $i \in \{2, \ldots, k\}$. In the conquer phase to construct J, we attach C and J_i to J_1. More precisely, we modify J_1 so that when the postman reaches c_1, he/she enters C to visit c_2 and moves on J_2. When the postman comes back to c_2, he/she continues to move on C to visit $c_3, \ldots, c_l$ similarly. The postman continues the rest of the traversal on J_1 at c_1. □

We prove by induction on $|V|$ that the penalty of J is at most $|V| - 1$. The basis is a case when G is a tree. Assume true for every connected graph with the number of vertices less than $|V|$, prove true for $|V|$.

Every edge in G belongs to either C or some S_i. The edges in C generate no penalty traversal. Thus, every edge with a penalty traversal must belongs to S_i, which implies

$$P(J) = P(J_1) + P(J_2) + \ldots + P(J_k),$$

where P is a function that returns the penalty of a postman tour. By induction hypothesis and the fact that J_i are vertex disjoint,

$$P(J) = \sum_{i=1}^{k} P(J_i) \leq \sum_{i=1}^{k} (V(J_i) - 1) \leq |V| - 1.$$

This proves the induction step. □

Algorithm *OfflinePostman* runs in linear time. For, we use the dfs algorithm. When we find a cycle C by detecting a back edge, we delete the edges in C.

Start the dfs algorithm at each vertex in C, marking an edge by C and the direction of its first time traversal. When the traversal of G is complete, recover the recursive structure of the obtained postman tours by the mark. Construct a desired postman tour for G in linear time.

The following theorem is our main claim.

Theorem 3.2 *There exists an online algorithm for computing The Corridor Model of The Postman Problem that generates a penalty at most* $2|V(G)|-2$ *for any connected graph* G.

Proof. The following is our algorithm. We think that one step of the algorithm is completed when an edge is traversed by the postman.

Algorithm *OnlinePostman*
Input: The postman initially locates at a vertex s in a connected graph $G = (V, E)$. He/she will explore the entire graph to return to s. At a current vertex v, every edge e from v and the other visited endpoint of e are known to the postman. The explored subgraph H is memorized by the postman.
Output: A postman tour of G with a penalty at most $2|V|-2$.

Run the dfs algorithm. We color edges and vertices in order to imply some traversal status. If we color an edge or a vertex in red, it is complete in the dfs part of the exploration.

If the postman finds an uncolored edge e from the current vertex v such that the other endpoint of e is not in H, he traverses and colors it in white. If he finds an uncolored e with the other endpoint in H, he does not traverse it at the moment and colors it in blue.

When every edge from v is colored, let G_v be the subgraph of G, which contains all the vertices reachable from v using only blue edges, and the blue edges. If every $w \in V(G_v) - \{v\}$ is colored in red, the postman explores G_v by *OfflinePostman*(G_v, v).

There exists a simple white path P_v that connects s and v. The postman colors v in red and moves back on the edge incident to v in P_v. He/she colors the edge in red.

Continue the above until the postman returns to s. □

Lemma 3.3 *At any step of the algorithm, the collection of the red and white edges forms a tree that spans over* $V(H)$*. Moreover, the set of the white edges forms a simple path between* s *and* v.

Proof. The edges form a connected component that spans over $V(H)$, since they are included in the trajectory of the walk so far. They do not contain a cycle, since if the other endpoint of an edge from v is already visited, the edge is colored in blue. The second statement is verified by induction on the number of steps executed. □

Lemma 3.4 *Algorithm OnlinePostman finishes when every edge in* G *is traversed and the postman returns to* s.

Proof. Suppose that it does not finish. Let G' be the subgraph of G induced by all the vertices visited by the postman in infinitely many steps or until the algorithm halts unexpectedly. The collection of the red edges forms a tree that spans over $V(G')$, because G' contains no white edges.

Since G is connected, there exists an unsaturated vertex v in G'. Let e be an untraversed edge incident to v. It is a blue edge so that the other endpoint belongs to G'. For, e must be colored due to $v \in V(G')$. It is colored in blue, else it is a red edge that is a witness of e being traversed.

The blue edges in G' form connected components. One of them C contains e. Every vertex in C is incident to both a blue edge and a red edge. There exists a step when the last white edge in C turns red. Immediately, *OfflinePostman* would have been issued for the postman to traverse e. Therefore, such e cannot exist.

Since there is no white edge in $G' = G$, the postman must return to s. This correctly finishes the algorithm. □

Lemma 3.5 *The generated penalty is bounded by* $2|V| - 2$.

Proof. The number of penalty traversals on the red edges is at at most $|V| - 1$ due to Lemma 3.3. We show that the the number of penalty traversals on the blue edges is at most $|V| - 1$. Let $G_{v_1}, G_{v_2}, \ldots, G_{v_k}$ be the collection of G_v for which *OfflinePostman* is issued. It suffices to show that they are vertex disjoint subgraphs of G.

Each G_{v_i} is a component connected by blue edges. Suppose there exists a vertex

$$w \in V\left(G_{v_1}\right) \cap V\left(G_{v_2}\right).$$

When *OfflinePostman*(G_{v_1}, v_1) is issued before *OfflinePostman*(G_{v_2}, v_2), w must be colored in red if $w \neq v_1$. Whether $w = v_1$ or not, the blue edges in G_{v_2} incident to w have been already colored at the step. Then, *OfflinePostman* at v_1 would not have been issued for G_{v_1}. Therefore, such w is impossible to exist. □

The above sequence of lemmas proves the theorem. □

4 Lowerbounds for the Corridor Model and the Labyrinth Model

In this section, we prove lower bounds of a penalty generated by any algorithms for the both models.

Theorem 4.1 *No algorithm for The Corridor Model of The Postman Problem guarantees a penalty less than* $\frac{5|V(G)|-6}{4}$ *for an arbitrary connected graph* G.

Proof. We construct a graph G_j for a given positive integer j, by the following procedure. Call the next graph a *unit.*

$$s = \Big(\{p(s), b(s), c_1(s), c_2(s), c_3(s)\}, \{(p(s), c_1(s)), (c_1(s), b(s)), (b(s), c_2(s)), \\ (c_2(s), p(s)), (c_1(s), c_3(s))\}\Big)$$

where $p(s)$, $b(s)$ and $c_i(s)$ are vertices. There are j units in total, say $s_1, s_2, \ldots, s_j$. s_i for each $i \in \{2, 3, \ldots, j-1\}$ is to be connected to the units s_{i-1} and s_{i+1} at $p(s_i)$ and $b(s_i)$, respectively. $p(s_i)$ is identified with $b(s_{i-1})$, and $b(s_i)$ with $p(s_{i+1})$. $b(s_j)$ is connected to a vertex w by an extra edge, and $p(s_1)$ is the start point of the exploration.

Lemma 4.2 *Any algorithm generates at least five penalty traversals in each s.*

Proof. The postman always enters each s at $p(s)$ for the first time. Due to the definition of The Corridor Model, it is possible that he/she reaches $c_2(s)$ without saturating $c_1(s)$ and $b(s)$. The postman must re-visit the two vertices.

If the postman traverses the edge $(c_2(s), p(s))$ at the next step, it takes at least five penalty traversal to visit $b(s)$, $c_1(s)$ and $c_3(s)$ to come back to $p(s)$ once again. If the postman traverses $(c_2(s), b(s))$ at the next step, he/she needs to visit every $c_i(s)$. This takes at least five penalty traversals also. □

$|V(G_j)|$ and $|E(G_j)|$ are $4j+2$, and $5j+1$, respectively. The number of penalty traversals is at least $|E(G_j)|$ due to the lemma. Thus, we have a penalty for G_j at least

$$\begin{aligned}\frac{5j+1}{4j+2} \cdot |V(G_j)| &= \frac{\frac{5}{4}(4j+2) - \frac{3}{2}}{4j+2} \cdot |V(G_j)| \\ &= \frac{5}{4}|V(G_j)| - \frac{3}{2}\end{aligned}$$

where $|V(G_j)|$ is arbitrary large. □

The lower bound $\frac{4|V|-5}{3}$ for The Labyrinth Model is obtained, by a similar way that changes a unit s into a 'square', *i.e.*, obtained by eliminating $c_3(s)$ from s. The only difference is that in The Labyrinth Model, the postman possibly explores every vertex in s at the first time visit to come back to $p(s)$ without saturating $b(s)$.

If a multiple edge is allowed for The Labyrinth Model, we can obtain a greater lower bound $2|V| - 1$. It is doable by further shrinking every s into a cycle of length two. In other words, if one comes up with an algorithm for The Labyrinth Model with a penalty less than $2|V| - 1$, it should use the fact that the given graph is free from multiple edges.

5 Conclusions

A $2|V| - 2$ penalty algorithm for The Corridor Model has been presented, and some penalty lower bounds for the both models are obtained. In addition, the tightest upper bound of a penalty for the offline problem is shown. The result is summarized in Table 1.

This also poses an open problem asking how close the both bounds can be in the two models. We doubt the existence of a $2|V|$ penalty algorithm for The Labyrinth Model, even in multiple edge free graph. What prevents us from

Table 1. Summarizing the Obtained Bounds

	Offline	The Labyrinth Model	The Corridor Model
Upper Bound	$\|V\|-1$	$3\|V\|$ *	$2\|V\|-2$
Lower Bound	$\|V\|-1$	$\frac{4}{3}\|V\|-\frac{5}{3}$ **	$\frac{5}{4}\|V\|-\frac{3}{2}$

* Due to [8] ** $2|V|-1$ if multiple edges are allowed,

matching the both bounds seems similar in the both models. We think that it is a kind of 'algorithm dependent bottlenecks', meaning that there are distinct types of lower bound graphs for distinct exploration strategies. Therefore, we feel that if one model is improved, so is the other.

References

1. S. Albers and M. R. Henzinger, "Exploring Unknown Environments", Proc. 29th Symp. on Theory of Computing (1997), pp. 416–425.
2. B. Awerbuch, M. Betke, R. Rivest, and M. Singh, "Piecemeal Graph Learning by a Mobile Robot", Proc. 8th Conf. on Comput. Learning Theory (1995),,pp. 321–328.
3. E. Bar-Eli, P. Berman, A. Fial and R. Yan, "On-line Navigation in a Room", Proc. 3rd ACM-SIAM Symp. on Discrete Algorithms (1992), pp. 237–249.
4. A. Blum, P. Raghavan and B. Shieber, "Navigating in Unfamiliar Geometric Terrain", Proc. 23rd Symp. on Theory of Computing (1991), pp. 494–504.
5. M. Betke, R. Rivest and M. Singh, "Piecemeal Learning of an unknown environment", Proc. 5th Conf. on Comput. Learning Theory (1993), pp. 277–286.
6. X. Deng, T. Kameda and C. H. Papadimitriou, "How to Learn an Unknown Environment", Proc. 32ns Symp. on Foundations of Computer Science (1991), pp. 298–303.
7. X. Deng and C. H. Papadimitriou, "Exploring an unkown graph", Proc. 31st Symp. on Foundations of Computer Science (1990), pp. 356–361.
8. P. Panaite and P. Pelc, "Exploring Unknown Graphs", SIAM J. Matrix and Appl., Vol. 11, 1990, pp. 83–88.
9. N. Rao, S. Hareti, W. Shi and S. Iyengar, "Robot Navitation in UnKnown Terrans: Introductory survey of Non-heuristic Algorithms", Tech.Report ORNL/TM12410, Oak Ridge National Laboratory, July 1993.
10. W. J. Cook, W. H. Cunningham, W. R. Pulleyblank and A. Schrijver, "Combinatorial Optimization", 1998, John Wiley and Sons Inc.

A Simple Dual Ascent Algorithm for the Multilevel Facility Location Problem

Adriana Bumb and Walter Kern

Faculty of Mathematical Sciences, University of Twente,
P.O. Box 217, 7500 AE Enschede, The Netherlands
a.f.bumb,kern@math.utwente.nl

Abstract. We present a simple dual ascent method for the multilevel facility location problem which finds a solution within 6 times the optimum for the uncapacitated case and within 12 times the optimum for the capacitated one. The algorithm is deterministic and based on the primal-dual technique.

1 Introduction

An important problem in facility location is to select a set of facilities, such as warehouses or plants, in order to minimize the total cost of opening facilities and of satisfying the demands for some commodity (see Cornuejols, Nemhauser & Wolsey [CNW90]).

In this paper we consider the multilevel facility location problem in which there are k types of facilities to be built: one type of depots and $(k-1)$ types of transit stations. For every type of facility the opening cost is given. Each unit of demand must be shipped from a depot through transit stations of type $k-1, \ldots, 1$ to the demand points. We assume that the shipping costs are positive, symmetric and satisfy the triangle inequality. The goal of the problem is to select facilities of each type to be opened and to connect each demand point to a path along open facilities such that the total cost of opening facilities and of shipping all the required demand from depots to demand points is minimized.

Being an extension of the uncapacitated facility location problem, which is known to be Max SNP-hard (see [GK98] and [S97]), this problem is Max SNP-hard as well. The first approximation algorithms for the multilevel facility location problem were developed by Shmoys, Tardos & Aardal [STA97] and Aardal, Chudak & Shmoys [ACS99] and were based on rounding of an LP solution to an integer one. The performance guarantees of these algorithms were 3.16, respectively 3. The first combinatorial algorithm for the multilevel facility location problem was developed by Meyerson, Munagala & Plotkin [MMP00], and finds a solution within $O(\log|D|)$ the optimum, where D is the set of demand points.

Using an idea from [JV99], we present a simple greedy (dual ascent) method for the multilevel facility location problem that finds a solution within 6 times the optimum. The algorithm extends to a capacitated variant of the problem, when each facility can serve only a certain number of demand points, with an increase of the performance guarantee to 12.

M. Goemans et al. (Eds.): APPROX-RANDOM 2001, LNCS 2129, pp. 55–63, 2001.

2 The Metric Multilevel Uncapacitated Facility Location Problem

Consider a complete $(k+1)-$partite graph $G = (V, E)$ with $V = V_0 \cup \ldots \cup V_k$ and $E = \bigcup_{l=1}^{k} V_{l-1} \times V_l$. The set $D = V_0$ is the set of *demand nodes* and $F = V_1 \cup \ldots \cup V_k$ is the set of possible *facility locations* (at *level* $1, \ldots, k$). We are given *edge costs* $c \in R_+^E$ and *opening costs* $f \in R_+^F$ (i.e., opening a facility at $i \in F$ incurs a cost $f_i \geq 0$). We assume that c is induced by a metric on V. Without loss of generality we can assume that there are no edges of cost 0.

Remark 1. Our results also hold in a slightly more general setting, where we require only for $e \in V_0 \times V_1$ that $c(e) \leq c(p)$ for any path p joining the endpoints of e.

Denote by P the set of paths of length $k-1$ joining some node in V_1 to some node in V_k. If $j \in D$ and $p = (v_1, \ldots, v_k) \in P$, we let jp denote the path $(j, v_1, \ldots, v_k)$. As usual $c(p)$ and $c(jp)$ denote the length of p resp. jp (with respect to c).

The corresponding facility location problem can now be stated as follows: Determine for each $j \in D$ a path $p_j \in P$ (along "open facilities") so as to minimize

$$\sum_{j \in D} c(jp_j) + f(\bigcup_{j \in D} p_j).$$

Remark 2. In this setting we assume that each $j \in D$ has a demand of one unit to be shipped along p_j. Our results easily extend to arbitrary positive demands.

To derive an integer programming formulation of the multilevel facility location problem, we introduce the $0-1$ variables y_i $(i \in F)$ to indicate whether $i \in F$ is open and the $0-1$ variables x_{jp} $(j \in D,\ p \in P)$ to indicate whether j is served along p.

We let

$$c(x) := \sum_{p \in P} \sum_{j \in D} c_{jp} x_{jp}$$

and

$$f(y) := \sum_{i \in F} f_i y_i \ .$$

The multilevel facility location problem is now equivalent to

$$
(P_{int}) \qquad
\begin{aligned}
&\text{minimize } c(x) + f(y) \\
&\text{subject to } \sum_{p \in P} x_{jp} = 1, && \text{for each } j \in D && (1)\\
&\sum_{p \ni i} x_{jp} \le y_i, && \text{for each } i \in F,\ j \in D && (2)\\
&x_{pj} \in \{0,1\}, && \text{for each } p \in P,\ j \in D \\
&y_i \in \{0,1\}, && \text{for each } i \in F
\end{aligned}
$$

Constraints (1) ensure that each j gets connected via some path and constraints (2) ensure that the paths only use open facilities.

The LP–relaxation of (P_{int}) is given by

$$
(P) \qquad
\begin{aligned}
&\text{minimize } c(x) + f(y) \\
&\text{subject to } (1), (2) \\
&x_{jp} \ge 0 \\
&y_i \ge 0 \ .
\end{aligned}
$$

Note that $x_{jp} \le 1$ is implied by (1) and $y_i \le 1$ holds automatically for any optimal solution (x, y) of (P).

The standard way of proving a $0-1$ solution (x, y) of (P_{int}) to be a ρ–approximation is to show that

$$
c(x) + f(y) \le \rho C_{LP} \tag{2.1}
$$

where C_{LP} is the optimum value of (P).

3 The Primal-Dual Algorithm

The basic idea of the primal-dual approach is to exhibit a primal $0-1$ solution (x, y) satisfying (2.1) by considering the dual of (P). Introducing dual variables v_j and t_{ij} corresponding to constraints (1) and (2) in (P), the dual becomes

$$
\begin{aligned}
&\text{maximize } \sum_{j \in D} v_j \\
&v_j - \sum_{i \in p} t_{ij} \le c(jp), \ \text{for each } p \in P,\ j \in D && (3)\\
&\sum_{j \in D} t_{ij} \le f_i, \quad \text{for each } i \in F && (4)\\
&t_{ij} \ge 0, \ \text{for each } i \in F,\ j \in D
\end{aligned}
$$

Intuitively, the dual variable v_j indicates how much $j \in D$ is willing to pay for getting connected. The value of t_{ij} indicates how much $j \in D$ is willing to contribute to the opening cost f_i (if he would be connected along a path through i).

We aim at constructing a primal feasible $0-1$ solution (x, y) and a feasible dual solution (v, t) such that

$$c(x) + f(y) \leq 6 \sum_{j \in D} v_j,$$

implying (2.1) for $\rho = 6$.

We first describe how to construct the dual solution (v, t). To this end, we introduce the following notation w.r.t. an arbitrary feasible solution (v, t) of (D):

A facility $i \in F$ is *fully paid* when

$$\sum_{j \in D} t_{ij} = f_i. \tag{3.1}$$

A demand point $j \in D$ *reaches* $i_l \in V_l$ if for some path $p = (i_1, \ldots, i_l)$ from $V_1$to i_l all facilities $i_1, \ldots i_{l-1}$ are *fully paid* and

$$v_j = c_{jp} + \sum_{i \in p} t_{ij}. \tag{3.2}$$

If, in addition, also i_l is fully paid, we say that j *leaves* i_l or, in case $l = k$, that j gets *connected* (along p to $i_k \in V_k$).

Our algorithm for constructing the dual solution is a dual ascent method, generalizing the approach in [JV99]. We start with $v \equiv t \equiv 0$ and increase all v_j uniformly ("with unit speed"). When some $j \in D$ reaches a not fully paid node $i \in F$, we start increasing t_{ij} with unit speed, until f_i is fully paid and j leaves i. We stop increasing v_j when j gets connected. The algorithm maintains the invariant that at time T the dual variables v_j that are still being raised are all equal to T. More precisely, we proceed as described below.

UNTIL all $j \in D$ are connected DO
- Increase v_j for all $j \in D$ not yet connected
- Increase t_{ij} for all $i \in F$, $j \in D$ satisfying $(i) - (iii)$,
  - (i) j has reached i
  - (ii) j is not yet connected
  - (iii) i is not yet fully paid.

Let (v, t) denote the final dual solution. Before constructing a corresponding primal solution (x, y), let us state a few simple facts about (v, t).

For each fully paid facility $i \in V_l$, $l \geq 2$, denote by T_i the time when facility i became fully paid. *The predecessor* of i will be the facility in the level $l-1$ via which i was for the first time reached by a demand point, i.e.,

$$Pred\,(i) = \left\{ i' \in V_{l-1} |\ i' \text{ is fully paid and } T_{i'} + c_{i'i} = \min_{\substack{i'' \in V_{l-1} \\ i'' \text{ fully paid}}} (T_{i''} + c_{i''i}) \right\}.$$

(Ties are broken arbitrarily.)

The predecessor of a fully paid facility $i \in V_1$ will be its closest demand point. We can define the time $T_{Pred(i)} = 0$.

For all fully paid facilities i in the $k - th$ level denote by j_i $p_i = (i_1, \ldots, i_k)$ the path through the following points:

- $i_k = i$
- $i_l = Pred(i_{l+1})$, for each $1 \leq l \leq k - 1$
- $j_i = Pred(i_1)$.

We will call the *neighborhood* of i the set of demand nodes *contributing* to p_i i.e.,

$$N_i = \{j \in D \mid t_{i'j} > 0 \text{ for some } i' \in p_i\} \ .$$

Since each $j \in D$ gets connected we may fix for each $j \in D$ a connecting path $\widetilde{p_j} \in P$ of fully paid facilities (ties are broken arbitrarily).

Lemma 1. *(i)* $c(j\widetilde{p_j}) \leq v_j$ *for all* $j \in D$

(ii) For all $j \in D$ *and* $i \in V_k$ *fully paid such that* $i \in \widetilde{p_j}$, *either* $v_j = T_i$ *and* $t_{ij} > 0$ *or* $v_j > T_i$ *and* $t_{ij} = 0$

(iii) For all fully paid facilities $i \in V_k$ *and corresponding paths* $p_i = (i_1, \ldots, i_k)$, *the following relation holds*

$$T_{i_1} \leq \ldots \leq T_{i_k}$$

(iv) Let $i \in V_k$ *be a fully paid facility and* $p_i = (i_1, \ldots, i_k)$ *its associated path. For all* $j \in D$ *and* $i_l \in p_i$ *with* $t_{i_l j} > 0$, *there exists a path* p *from* V_1 *to* i_l *such that*

$$c(jp) + \sum_{s=l}^{k-1} c_{i_s i_{s+1}} \leq T_i \ .$$

In particular, $c(j_i p_i) \leq T_i$

(v) If i, i' *are two fully paid facilities in* V_k *with intersecting neighborhoods then for each* $j' \in D$, *such that* $i' \in \widetilde{p_{j'}}$, $c_{j_i j'} \leq 4 \max\{T_i, v_{j'}\}$

(vi) $\sum_{i' \in p_i} t_{i'j} \leq v_j$ *for all* $j \in D$

Proof. The first claim is straightforward from (3.2) and the definition of $\widetilde{p_j}$.

The second claim is based on the observation that at time T all the $v-$values that can be increased are equal with T and that the final $v-$values reflect the times when the demand points get connected. There are two possibilities that a fully paid facility $i \in V_k$ is on the connecting path of a demand point j. One is that j reached i before T_i and got connected when i became fully paid. In this case $t_{ij} > 0$ and $v_j = T_i$. The other possibility is that j reached i after i was fully paid, which means that $t_{ij} = 0$ and $v_j > T_i$.

The definition of a predecessor implies that for each fully paid $i \in F$

$$c_{Pred(i)i} + T_{Pred(i)} \leq T_i \ . \tag{3.3}$$

The third claim follows immediately.

For the forth claim, by adding the inequalities (3.3) for $i_{l+1}, \ldots, i_{k-1}$ one obtains

$$\sum_{s=l}^{k-1} c_{i_s i_{s+1}} + T_{i_l} \leq T_{i_k} \quad .$$

Since $t_{i_l j} > 0$, there is a path p along which j reached i_l before T_{i_l}. Clearly, $c(jp) \leq T_{i_l}$, which implies (iv).

For proving (v), let $j \in N_i \cap N_{i'}$. Since $j \in N_i$, there is an $i_l \in p_i$ such that $t_{i_l j} > 0$. Then by (iv), there exists a path q from V_1 to i_l such that $c\,(jq) \leq T_i$.

Suppose $p_{i'} = (i'_1, \ldots, i'_k)$. Similarly, there is an $i'_r \in p_{i'}$ and a path q' from V_1 to i'_r such that $c\,(jq') + \sum_{s=r}^{k-1} c_{i'_s i'_{s+1}} \leq T_{i'}$.

Using the triangle inequality and (ii), we obtain

$$\begin{aligned}
c_{j_i j'} &\leq c(j_i p_i) + c(jq) + c(jq') + \sum_{s=r}^{k-1} c_{i'_s i'_{s+1}} + c(j' \widetilde{p_{j'}}) \\
&\leq 2T_i + T_{i'} + v_{j'} \\
&\leq 2T_i + 2v_{j'} \\
&\leq 4 \max \{T_i, v_{j'}\} \quad .
\end{aligned}$$

Finally, for proving the statement in the last claim is enough to show that no demand point j could increase simultaneously two values $t_{i_l j}, t_{i_s j}$, for $i_l \neq i_s$ and $i_l, i_s \in p_i$. This follows from the definition of p_i, which implies that whenever a demand point reaches a facility on p_i, the predecessor of that facility should have been already paid, and subsequently all the facilities of p_i situated on inferior levels. □

We now describe how to construct a corresponding primal solution (x, y). Suppose there are r fully paid facilities in the last level. Order them according to nondecreasing $T-$values, say

$$T_1 \leq \ldots \leq T_r \quad .$$

Construct greedily a set $C \subseteq V_k$ of *centers* which have parewise disjoint neighborhoods and assign each $j \in D$ to some center $i_0 \in C$ as follows:

INITIALIZE $C = \emptyset$

FOR $i = 1, \ldots, r$ DO

IF $N_i \cap N_{i_0} \neq \oslash$ for some $i_0 \leq i$, assign to p_{i_0} all demand nodes $j \in D$ with $i \in \widetilde{p_j}$

ELSE $C = C \cup \{i\}$ and assign to p_i all the demand nodes $j \in D$ with the property that $i \in \widetilde{p_j}$

The paths p_i $(i \in C)$ are called *central paths*.

Remark 3. Note that each demand point j is assigned to one center. Furthermore, by construction of C, j "contributes" to at most one central path (not necessarily the one to which it is assigned).

The primal solution (x, y) is obtained by connecting all demand nodes along their corresponding central paths:

$$x_{jp} := \begin{cases} 1 \text{ if } p = p_i \text{ and } j \text{ was assigned to } p_i \\ 0 \text{ otherwise} \end{cases}$$

and

$$y_i := \begin{cases} 1 \text{ if } i \text{ is on a central path} \\ 0 \text{ otherwise} \end{cases} .$$

The shipping cost $c(x)$ is easily bounded as follows.

If $j \in D$ is assigned to p_{i_0} then $T_{i_0} \leq T_i$,where $\{i\} = \widetilde{p_j} \cap V_k$. Due to Lemma 1 (ii) and (v), we get $T_{i_0} \leq v_j$ and

$$c_{jp_{i_0}} \leq c_{j_{i_0}j} + c_{j_{i_0}p_{i_0}} \leq 4v_j + T_{i_0} \leq 5v_j .$$

The cost of opening facilities along a central path p_{i_0} can be also bounded with the help of Lemma 1(vi)

$$\sum_{i \in p_{i_0}} f_i = \sum_{i \in p_{i_0}} \sum_{j \in N_i} t_{ij} \leq \sum_{j \in N_i} v_j .$$

Since the centers have pairwise disjoint neighborhoods, we further conclude that

$$f(y) = \sum_{i_0 \in C} \sum_{i \in p_{i_0}} f_i \leq \sum_{j \in D} v_j .$$

We have proved

Theorem 1. *The above primal solution (x, y) satisfies*

$$c(x) + f(y) \leq 6 \sum_{j \in D} \nu_j .$$

4 A Capacitated Version

The following capacitated version has been considered in the literature: Each $i \in F$ has an associated *node capacity* $u_i \in N$ which is an upper bound on the number of paths using i. On the other hand, we are allowed to open as many copies of i (at cost f_i each) as needed.

To formulate this, we replace the $0-1$ variables y_i in (P_{int}) by nonnegative integer variables $y_i \in Z_+$, indicating the number of open copies of $i \in F$. Furthermore, we add *capacity constraints*

$$\sum_{j \in D} \sum_{p \ni i} x_{jp} \leq u_i y_i, \text{ for each } i \in F \ . \tag{4.1}$$

Again, we let C_{LP} denote the optimum value of the corresponding LP-relaxation.

The idea to approach the capacitated case (also implicit in [JV99] for the 1-level case) is to move the capacity constraints to the objective using Lagrangian multipliers $\lambda_i \geq 0$, for each $i \in F$. This results in an uncapacitated problem

$$\begin{aligned} C(\lambda) &:= \text{minimize } c(x) + f(y) + \sum_{i \in F} \lambda_i \left(\sum_{j \in D} \sum_{p \ni i} x_{jp} - u_i y_i \right) \\ &= \text{minimize } \widetilde{c}(x) + \widetilde{f}(y) \end{aligned}$$

with $\widetilde{f}_i = f_i - \lambda_i u_i$, for each $i \in F$ and $\widetilde{c}(e) = c(e) + \lambda_i$ if i is the endpoint of $e \in E$. Note that each $\lambda \geq 0$ gives $C(\lambda) \leq C_{LP}$.

As in section 3. we compute a primal $0-1$ solution (x, y) of $C(\lambda)$ with

$$\widetilde{c}(x) + \widetilde{f}(y) \leq 6C(\lambda) \ .$$

Note that this does not necessarily satisfy the capacity constraints (4.1). However, a clever choice of the Lagrangian multipliers $\lambda_i = \frac{1}{2} \frac{f_i}{u_i}$ $(i \in F)$ yields

$$\begin{aligned} \widetilde{c}(x) + \widetilde{f}(y) &= c(x) + \frac{1}{2} \sum_{i \in F} \frac{f_i}{u_i} \sum_{p \ni i} \sum_{j \in D} x_{jp} + \frac{1}{2} \sum_{i \in F} f_i y_i \\ &\geq c(x) + \frac{1}{2} \sum_{i \in F} f_i \overline{y_i} \ , \end{aligned}$$

where $\overline{y_i} := \left\lceil \frac{1}{u_i} \sum_{p \ni i} \sum_{j \in D} x_{jp} \right\rceil$ opens each facility $i \in F$ sufficiently many times. Hence $(x, \overline{y})$ is indeed a feasible solution of the capacitated problem satisfying

$$c(x) + \frac{1}{2} f(\overline{y}) \leq 6C(\lambda) \leq 6C_{LP} \ ,$$

hence

$$c(x) + f(\overline{y}) \leq 12 C_{LP} \ .$$

Theorem 2. *Our greedy dual ascent method yields a $12-$approximation of the multilevel capacitated facility location problem.*

References

ACS99. Aardal,K.I., Chudak, F., Shmoys, D.B.: A 3-approximation algorithm for the k-level uncapacitated facility location problem. Information Processing Letters, **72**, (1999), 161–167

CNW90. Cornuejols, G., Nemhauser, G. L., Wolsey, L. A.: The uncapacitated facility location problem. In P. Mirchandani and R. Francis, editors, Discrete Location Theory, John Wiley and Sons, New York, (1990), 119–171

GK98. Guha, S., Khuller,S.: Greedy strikes back: improved facility algorithms. Proceedings of the 9th Annual ACM-SIAM Symposium on Discrete Algorithms, (1998), 649–657

JV99. Jain, K., Vazirani,V.V.: Primal-dual approximation algorithms for metric facility location and $k-$median problems. Proceedings of the 40th Annual IEEE Symposium on Foundations of Computer Science, (1999)

MMP00. Meyerson, A., Munagala, K., Plotkin S.: Cost distance: Two metric network design. Proceedings of the 41th IEEE Symposium on Foundation of Computer Science, (2000)

STA97. Shmoys, D. B., Tardos, E. Aardal, K. I.: Approximation algorithms for facility location problems. Proceedings of the 29th Annual ACM Symposium on Theory of Computing, (1997), 265–274

S97. Sviridenko, M.: Personal communication, (1997)

Approximation Schemes for Ordered Vector Packing Problems

Alberto Caprara[1,⋆], Hans Kellerer[2], and Ulrich Pferschy[2]

[1] DEIS, University of Bologna, viale Risorgimento 2, I-40136 Bologna, Italy
acaprara@deis.unibo.it
[2] University of Graz, Department of Statistics and Operations Research,
Universitätsstr. 15, A-8010 Graz, Austria
{hans.kellerer,pferschy}@uni-graz.at

Abstract. In this paper we deal with the d-dimensional vector packing problem, which is a generalization of the classical bin packing problem in which each item has d distinct weights and each bin has d corresponding capacities. We address the case in which the vectors of weights associated with the items are totally ordered, i.e., given any two weight vectors a_i, a_j, either a_i is componentwise not smaller than a_j or a_j is componentwise not smaller than a_i, and construct an asymptotic polynomial-time approximation scheme for this case. As a corollary, we also obtain such a scheme for the bin packing problem with cardinality constraint, whose existence was an open question to the best of our knowledge.
We also extend the result to instances with constant Dilworth number, i.e. instances where the set of items can be partitioned into a constant number of totally ordered subsets. We use ideas from classical and recent approximation schemes for related problems, as well as a nontrivial procedure to round an LP solution associated with the packing of the small items.

1 Introduction

In the classical one dimensional *bin packing problem* (BP) we are given n items, the i-th item having a weight $a_i \in (0, 1]$ for $i = 1, \ldots, n$, and an infinite number of unit capacity bins. The goal is to assign each item to a bin so that the sum of the weights in each bin does not exceed the capacity and the total number of nonempty bins is minimized. The *d-dimensional vector packing problem* (VP) is the generalization of BP in which every item i instead of just one weight has d weights $a_i^1, a_i^2, \ldots, a_i^d$ and every bin has d corresponding capacities all equal to 1 (possibly after scaling). For notational convenience we simply use a_i^j which should not be confused with the j-th power of a_i. We let $a_i := (a_i^1, a_i^2, \ldots, a_i^d)$ denote the *weight vector* associated with item i. This generalization was first introduced in [4] and then studied by various authors, being both of practical and theoretical interest.

⋆ The work of the first author was partially supported by CNR and MURST, Italy.

M. Goemans et al. (Eds.): APPROX-RANDOM 2001, LNCS 2129, pp. 63–75, 2001.

A relevant special case of VP arises when $d = 2$ and the item weights on the second dimension are identical, say $a_i^2 = 1/k$ for all i. In this case, we have a BP with an upper bound k on the *number of items* which can be put together into one bin, called the *k-item bin packing problem* (kBP). This problem was first treated in 1975 by Krause, Shen and Schwetman [7]. In their paper kBP appears as a formulation of a task–scheduling problem on a multiprogramming computer system, which is still an important part of every operating system for multiple processor machines 25 years later. Further details are given in [6] and [7].

Also the general 2-dimensional vector packing problem has many practical applications, some of which are mentioned in [8], [9] and [1], where heuristics and exact algorithms are proposed.

An illustrative example of an important application of 2-dimensional vector packing arises in the logistics of cargo airplanes. In a real-world problem from a major cargo airline we are given a set of n transportation requests for a certain very busy flight route, e.g. Frankfurt – New York. Each such request basically consists of a single freight piece or package. The obvious objective of the freight disponent is to find an assignment of the packages to planes such that all packages can be transported on a minimal number of planes. Assuming that all available planes have the same capacity is quite common, since many airlines try to use homogenuos fleets for certain routes. Moreover, each plane usually makes the round-trip several times such that a demand of e.g. 8 planes may be realized by using only 2 planes each making the trip four times.

For each package i we are given two parameters, its *weight* a_i^1 and its *volume* a_i^2. Although an optimal packing of the shapes of the items to be transported would be a three-dimensional packing problem, in practice the actual loading of the items assigned to each plane is done by experienced workers without mathematical optimization of the loading pattern. For practical purposes it is sufficient to ensure that the volume of each (usually rectangular) package does not exceed a given upper bound, which is naturally smaller than the actual volume available on the plane. The upper bound on the total weight in each plane is given by its pay load. Clearly, the resulting optimization problem is a special case of VP with $d = 2$. In practice, packages from the same customer share the main characteristics of weight and volume, in the sense that a heavier package also requires more volume.

We will first consider the so-called *ordered* VP. To this end, we define a natural order relation on the vectors in $\mathbb{R}^d$ and write $u \succeq v$ for any two vectors u, v in the same space if u is at least as large as v in *every component*. The ordered VP is the special case of VP in which, for any two items i, j either $a_i \succeq a_j$ or $a_j \succeq a_i$ holds, i.e. the relation "$\succeq$" defines a total ordering on the set of items. Note that kBP is a special example of the ordered 2-dimensional VP. Furthermore, we can introduce a *cardinality constraint* for VP analogous to kBP by requiring that at most k items may be packed into every bin. This can be modeled by extending VP to a $(d+1)$-dimensional vector packing problem where $a_i^{d+1} = 1/k$ for all i. Clearly, an ordered VP is still ordered also after adding this additional dimension. Hence, all results derived in this paper for the ordered VP also hold for the corresponding cardinality constrained problem.

Clearly, the item set N of every general VP can be partitioned into subsets $N^1, \ldots, N^c$ such that, for every subset N^j and every two items $i, h \in N^j$, either $a_i \succeq a_h$ or $a_h \succeq a_i$ holds, i.e. "$\succeq$" defines a total ordering on each subset N^j. The *Dilworth number* of a VP instance is the minimum c for which such a partitioning exists. Of course, an ordered VP instance has Dilworth number 1. The Dilworth number is equal to the "classical" Dilworth number for the partially ordered set given by the item weight vectors and the order relation "$\succeq$". Hence, the Dilworth number of a VP instance can be computed efficiently by max-flow techniques (see e.g. [5]). We will also show in this paper how to extend our results to VP with *constant* Dilworth number. The discussion above points out that VP instances arising from real-world applications often have a relatively small Dilworth number.

All the problems mentioned above are generalizations of BP and hence NP-hard in the strong sense. Moreover, it is well-known that no polynomial time approximation algorithm with a worst-case performance ratio better than 3/2 can exist for BP, unless P=NP. However, Fernandez de la Vega and Lueker [3] gave an *asymptotic polynomial time approximation scheme* (APTAS) for BP. Their method can easily be extended to an asymptotic $(d+\varepsilon)$-approximation algorithm for VP. Recently, Chekuri and Khanna [2] improved this result proposing a polynomial-time algorithm with worst-case performance ratio $1+\varepsilon d+O(\ln \varepsilon^{-1})$, for any $\varepsilon > 0$, which leads to a polynomial-time algorithm with worst-case performance ratio $O(\ln d)$ when the dimension d is fixed. The existence of an APTAS, even for the case $d = 2$, was recently ruled out by Woeginger [10] (on the assumption P$\neq$NP). For kBP, to the best of our knowledge, the best known asymptotic approximation ratio achievable is 3/2, as given in [6].

Our main result is showing the existence of an APTAS for the ordered VP in Section 2. As a corollary, we get an APTAS for kBP. Our scheme is based on standard techniques such as small items elimination and item grouping (cf. [3]), but also on the enumeration of the solutions for the instance obtained from grouping and on the use of an LP to include the small items for each solution. These latter techniques are similar to those used in [2] for multiprocessor vector scheduling (a problem related to VP in which the number of bins is fixed, all bins must have the same capacity b on all dimensions, and one would like to minimize the value of b so as to pack all items). Nevertheless, while the rounding of the LP solution is trivial for the case of the *constant* dimension, see [2], when the dimension is a part of the input rounding has to be done carefully to achieve an APTAS, as we will show in Section 2. In Section 3 we will show how our approach can be extended to VP with *constant* Dilworth number. An immediate corollary is an APTAS for the special case of a 2-dimensional VP where the number of different item weights in one dimension is bounded by a constant.

In the conclusions of [10] it is mentioned that "a slight modification of the method of Fernandez de la Vega and Lueker [3] yields asymptotic polynomial time approximation schemes for the subproblems of d-dimensional vector packing with constant Dilworth number". This would cover the results presented in this paper (at least for a constant dimension – it is not clear whether the above sentence refers to a constant dimension or not). Nevertheless, to the best of our understanding, in order to achieve our results a *substantial* modification of the method in [3] is necessary, along the lines presented in this paper. In particular, even if a near-optimal solution for the *large* items (see the next section) in an ordered VP instance is easy to achieve following the approach in [3], it is not always possible to extend this solution to a near-optimal one for the overall instance, including the *small* items. (Certainly a greedy approach does not work, as it may end up with bins almost filled in different dimensions.) In fact, the packing of the large items can hardly be done independently of the small items, as it is done in [3] and other similar approaches, to achieve near-optimality.

Consider for example the kBP instance with $k = 5$ and $n = 2s + 8s$ items (where s is a positive integer), $2s$ with weight $1/2$ and $8s$ with weight $1/8$. Clearly, the optimal solution packs one item of weight $1/2$ and four items of weight $1/8$ per bin, requiring a total of $2s$ bins. If the small items are those with weight $1/8$, the optimal packing of the large items, requiring s bins, is obtained by packing two items of weight $1/2$ in each bin. Then, the remaining small items must be packed into separate bins, requiring $\lceil \frac{8s}{5} \rceil$ additional bins. If we replace 8 by 2^p for some integer $p \geq 3$ and let $k = 2^{p-1} + 1$, assuming the small items are those with weight $1/2^p$, the optimal solution still needs $2s$ bins, whereas the solution obtained by optimally packing the large items first uses $s + \left\lceil \frac{2^p\, s}{2^{p-1}+1} \right\rceil$ bins, which goes to $3s$ as p goes to infinity. Accordingly, the worst-case approximation guarantee cannot be better than $3/2$ (even for kBP) if the large items are packed first.

Although we have already used the notion of approximation algorithm throughout the Introduction, we will conclude this section with a precise definition of approximation algorithms used in the paper. These algorithms are generally classified by their *worst-case ratio*: For any VP algorithm A let $C^A(I)$ denote the number of bins used by algorithm A, and let $C^{OPT}(I)$ denote the minimum (optimum) number of bins required to pack the items of a given instance I. Then the *asymptotic* worst-case performance ratio is defined as

$$R_A = \lim_{C^{OPT}(I)\to\infty} \sup_I \frac{C^A(I)}{C^{OPT}(I)} .$$

Observe that $R_A \leq K_1$ if there are two constants K_1 and K_2 such that for every problem instance I

$$C^A(I) \leq K_1 C^{OPT}(I) + K_2 .$$

Moreover, R_A is an *absolute* worst-case ratio if $K_2 = 0$. Consider a value $\varepsilon \in (0, 1)$. We say that A is an *asymptotic* $(1 + \varepsilon)$*-approximation algorithm* if $R_A \leq (1 + \varepsilon)$. An APTAS is an asymptotic $(1 + \varepsilon)$-approximation algorithm for any $\varepsilon > 0$ with running time polynomial in the size of the encoded instance.

2 An APTAS for the Ordered VP

This main section contains the description of an APTAS for the ordered VP both for the fixed dimension and for the case where d is a part of the input. Let ε be the required accuracy. We will assume $\varepsilon < 1/2$. As it is usually the case for packing problems we will distinguish between *small* and *large items*. Let us define for the given value $\varepsilon \in (0, 1/2)$

$$L := \{i \mid a_i^k \geq \varepsilon \text{ for some } k \in \{1, \dots, d\}\},\ \ell := |L|, \qquad \textit{large items}$$
$$S := \{i \mid a_i^k < \varepsilon \text{ for all } k \in \{1, \dots, d\}\},\ s := |S|, \qquad \textit{small items.}$$

For simplicity we assume that $L = \{1, \dots, \ell\}$ with $a_1 \succeq a_2 \succeq \dots \succeq a_\ell$.

2.1 Enumerating Packings for the Large Items

Let us first assume that $\ell > 2/\varepsilon^2$. Later on we will also illustrate how to handle the (simpler) case $\ell \leq 2/\varepsilon^2$. We introduce *item grouping* to attain a simplified structure of packings for the large items, so that only a polynomial number of different packings have to be considered. In particular, we group the items in L as follows. Define

$$h := \lfloor \ell \varepsilon^2 \rfloor$$

and let p and q be such that $\ell = ph + q$ and $1 \leq q \leq h$. Note that $h \geq 2$. We now show that

$$p \leq \frac{2}{\varepsilon^2} \tag{1}$$

and hence p is bounded by a constant. If (1) were not true we would have the contradiction

$$\ell = ph + q > \frac{2}{\varepsilon^2} \lfloor \ell \varepsilon^2 \rfloor \geq \frac{2}{\varepsilon^2}(\ell \varepsilon^2 - 1) = \ell + \left(\ell - \frac{2}{\varepsilon^2} \right) > \ell,$$

because $\ell > 2/\varepsilon^2$.

We partition L into the $p+1$ subsets $L_i := \{ih + 1, \dots, (i+1)h\}$, $i = 0, \dots, p-1$, each containing h items and $L_p := \{ph + 1, \dots, \ell\}$. Note that each bin in a feasible solution contains at most $\lfloor \frac{1}{\varepsilon} \rfloor$ items from L, since a_ℓ has a weight of at least ε in at least one dimension and all other items are not smaller in this dimension. Accordingly, we let a *packing vector* t be a $(p+1)$-dimensional vector with $t_i \leq \lfloor \frac{1}{\varepsilon} \rfloor$ for $i = 0, \dots, p$, where t_i denotes the number of items from L_i. We observe that, while all sets of large items that fit in a bin correspond to a packing vector, the converse does not hold.

The total number f of all possible packing vectors t with $t_i \leq \lfloor \frac{1}{\varepsilon} \rfloor$ for all i is clearly bounded by

$$f \leq \left(\left\lfloor \frac{1}{\varepsilon} \right\rfloor + 1 \right)^{p+1}$$

since each of the $p+1$ components of t can be chosen between 0 and $\lfloor \frac{1}{\varepsilon} \rfloor$. Hence, f is constant for any fixed ε.

In our approximation scheme we compute all possible combinations of packing vectors to the bins. To this end, we need to know the number $m = C^{OPT}(I)$ of bins used by the optimal solution. This can be done either by trying all the possible values of m, between 1 and n, or by performing binary search. For the given m, we consider all possible *assignments* of packing vectors to the m bins. To get a very rough upper bound on the number of such assignments, consider that each of the f packing vectors can be selected up to m times. Hence, we can count for every packing vector the number of times it is assigned to a bin. Generating f such numbers between 0 and m yields a total of $(m+1)^f$ possible assignments which is polynomial in m (and n). Note that we do not have to identify the particular bin a vector is assigned to.

An assignment is called *feasible* if the sum over the ith components of all m packing vectors is equal to $|L_i|$ for $i = 0, 1, \ldots, p$. Note that in this way we will also consider the *optimal* assignment, i.e. the assignment that associates with each bin exactly the packing vector defined by this bin in the optimal solution.

We can define for every packing vector t an *induced packing* of items in L into a bin (not necessarily feasible) by packing *up to* t_i arbitrary items from set L_{i+1} for $i < p$. The entry t_p of the packing vector is ignored in the induced packing. If the packing vector t corresponds to a feasible packing then also the packing induced by t is feasible, since every item $k \in L_i$ in the feasible packing is replaced in the induced packing either by no item or by an arbitrary item $j \in L_{i+1}$ such that $a_j \preceq a_k$. Note that a feasible assignment may involve a packing vector corresponding to an infeasible bin. However, the crucial point that guarantees that our approach is correct is that a packing vector corresponding to a feasible bin gives rise to an induced packing which is always feasible.

For every feasible assignment considered, we generate the corresponding induced packing for all bins. Recall that this means that for every bin with an assigned packing vector t we pack up to t_i arbitrary items from L_{i+1} for $i = 0, \ldots, p-1$. Since the assignment is feasible, it is easy to generate the induced packing so that all items in $L_1, L_2, \ldots, L_p$ are packed, noting that $|L_i| = |L_{i+1}|$ for $i = 0, \ldots, p-2$ and $|L_{p-1}| \geq |L_p|$. If in the thereby generated packing the capacity of some bin is exceeded, the feasible assignment we started from cannot be the optimal one, and we simply discard it. Otherwise, i.e. if all the m d-dimensional capacity constraints are fulfilled, let $b_j = (b_j^1, \ldots, b_j^d)$ be the residual capacity vector of bin j for $j = 1, \ldots, m$. By construction of the packing induced by the packing vectors of the optimal assignment, the values b_j of this packing are componentwise at least as large as the residual capacities in the optimal solution for each bin j.

The items in L_0 are left unpacked and are put each into a separate bin, which requires h extra bins. A trivial bound on the total weight of large items yields $m \geq \lceil \ell\varepsilon \rceil$ since, as mentioned before, in at least one dimension item a_ℓ and also all other large items have a weight at least ε. Hence, we get

$$h = \lfloor \ell\varepsilon^2 \rfloor \leq \lceil \ell\varepsilon \rceil \varepsilon \leq m\varepsilon.$$

Summarizing, since one of all the assignments derived from the considered assignments of packing vectors to bins corresponds to the optimal packing, we managed to pack all large items into $m + m\varepsilon$ bins while leaving in every bin a residual capacity for the small items at least as large as in the optimal solution.

We now consider the case $\ell \leq 2/\varepsilon^2$. In this case, the number of large items is constant and, assuming we know the number m of bins in an optimal solution, we enumerate all the $O(m^\ell)$ feasible packings of the large items into the bins, i.e. packings fulfilling the capacity constraints in every dimension, among which is the optimal assignment.

Summarizing, we have proved

Theorem 1. *For any fixed $\varepsilon \in (0, 1/2)$, there is a $(1+\varepsilon)$-approximation algorithm for the ordered VP if all items are large with respect to ε.* □

2.2 Packing the Small Items

For each feasible packing of large items generated above in either case, we pack the small items. First we solve the following LP.

For each small item $i \in S$ and bin $j = 1, \ldots, m$ let $x_{ij} \in [0, 1]$ denote the (fractional) part of item i which is packed into bin j.

$$\sum_{j=1}^{m} x_{ij} = 1, \quad i \in S, \tag{2}$$

$$\sum_{i \in S} a_i^k x_{ij} \leq b_j^k, \quad j = 1, \ldots, m, k = 1, \ldots, d, \tag{3}$$

$$x_{ij} \geq 0, \quad i \in S, \; j = 1, \ldots, m. \tag{4}$$

Note that we are only interested in a feasible solution to this set of linear inequalities. To formally define an LP any objective function may be added.

If the LP has no feasible solution, the feasible assignment cannot correspond to the optimal assignment by the above discussion, and this assignment is discarded. Otherwise, a *basic* feasible solution of the LP contains at most $s + dm$ nonzero variables, and a straightforward counting argument (also given in [2]) shows that the number of items that are not packed into a single bin is at most dm. These items can be put into separate bins. Recalling that $a_i^k < \varepsilon$ for each $i \in S$ and $k = 1, \ldots, d$, this requires no more than $\left\lceil \frac{dm}{\lfloor \frac{1}{\varepsilon} \rfloor} \right\rceil$ bins, which can be bounded from above with a simple calculation by $2dm\varepsilon + 1$ for $\varepsilon < 1/2$.

Accordingly, the solution found for the correct value of m and the optimal assignment has a value bounded by

$$m + m\varepsilon + \left\lceil \frac{dm}{\lfloor \frac{1}{\varepsilon} \rfloor} \right\rceil \leq m + (2d + 1)m\varepsilon + 1,$$

which proves

Theorem 2. *There is an APTAS for the ordered VP if the dimension d is constant.* □

Corollary 1. *There is an APTAS for kBP.* □

2.3 Rounding the LP Solution for the General Dimension

The case in which d is a part of the input needs a more careful approach. Let x^* be an optimal solution of the above LP. Our final aim is to round this solution in a way such that only $2m$ items remain unpacked in bins $1, \ldots, m$. By the discussion above, packing these items in additional bins in a greedy way will require no more than $4m\varepsilon + 1$ additional bins, yielding an APTAS also for the case of the general dimension d.

To simplify the notation, we will assume that no small item is entirely packed into one bin, i.e. that no entry of x^* is equal to one. The generalization to the case in which $x^*_{ij} = 1$ for some i, j is obvious by fixing item i into bin j and considering only the packing of the fractional items.

Let $S = \{1, \ldots, s\}$ with $a_1 \preceq a_2 \preceq \ldots \preceq a_s$. We will pack the items in S in increasing order into the m given bins until we "get stuck". Lemma 1 will state that at this point there are at most $2m$ small items left unpacked. In principle, in order to pack some item i we will try to pick a bin (which possibly already contains a fractional part of i) and pack item i completely into that bin. To make room for item i we will "throw out" the fractional parts of larger items $i+1, i+2, \ldots$ as far as necessary.

More formally, we will derive step by step a rounded solution $\tilde{x}$ from the original LP solution x^*. Initially, we set $\tilde{x} := x^*$. We pack the smallest item 1 as follows. If there are bins j such that $\sum_{i \in S} \tilde{x}_{ij} = \sum_{i \in S} x^*_{ij} < 1$, we set $\tilde{x}_{ij} := 0$ for $i \in S$, i.e. we do not consider these bins to pack any item because it may happen that no item fits completely into such a bin. Then we find an (arbitrarily chosen) bin j such that $\tilde{x}_{1j} > 0$, letting $f_1 := 1$. If no such bin exists (which is possible due to the fact that some bins are not considered) we choose a bin j such that $\tilde{x}_{f_1 j} > 0$ and f_1 is as small as possible. (If all entries in $\tilde{x}$ are 0, item 1 as well as all other small items are unpacked, see below.) We pack item 1 in bin j, using the capacity of bin j allocated by the LP solution for items $f_1, \ldots, e_1$, where e_1 is the smallest index such that $y := \sum_{h=f_1}^{e_1} \tilde{x}_{hj} \geq 1$. In our rounded solution, this corresponds to setting $\tilde{x}_{1j} := 1$, $\tilde{x}_{1\ell} := 0$ for $\ell \neq j$, $\tilde{x}_{hj} := 0$ for $1 < h < e_1$, and $\tilde{x}_{e_1 j} := y - 1$. We say that the packing of item 1 *starts at level* f_1 and *ends at level* e_1. Note that, if $y > 1$, not all capacity allocated for item e_1 is used for item 1; in other words we still have a fraction $\tilde{x}_{e_1 j}$ that may be used to pack into bin j items $2, 3, \ldots, e_1$. To be more precise, we formally state our rounding procedure in Figure 1.

If after the bin closing step no bin is left for item i, the remaining items $i, \ldots, s$ will be packed into separate bins as above.

The procedure is illustrated in Figure 2 on an example, showing the rounded solution $\tilde{x}$ at the beginning as well as after each solution updating step. Among

Procedure LP-round

For each item $i = 1, \ldots, s$ perform the following steps:
Bin closing:
For each bin j such that $\sum_{h \geq i} \tilde{x}_{hj} < 1$, let $\tilde{x}_{hj} := 0$ for $h \geq i$
(close bin j)
If all bins are closed then terminate
Bin selection:
If $\tilde{x}_{ij} > 0$ for some bin j then
Let $\tilde{x}_{i\ell} := 0$ for $\ell \neq j$ and $f_i := i$
Else
Let $f_i := \min\{r : r > i \text{ and } \tilde{x}_{rk} > 0 \text{ for some bin } k\}$
Let j be a bin such that $\tilde{x}_{f_i j} > 0$
Solution updating:
Let $\tilde{x}_{ij} := 1$ (pack item i in bin j)
Let $e_i := \min\{r : \sum_{h=i}^{r} \tilde{x}_{hj} \geq 1\}$ and $y := \sum_{h=i}^{e_i} \tilde{x}_{hj}$
Let $\tilde{x}_{hj} := 0$ for $h = 1, \ldots, e_i - 1$ and $\tilde{x}_{e_i j} := y - 1$

Fig. 1. The LP rounding procedure.

6 items, items $1, 2, 3, 4$ are packed, respectively, into bins $1, 2, 3, 3$, with $f_1 = 1, e_1 = 4; f_2 = 2, e_2 = 6; f_3 = 3, e_3 = 4; f_4 = 5, e_4 = 6$.

Lemma 1. *The number of small items not packed into the first m bins at the end of the rounding procedure is at most $2m$.*

Proof. The number of small items packed by the rounded solution into bins $1, \ldots, m$, is equal to $\sigma := \sum_{i \in S} \sum_{j=1}^{m} \tilde{x}_{ij}$. We will compare this value to $s = \sum_{i \in S} \sum_{j=1}^{m} x^*_{ij}$. Note that σ is initially equal to s and may be decreased only in the bin closing step and in the bin selection step, whereas it is unchanged in the solution updating step. It is immediate to see that the overall decrease in all bin closing steps is at most m. In the following we will bound the overall decrease in all bin selection steps.

Noting that there may be some decrease in the bin selection step for item i only if $f_i = i$, let k be the last such that $f_k = k$, i.e. $f_i > i$ for $i > k$. If $k \leq m$, then the overall decrease in all bin selection steps is not larger than m, as the decrease in each step is not larger than 1.

We now consider the case $k > m$. The definition of f ensures $f_1 \leq f_2 \leq \ldots \leq f_k$. Hence, since $f_k = k$, the packing of all items in $1, \ldots, k$ starts at a level not larger than k, i.e. $f_i \leq k$ for $i = 1, \ldots, k$. Clearly, the number of items i with $f_i \leq k$ whose packing ends at a level larger than k, i.e. $e_i > k$, cannot be larger than m, the number of bins available. This means that $e_i \leq k$ for at least $k - m$ items among $1, \ldots, k$. Now, since the decrease of σ during the bin selection steps in the first k iterations is only due to changing entries $\tilde{x}_{ij}$ with $i \leq k$, this decrease is at most m since $\sum_{i=1}^{k} \sum_{j=1}^{m} x^*_{ij} = k$, and $\sum_{i=1}^{k} \sum_{j=1}^{m} \tilde{x}_{ij} \geq k - m$ (at least $k - m$ items among $1, \ldots, k$ are packed by the rounding procedure). After

$$\tilde{x} = \begin{bmatrix} 1/2 & 1/2 & 0 \\ 1/8 & 3/8 & 1/2 \\ 1/8 & 1/8 & 3/4 \\ 1/2 & 1/4 & 1/4 \\ 1/4 & 0 & 3/4 \\ 0 & 1/2 & 1/2 \end{bmatrix} \rightarrow \begin{bmatrix} 1 & 0 & 0 \\ 0 & 3/8 & 1/2 \\ 0 & 1/8 & 3/4 \\ 1/4 & 1/4 & 1/4 \\ 1/4 & 0 & 3/4 \\ 0 & 1/2 & 1/2 \end{bmatrix} \rightarrow \begin{bmatrix} 1 & 0 & 0 \\ 0 & 1 & 0 \\ 0 & 0 & 3/4 \\ 0 & 0 & 1/4 \\ 0 & 0 & 3/4 \\ 0 & 1/4 & 1/2 \end{bmatrix} \rightarrow$$

$$\rightarrow \begin{bmatrix} 1 & 0 & 0 \\ 0 & 1 & 0 \\ 0 & 0 & 1 \\ 0 & 0 & 0 \\ 0 & 0 & 3/4 \\ 0 & 0 & 1/2 \end{bmatrix} \rightarrow \begin{bmatrix} 1 & 0 & 0 \\ 0 & 1 & 0 \\ 0 & 0 & 1 \\ 0 & 0 & 1 \\ 0 & 0 & 0 \\ 0 & 0 & 1/4 \end{bmatrix} \rightarrow \begin{bmatrix} 1 & 0 & 0 \\ 0 & 1 & 0 \\ 0 & 0 & 1 \\ 0 & 0 & 1 \\ 0 & 0 & 0 \\ 0 & 0 & 0 \end{bmatrix}$$

Fig. 2. Illustration of the rounding procedure.

packing k no further decrease takes place in the bin selection step, so the total decrease in this step is at most m.

In both cases, we have an overall decrease not larger than $2m$, i.e. the number of small items unpacked at the end of the rounding step does not exceed $2m$. □
The discussion above shows

Theorem 3. *There is an APTAS for the ordered VP.* □

As already mentioned, adding a cardinality constraint to an ordered VP instance yields another ordered VP instance. This shows

Corollary 2. *There is an APTAS for the k-item ordered VP.* □

We outline our APTAS in Figure 3. It should be noted that efficiency was not an objective in the design of the algorithm. Its aim is rather to illustrate the existence of an APTAS for the ordered VP.

In particular, the practical inefficiency is due to the complete enumeration of the feasible assignments of large items to the bins. Although the inefficiency in handling large items is common to all approximation schemes for packing problems, in our scheme the number of feasible assignments is (roughly) $n^{\frac{1}{\varepsilon}^{\frac{1}{\varepsilon}^2}}$, for each of them we solve an LP, etc., whereas most of the schemes for easier packing problems have linear running time for fixed ε.

Although this makes the practical application of our approximation scheme very unlikely, some of the ideas may be useful within the design of practical heurstics. For instance, the separation between "large" and "small" items, packing the large ones first (by some reasonable heuristic instead of almost complete enumeration) and then the small ones (possibly by LP and rounding) may turn out to be effective in practice.

Algorithm APTAS

Partition the item set into S and L
For all $m = 1, \ldots, n$ (possible number of bins in the optimal solution)
  If $\ell > 2/\varepsilon^2$ then
    Partition L into $L_0, \ldots, L_{p-1}$ (each containing $\lfloor |L|\varepsilon^2 \rfloor$ items) and L_p
      (containing $|L| - p \lfloor |L|\varepsilon^2 \rfloor$ items)
    Pack the items in L_0 into $|L_0|$ bins
    For each feasible assignment of packing vectors to the m bins
      Pack the items in $L_0, \ldots, L_{p-1}$ into the m bins
        as induced by the packing vectors
      If the packing generated is feasible then
        Pack the items in L_p into $|L_p|$ extra bins
        Solve the LP corresponding to the packing of the items
          in S into the m bins
        If the LP has a feasible solution then
          Pack the items in S according to the LP solution,
            using extra bins for the unpacked small items
          Possibly update the best solution found so far
  Else ($\ell \leq 2/\varepsilon^2$)
    For each feasible packing of the large items to the m bins
      Solve the LP corresponding to the packing of the items in S
        into the m bins
      If the LP has a feasible solution then
        Pack the items in S according to the LP solution,
          using extra bins for the unpacked small items
        Possibly update the best solution found so far

Fig. 3. Outline of the APTAS for the ordered VP.

3 Extension to Bounded Dilworth Number

Let $N^1, \ldots, N^c$ be the partitioning of the item set N into the minimum number of ordered sets as described in the Introduction. Throughout the section, we will consider the case in which c is a constant.

The APTAS for this problem is derived along the same lines as the approximation scheme in Section 2. We treat the items in each set N^j in the same way as we treated the whole item set N in the previous section. Namely, we consider the set L^j of the large items in N^j, i.e. items with $a_i^k \geq \varepsilon$ for some $k \in \{1, \ldots, d\}$. Let $\ell^j := |L^j|$. If $\ell^j \leq 2/\varepsilon^2$, there is only a constant number of large weight items in L^j. Otherwise, we perform item grouping again and partition L^j into a constant number $p^j + 1$ of subsets containing (again except the last one) $\lfloor \ell^j \varepsilon^2 \rfloor$ consecutive items. The small items from all sets N^j will be considered later.

Now we consider all packing vectors corresponding to the packing of large items into bins. Since the number of sets L^j is at most c and each of them is either partitioned into a constant number $p^j + 1$ of subsets with consecutive items, or contains only a constant number of items, a packing vector has constant length.

More precisely, now a packing vector has $p^j + 1$ entries for each set N^j which is partitioned into $p^j + 1$ subsets, each telling the number of large items packed from each subset. Moreover, a packing vector has ℓ^j entries for each set N^j with a constant number of large items, and has value 0 or 1 indicating whether the corresponding item is packed or not. Obviously, every entry of a packing vector is again at most $\lfloor \frac{1}{\varepsilon} \rfloor$. Therefore, the overall number of packing vectors is again bounded by a constant and the number of possible assignments of packing vectors to the m bins is bounded by a polynomial in m.

The induced packing can be constructed from each generated packing vector in the same way as described in Section 2.1 (of course only for the components corresponding to sets N^j with $\ell^j > 2/\varepsilon^2$). As before we consider only feasible assignments of packing vectors to bins which leaves again at most $\lfloor \ell^j \varepsilon^2 \rfloor$ items unpacked for every set N^j. Packing all these items into separate bins requires at most $cm\varepsilon$ extra bins by employing a trivial weight bound separately on each set N^j.

There remain the small items to be packed for every generated feasible packing. This is done again by the same LP as in the previous section. The fractional solution values are rounded separately for each set N^j. By the argument of Section 2.3 we now have at most $2cm$ items which are not packed into the first m bins, and we can pack any $\lfloor \frac{1}{\varepsilon} \rfloor$ of them together in one bin thus requiring as before at most $\left\lceil \frac{2cm}{\lfloor \frac{1}{\varepsilon} \rfloor} \right\rceil$ bins. Overall, the number of bins used is at most $m + cm\varepsilon + 4cm\varepsilon + 1$, which shows the following

Theorem 4. *There is an APTAS for VP with constant Dilworth number.* □

References

1. A. Caprara, P. Toth, "Lower Bounds and Algorithms for the 2-Dimensional Vector Packing Problem", to appear in *Discrete Applied Mathematics* (2001).
2. C. Chekuri, S. Khanna, "On Multi-dimensional Packing Problems", in *Proceedings of the Tenth Annual ACM-SIAM Symposium on Discrete Algorithms (SODA'99)*, ACM Press (1999).
3. W. Fernandez de la Vega, G.S. Luecker, "Bin packing can be solved within $1 + \varepsilon$ in linear time", *Combinatorica* **1** (1981), 349–355.
4. M.R. Garey, R.L. Graham, D.S. Johnson, A.C. Yao, "Resource constrained scheduling as generalized bin packing", *J. Combinatorial Theory Ser. A* **21** (1976), 257–298.
5. M. Grötschel, L. Lovász, A. Schrijver, *Geometric Algorithms and Combinatorial Optimization*, Springer-Verlag, 1988.
6. H. Kellerer, U. Pferschy, "Cardinality Constrained Bin-Packing Problems", *Annals of Operations Research* **92** (1999), 335–348.
7. K.L. Krause, V.Y. Shen, H.D. Schwetman, "Analysis of several task-scheduling algorithms for a model of multiprogramming computer systems", *Journal of the ACM* **22** (1975), 522–550.

8. K. Maruyama, S.K. Chang, D.T. Tang, "A General Packing Algorithm for Multidimensional Resource Requirements", *International Journal Comput. Inform. Sci.* **6** (1977), 131–149.
9. F.C.R. Spieksma, "A Branch-and-Bound Algorithm for the Two-Dimensional Vector Packing Problem", *Computers and Operations Research* **21** (1994), 19–25.
10. G.J. Woeginger, "There is no Asymptotic PTAS for Two-Dimensional Vector Packing", *Information Processing Letters* **64** (1997), 293–297.

Incremental Codes

Yevgeniy Dodis[1] and Shai Halevi[2]

[1] Department of Computer Science, New York University, 251 Mercer St, New York, NY 10012, USA
dodis@cs.nyu.edu

[2] IBM T.J. Watson Research Center, P.O. Box 704, Yorktown Heights, New York 10598, USA
shaih@watson.ibm.com

Abstract. We introduce the notion of *incremental* codes. Unlike a regular code of a given rate, which is an unordered set of elements with a large minimum distance, an incremental code is an ordered vector of elements each of whose prefixes is a good regular code (of the corresponding rate). Additionally, while the quality of a regular code is measured by its minimum distance, we measure the quality of an incremental code $\mathcal{C}$ by its *competitive ratio* A: the minimum distance of *each* prefix of $\mathcal{C}$ has to be at most a factor of A smaller than the minimum distance of the *best* regular code of the same rate.

We first consider incremental codes over an arbitrary compact metric space M, and construct a 2-competitive code for M. When M is finite, the construction takes time $O(|M|^2)$, exhausts the entire space, and is NP-hard to improve in general. We then concentrate on 2 specific spaces: the real interval $[0,1]$ and, most importantly, the Hamming space F^n. For the interval $[0,1]$ we construct an optimal (infinite) code of competitive ratio $\ln 4 \approx 1.386$. For the Hamming space F^n (where the generic 2-competitive constructive is not efficient), we show the following. If $|F| \geq q$, we construct optimal (and efficient) 1-competitive code that exhausts F^n (has rate 1). For small alphabets ($|F| < q$), we show that 1-competitive codes do not exist and provide several efficient constructions of codes achieving constant competitive ratios. In particular, our best construction has rate $(1 - o(1))$ and competitive ratio $(2 + o(1))$, essentially matching the bounds in the generic construction.

1 Motivating Example

Imagine the following problem which was actually given to one of the authors. An Internet company wants to assign account numbers to its customers when the latter shop on-line. An account number allows the customer to check the status of the order, get customer support, etc. In particular, the customer can enter it over the phone. Because of that and several other reasons, account numbers should not be too long. On the other hand, we would like account numbers to be somewhat far from each other, so that it is unlikely for the customer to access a valid number by mis-entering few digits. One way to achieve this would be to use an error-correcting code of reasonable minimum distance (for example, a random account number might work for a while). This has two problems, however. First,

M. Goemans et al. (Eds.): APPROX-RANDOM 2001, LNCS 2129, pp. 75–90, 2001.

good distance implies not very good rate, and since the account numbers are quite short, we "waste" a lot possible account numbers, and exhaust our small account space too quickly (thus, losing customers). Secondly, when the number of customers is small, the corresponding prefix of our code is not as good as we could have made it with so few account numbers.

We propose a much better solution to this problem, namely, to use an *incremental* code. Such a code will eventually exhaust (or nearly exhaust) the whole space. Indeed, when the number of customers is huge, we prefer to have close accounts numbers rather than to lose customers. On the other hand, when the number of customers is i, incremental code *guarantees* that the first i account numbers we assign will be almost as far from each other as *any* possible i account numbers could be! In other words, the minimal distance of larger and larger prefixes of the code slowly decreases at an almost optimal pace.

Notice that while it is customary to measure regular codes of a given rate in terms of their minimum distance, a more relevant measure of incremental codes is the relative behavior of minimal distance on larger and larger prefixes. This leads to the notion of a *competitive ratio* of an incremental code $\mathcal{C}$. Namely, $\mathcal{C}$ is A-competitive if the minimum distance of each prefix of $\mathcal{C}$ is at most A times smaller than that of the *optimum* code of the same rate.

ORGANIZATION. While the main motivation for incremental codes comes from the Hamming spaces, we start in Section 2 by defining and studying the corresponding notion on arbitrary metric spaces. In particular, we obtain a 2-competitive codes for any such metric space, and show that it is NP-hard (in general) to beat this competitive ratio. We then study two specific spaces. In Section 3 we construct an optimal 1.386-competitive incremental code for the real interval $[0,1]$. Finally, in Section 4 we study the most important Hamming space F^n (where the generic construction is inefficient). In Section 4.1 we give a simple and efficient 1-competitive code for the Hamming space over moderate alphabets (in particular solving the "account problem" above). In the remaining subsections we concentrate on the intricacies of the Hamming space over small alphabets. While it is much harder to obtain competitive codes in this case (since our understanding of optimal codes is somewhat limited), we give several efficient constructions achieving constant competitive ratios.

2 General Notion and Construction

For simplicity, we restrict our attention to finite metric spaces, even though most results extend to compact metric spaces as well. So let $\mathcal{M} = (M, D)$ be any finite metric space on point set M with metric function D. A (regular) *code* on $\mathcal{M}$ is simply a subset of points $S \subseteq M$. The *minimum distance* $d_{\mathcal{M}}(S)$ of S is the smallest pairwise distance between distinct points in S. For an integer i we define the quantity $\mathsf{opt}\text{-}d_{\mathcal{M}}(i)$ to be the largest minimal distance of a code of cardinality i: $\mathsf{opt}\text{-}d_{\mathcal{M}}(i) = \max_{|S|=i} d_{\mathcal{M}}(S)$.

An *incremental code* $\mathcal{C} = \langle c_1 \dots c_k \rangle$ is an *ordered* sequence of distinct points of M. $\mathcal{C}$ is *exhaustive* if $k = |M|$, i.e. the code eventually runs through the entire space. For every $i \in [k]$ we define the i-th prefix of $\mathcal{C}$, $\mathcal{C}_i = \{c_1 \dots c_i\}$, and view it as a regular (unordered) code of cardinality i. We say that $\mathcal{C}$ is A-*competitive*, if for every $i \in [k]$, the i-th prefix $\mathcal{C}_i$ of $\mathcal{C}$ forms a code of distance at least $\mathsf{opt}\text{-}d_{\mathcal{M}}(i)/A$, i.e. $\mathsf{opt}\text{-}d_{\mathcal{M}}(i) \leq A \cdot d_{\mathcal{M}}(\mathcal{C}_i)$. We denote by $r_{\mathcal{M}}(\mathcal{C})$ the (best) competitive ratio of $\mathcal{C}$, and by $\mathsf{opt}\text{-}r_{\mathcal{M}}(k)$ the smallest competitive ratio of any incremental code of cardinality k: $\mathsf{opt}\text{-}r_{\mathcal{M}}(k) = \min_{|\mathcal{C}|=k} r_{\mathcal{M}}(\mathcal{C})$. We define $\mathsf{opt}\text{-}r_{\mathcal{M}} = \mathsf{opt}\text{-}r_{\mathcal{M}}(|M|)$, and call it the *competitive ratio* of $\mathcal{M}$. (We notice that since the prefix an A-competitive incremental code is also A-competitive, we have that $\mathsf{opt}\text{-}r_{\mathcal{M}}(k)$ is a non-decreasing function of k.) We say that an incremental code $\mathcal{C}$ is *perfect* if $\mathcal{C}$ is 1-competitive, and that the space $\mathcal{M}$ is *incrementally perfect* if it has an exhaustive 1-competitive code ($\mathsf{opt}\text{-}r_{\mathcal{M}} = 1$).

Theorem 1.

1. *The competitive ratio of any $\mathcal{M}$ is at most 2: $\mathsf{opt}\text{-}r_{\mathcal{M}} \leq 2$. Moreover, given $\mathcal{M}$ as an input, one can construct an exhaustive 2-competitive incremental code $\mathcal{C}$ for $\mathcal{M}$ in time $O(|M|^2)$. In fact, constructing k-prefix of $\mathcal{C}$ can be done in time $O(k \cdot |M|)$.*
2. *There exist $\mathcal{M}$ with competitive ratio 2.*
3. *For any $A < 2$ and given $\mathcal{M}$ as an input, it is* NP*-hard to construct A-competitive incremental code for $\mathcal{M}$, even when the competitive ratio of $\mathcal{M}$ is known to be 1. In particular, it is* NP*-hard to approximate the competitive ratio of $\mathcal{M}$ to within a factor less than 2.*

Proof. Given a point p and a finite set of points S, define the distance from p to S to be $D(p, S) = \min_{q \in S} D(p, q)$. We use the following simple greedy algorithm for constructing $\mathcal{C}$.

1. Let c_1 be any point of M, and let $\mathcal{C}_1 = \{c_1\}$.
2. For all $p \in M$: set $\mathsf{closest}(p) = c_1$. `% Note,` $D(p, \mathsf{closest}(p)) = D(p, \mathcal{C}_1)$
3. For $k = 2$ to $|M|$:
   - Let c_k be the furthest point from $\mathcal{C}_{k-1}$, i.e. maximizing $D(c_k, \mathcal{C}_{k-1})$. `% Done in linear time by using` $D(c_k, \mathcal{C}_{k-1}) = D(c_k, \mathsf{closest}(c_k))$
   - Set $\mathcal{C}_k = \{c_k\} \cup \mathcal{C}_{k-1}$.
   - For all $p \in M$: if $D(p, c_k) < D(p, \mathsf{closest}(p))$, update $\mathsf{closest}(p) = c_k$. `% This insures that` $D(p, \mathcal{C}_k) = D(p, \mathsf{closest}(p))$ `indeed`
4. Output $\mathcal{C} = \langle c_1 \dots c_{|M|} \rangle$.

It is easy to see that each iteration of the greedy algorithm is implemented in linear time $O(|M|)$, justifying the running time.

Now, take any $2 \leq k \leq |M|$. The 2-competitiveness of $\mathcal{C}$ follows from the two claims below.

Claim 1: $d_{\mathcal{M}}(\mathcal{C}_k) = D(c_k, \mathcal{C}_{k-1})$, i.e. the closest pair of points in $\mathcal{C}_k$ includes c_k.
Proof: Assume $d_{\mathcal{M}}(\mathcal{C}_k) = D(c_i, c_j) < D(c_k, \mathcal{C}_{k-1})$, where $i < j < k$. Then $D(c_j, \mathcal{C}_{j-1}) = D(c_i, c_j) < D(c_k, \mathcal{C}_{k-1}) \leq D(c_k, \mathcal{C}_{j-1})$, i.e. c_k should have been added before c_j, a contradiction. $\square$

Claim 2: $D(c_k, \mathcal{C}_{k-1}) \geq \frac{1}{2} \cdot \mathsf{opt}\text{-}d_{\mathcal{M}}(k)$.
Proof: Let $b_1 \dots b_k$ be the optimum code of cardinality k, i.e. $D(b_i, b_j) \geq$

$\mathsf{opt}\text{-}d_{\mathcal{M}}(k)$ for some $i \neq j$. Then the k open balls of radius $R = \frac{1}{2} \cdot \mathsf{opt}\text{-}d_{\mathcal{M}}(k)$ around the b_i's are all disjoint. Hence, at least one of these k balls does not contain any of the first $(k-1)$ selected points $c_1 \ldots c_{k-1}$. Say this is the ball around b_j. Hence, $D(b_j, \mathcal{C}_{k-1}) \geq R$. But c_k is the *furthest* point from $\mathcal{C}_{k-1}$, and, therefore, $D(c_k, \mathcal{C}_{k-1}) \geq D(b_j, \mathcal{C}_{k-1}) \geq R$. □

We next give an example of $\mathcal{M}$ with $\mathsf{opt}\text{-}r_{\mathcal{M}} = 2$. Let $M = \{w, x_1, x_2, y_1, y_2, z\}$, where $D(w, x_i) = 1$, $D(x_i, y_j) = 2$, $D(y_j, z) = 1$, $i, j = 1, 2$, and the other distances are the length of the shortest paths induced by the above assignments. In particular, the two furthest points are w and z of distance 4, and the best 4-code is $\{x_1, x_2, y_1, y_2\}$ of minimum distance 2. In other words, $\mathsf{opt}\text{-}d_{\mathcal{M}}(2) = 4 = D(w, z)$ and $\mathsf{opt}\text{-}d_{\mathcal{M}}(4) = 2 = D(x_i, y_j)$, $i, j = 1, 2$. Now, for any incremental code $\mathcal{C} = \langle c_1, c_2, c_3, c_4 \rangle$, unless $\mathcal{C}_4 = \{x_1, x_2, y_1, y_2\}$, one of the pairwise distances in $\mathcal{C}_4$ will be 1, giving a gap of $2/1 = 2$. On the other hand, if $\mathcal{C}_4 = \{x_1, x_2, y_1, y_2\}$, then $D(c_1, c_2) = 2$, giving again a gap of $4/2 = 2$ for the 2-prefix of $\mathcal{C}$.

Finally, we show that it is NP-hard to construct an A-competitive code for $A < 2$ when given $\mathcal{M}$ as an input, even if $\mathsf{opt}\text{-}r_{\mathcal{M}} = 1$. We make a reduction from the Maximum Independent Set problem, which is known to be NP-complete [GJ79]. Given a graph $G = (V, E)$, we define a metric space $\mathcal{M} = (M, d)$, where $M = V$ and $D(i, j) = 1$ iff $(i, j) \in E$, and $D(i, j) = 2$ otherwise. Let k be the (unknown) value of the maximum independent set in G. It is easy to see that the an optimal incremental code for $\mathcal{M}$ is 1-competitive, and should first list the k elements of any maximum independent set of G (in any order), followed by the other elements (in any order). In particular, $\mathsf{opt}\text{-}d_{\mathcal{M}}(i) = 2$ for $i \leq k$ and $\mathsf{opt}\text{-}d_{\mathcal{M}}(i) = 1$ otherwise. Moreover, any such 1-competitive code induces the maximum independent set by looking at the largest i-prefix with minimal distance 2.

On the other hand, any code for $\mathcal{M}$ which is not 1-competitive for $\mathcal{M}$ must be 2-competitive. Hence, if we have a procedure that can produce an A-competitive code for $\mathcal{M}$, where $A < 2$, this procedure must in fact produce an optimal 1-competitive code. But we just argued that in this case we can compute the largest independent set of G, which is NP-hard.

Remark 1. Notice that the greedy algorithm above is exactly the same as that of Gonzalez [G85] for the so called k-center problem. This is a just a coincidence, since our problems and the analysis are quite different.

Remark 2. While the greedy algorithm is extremely efficient for generic metric spaces, we are mainly interested in the Hamming space F^n. In this case we cannot afford to scan the entire (exponential in n) space to add every new point. Therefore, the greedy algorithm is inefficient for the Hamming space. See Section 4 for the discussion of efficiency and efficient constructions for F^n.

Extending to compact spaces. Aside from the complexity considerations, most of the discussion above (in particular, the greedy algorithm) extends to infinite *compact* metric spaces, like "nice" compact subsets in $\mathbb{R}^n$. The main difference is that "exhaustive" codes now become countably infinite codes, and we require every finite prefix of such a code to be "A-compatible" w.r.t. the best finite code of the same cardinality. We illustrate it in detail in the next section, when talking about the real interval $[0, 1]$.

3 Optimal Code for $[0, 1]$

An incremental code over $[0, 1]$ is simply a sequence of points $\mathcal{C} = \langle p_1, p_2, \ldots \rangle$. If we let $q_1^i \ldots q_i^i$ denote $p_1 \ldots p_i$ in the increasing order, then after i steps $[0, 1]$ is split into $(i + 1)$ intervals $I_0 = [0, q_1^i]$, $I_2 = [q_1^i, q_2^i], \ldots, I_i = [q_i^i, 1]$. Clearly, the minimal distance of $\mathcal{C}_i$ is $d(\mathcal{C}_i) = \min(|I_1|, \ldots, |I_{i-1}|)$, while the optimum distance is $\mathsf{opt}\text{-}d(i) = 1/(i-1)$ (by spreading the points uniformly). When adding p_{i+1} we simply subdivide one of the I_j's into two subintervals. If we assume that $p_1 = 0$ and $p_2 = 1$ (which will happen in our solution and will be the "worst case" in the lower bound proof), then the "border" intervals I_0 and I_{i+1} disappear, and our objective is to place the points $p_3, p_4, \ldots$ on $[0, 1]$ in such a manner that the length of the *smallest* interval after each p_i is as close to $1/(i-1)$ as possible. We notice that the dual "maximal interval" version of the latter problem – make the *largest* interval as close to $1/(i - 1)$ as possible – is a well known *dispersion problem* (see [DT97,C00,M99]). While our lower bound and its proof will be somewhat different for our "minimal interval" version, it will turn out that the optimal sequence for both versions will be the *same*, which is not at all clear a-priori.

As a warm-up, let us examine the behavior of the greedy algorithm in Theorem 1, which is guaranteed to give at least a 2-competitive code. Assume we start with $c_1 = 0$. The next step makes $c_2 = 1$, and then we simply keep subdividing the largest interval in half. Thus, after 2^k points, all but one interval will be of size $1/2^k$, but that last interval will be of size $1/2^{k-1}$. Since the optimum for 2^k points is $1/(2^k - 1)$, we get the ratio at least $(2^k - 1)/2^{k-1} \to 2$. Thus, greedy actually gives a ratio of 2. As we show now, the best (infinite) incremental code for $[0, 1]$ actually achieves the ratio of $\ln 4 = \log_e 4 \approx 1.386 < 2$.

Let $H(k) = (1 + \frac{1}{2} + \ldots + \frac{1}{k})$ denote the k-th harmonic series.

Lemma 1. *Unless* $A \geq 2 \cdot [H(2i) - H(i + 1)]$, *no incremental code of* $(2i + 1)$ *points in* $[0, 1]$ *is* A*-competitive.*

Proof. Consider a code of $2i + 1$ points in $[0, 1]$ with competitive ratio A. For every $j \leq 2i + 1$, consider the distances between adjacent points after placing the first j points. Let $\ell_1^j \leq \ell_2^j \leq \ldots \ell_{j-1}^j$ be these distances, sorted in increasing order. We need the following claim:

Claim: $\ell_k^j \leq \ell_{k+2}^{j+1}$ for $1 \leq k \leq j - 2$.
Proof: Adding a point (in this case, $(j + 1)$-st point) can either add one more distance to the list of interval distances (if the new point is the rightmost or the leftmost), or it can remove one length from the list, replacing it with two others (if the new point lies between two old points). In either case, there are at most two new lengths that are added to list. This means that among the first $k + 2$

lengths on the new list, there are at least k lengths that were already on the old list before we added the last point. Hence, the $(k+2)$-nd smallest length on the new list cannot be smaller than the k-th smallest length on the old list. □

By iterating the above claim, we get for all $0 \le j \le i-1$, $\ell_1^{2i+1-j} \le \ell_{1+2j}^{2i+1}$. Notice, ℓ_1^{2i+1-j} is the length of the smallest interval after adding $(2i+1-j)$ points. Since our code is A-competitive (and since the optimal arrangement of $2i+1-j$ points has distance $1/(2i-j)$), we must have $\frac{1/(2i-j)}{\ell_1^{2i+1-j}} \le A$, which means that $\ell_{1+2j}^{2i+1} \ge \ell_1^{2i+1-j} \ge 1/(A(2i-j))$. Summing the last inequality for $j = 0 \ldots i-1$, we get

$$\sum_{j=0}^{i-1} \ell_{1+2j}^{2i+1} \ge \frac{1}{A} \cdot \left(\frac{1}{2i} + \frac{1}{2i-1} + \ldots + \frac{1}{i+2} \right) = \frac{1}{A} \cdot [H(2i) - H(i+1)] \qquad (1)$$

On the other hand, since $\ell_{1+2j}^{2i+1} \le \ell_{2+2j}^{2i+1}$ and all the $2i$ intervals sum to at most 1, so we get

$$\sum_{j=0}^{i-1} \ell_{1+2j}^{2i+1} \le \sum_{j=0}^{i-1} \left(\frac{\ell_{1+2j}^{2i+1} + \ell_{2+2j}^{2i+1}}{2} \right) \le \frac{1}{2} \qquad (2)$$

Combining Equation (1) and Equation (2), we get $A \ge 2 \cdot [H(2i) - H(i+1)]$. □

Since $2 \cdot [H(2i) - H(i+1)] \approx 2 \ln \left(\frac{2i}{i+1} \right) \stackrel{i \to \infty}{\longrightarrow} \ln 4$, we get

Corollary 1. *If C is an infinite A-competitive code, then $A \ge \ln 4 \approx 1.386$.*

We now show an incremental code achieving the bound above. We let $p_0 = 0$, $p_1 = 1$, and explicitly tell the lengths of the i intervals after the first $(i+1)$ points. They are (in increasing order): $\log_2(1 + \frac{1}{2i-1}), \log_2(1 + \frac{1}{2i-2}), \ldots, \log_2(1 + \frac{1}{i})$. Notice, $\sum_{j=1}^{i} \log_2(1 + \frac{1}{2i-j}) = \log_2(\prod_{j=1}^{i} \frac{2i-j+1}{2i-j}) = \log_2(\frac{2i}{i}) = 1$, indeed. Also, for $i = 1$ our only interval is indeed of size $1 = \log_2(1 + \frac{1}{1})$. To add the $(i+2)$-nd point, we subdivide the currently largest interval of size $\log_2(1 + \frac{1}{i})$ into two intervals of sizes $\log_2(1 + \frac{1}{2i})$ and $\log_2(1 + \frac{1}{2i+1})$ (again, arithmetic works), as claimed. We see that the length of the smallest interval after $(i+1)$ points is $\log_2(1 + \frac{1}{2i-1}) \ge \frac{1}{i \ln 4}$ (the latter is easy to check), proving that this sequence has competitive ratio $\ln 4$.

Remark 3. The above argument (construction and the lower bound) can be adjusted to the case of the closed circle S^1 (where distance is the shortest arc of the circle). Any point p on the circle is identified with both 0 and 1 on the interval, and then the same code as above is used. Thus, the i intervals we get after i points have the same lengths as the i intervals we get after $(i+1)$ points in the $[0,1]$-construction, but both optima are still $1/i$, so the ratio is unchanged.

4 Hamming Space (Error-Correcting Codes)

We now turn our attention to the most practically important space – the Hamming space F^n. For that, we first recall some terminology and address some of the issues specific to this space.

TERMINOLOGY. Recall, the Hamming distance between $x, y \in F^n$ is the number of positions they disagree. Unless otherwise stated, we will assume that $F = [q]$ is a field of size q. A code with K codewords and minimal distance d over F is said to have *rate* $\tau = \log_q K/n$, *dimension* $k = \log_q K$ and *relative distance* $\delta = d/n$. We omit the subscript to the space from the quantities $\mathsf{opt}\text{-}r$ and $\mathsf{opt}\text{-}d$ when the Hamming space is clear from the context, and otherwise write $\mathsf{opt}\text{-}r(K;\ q, n)$ and $\mathsf{opt}\text{-}d(K;\ q, n)$ to emphasize the space. We let $\mathsf{opt}\text{-}\delta(\tau;\ q, n) = \frac{1}{n} \cdot \mathsf{opt}\text{-}d(q^{\tau n};\ q, n)$ be the largest possible relative distance of a code of rate τ over F. We let $V_q(R;\ n)$ denote the volume of an n-dimensional sphere of radius R in F^n, and notice that asymptotically, we have $\frac{1}{n} \cdot \log_q V_q(\alpha n;\ n) \approx H_q(\alpha) = \alpha \log_q(q-1) - \alpha \log_q \alpha - (1-\alpha)\log_q(1-\alpha)$, where $H_q(\cdot)$ above is the q-ary entropy function (in particular, $V_q(\alpha n;\ n) \le q^{nH_q(\alpha)}$). Finally, a *linear* code of dimension k is given by the $k \times n$ generator matrix G, where the codewords are of the form xG for $x \in F^k$. Most codes we will use below are linear.

WHAT IS AN "EFFICIENT" CONSTRUCTION? As opposed to the generic case, where we are given the entire metric space as input and need to produce as output the "code" itself (as a list of points), in the case of the Hamming space we usually think of entire space as being exponential in the relevant parameters, and we think of the code as having some implicit small representation. What we may require in terms of efficiency is to have a representation of the code whose length is polynomial in n and $\log q$, and an efficient procedure that given this representation and an index i, produces c_i, the i-th codeword. For example, viewed in this light, a random code is not an "efficient construction", but a random linear code is.

HOW GOOD IS THE OPTIMAL CODE? To compute competitive ratios, we need to understand the behavior of the optimal code of a given rate. This is well known for codes over "large" alphabets ($q \ge n$, see Section 4.1). For codes over small alphabets, we only have bounds on the minimal distance of the optimal code, rather than a closed-form formula. Hence, the competitive ratio that we can prove depends not only on the performance of the code in question, but also on the quality of these bounds.

On a brighter side, the discrepancy between the known upper- and lower-bounds on the optimal distance as a function of rate is at most a small constant factor. In fact, the ratio between the Hamming bound (an upper-bound) and the Gilbert-Varshamov bound (a lower-bound) is a factor of 2 "in spirit". To see that, recall that the Hamming bound says that for any code with K codewords and minimum distance d over $[q]^n$, it holds that $K \cdot V_q(d/2;\ n) \le q^n$, i.e. $\mathsf{opt}\text{-}d(K;\ q, n) \le 2V_q^{-1}(q^n/K;\ n)$. The Gilbert-Varshamov bound, on the other hand, says that there exists a K-word code with minimum distance d satisfying $K \cdot V_q(d;\ n) \ge q^n$, i.e. $\mathsf{opt}\text{-}d(K;\ q, n) \ge V_q^{-1}(q^n/K;\ n)$.

In principle, this means that when we use the Hamming bound as our estimate for the performance of the optimal code, we only lose a factor of two (or less). However, notice that when we use that bound, we usually use some estimate for V_q^{-1} (since working with V_q^{-1} itself is too hard), so we may lose some small additional factor there (see Section 4.5 for an example).

ORGANIZATION. The rest of this large section is organized as follows. In Section 4.1 we consider the case $q \geq n$, and show that F^n is incrementally perfect, i.e. has a 1-competitive exhaustive code. The remaining sections deal with a more difficult case of small alphabets ($q < n$). Section 4.2 shows that F^n is not incrementally perfect, while the last three sections deal with efficient constructions achieving constant competitive ratios for various settings of parameters.

4.1 Optimal 1-Competitive Code for Hamming Space F^n, $|F| \geq n$

Let $|F| = q \geq n$. In this case, the well known *Reed-Solomon* codes are known to be the optimal codes for any given dimension k. These linear codes first view the elements of F^k as polynomials of degree at most $(k-1)$, and then output the values of these polynomial at n arbitrary points of F (here we use $q \geq n$). We make the following simple observation to turn them into (optimal) incremental codes: polynomials of degree at most 0 are special cases of polynomials of degree 1, which in turn are special cases of polynomials of degree at most 2, and so on. We now formalize this idea to get the claimed 1-competitive code.

Let $\alpha_1 \dots \alpha_n$ be arbitrary distinct elements of F. Given $a = (a_0 \dots a_{k-1}) \in F^k$, assign a polynomial $p_a(x) = \sum_{i=0}^{k-1} a_i x^i$ of degree at most $(k-1)$, and output the codeword $(p(\alpha_1) \dots p(\alpha_n)) \in F^n$. Since any two distinct polynomials of degree at most $(k-1)$ can agree on at most $(k-1)$ points in F, the distance of the RS-code at least (in fact, exactly) $d = n - k + 1$. On the other hand, the classical singleton bound says that *any* code of dimension k must have minimal distance at most $(n - k + 1)$, achieved by the corresponding RS-code.

Now, if we first encode (i.e., evaluate at n points) the polynomials of degree 0, followed by the polynomials of degree 1 and so on, we see that the minimal distance of our now *incremental* code slowly decreases from n to $(n-1), \dots$, all the way to 1. Moreover, at the time we are encoding polynomials of degree *exactly* k, our current code's minimal distance $(n-k+1)$ is optimal by the singleton bound (as we already listed q^{k-1} polynomials of degree at most $(k-1)$). Thus, we showed that the Hamming space F^n has a 1-competitive exhaustive code $\mathcal{C} = \langle c_0, \dots, c_{N-1} \rangle$, where $N = q^n$. Additionally, it is easy to compute c_i given i. Indeed, if we write the elements of F as numbers $0, \dots, (q-1)$ (0 being the "zero" of F), and then interpret the representation of an integer $i \in \{0, \dots, N-1\}$ base $|F|$ as a string $a(i) \in F^n$, then listing i in the increasing order corresponds to the lexicographic order of the $a(i)$'s, which also lists the polynomials $p_{a(i)}$ in the order of increasing degrees, as needed. To summarize, we get

Theorem 2. *F^n is incrementally perfect when $|F| = q \geq n$ and F is a field. Specifically, letting $c_i = (p_{a(i)}(\alpha_1), \dots, p_{a(i)}(\alpha_n)) \in F^n$, the incremental code*

$\mathcal{C} = \langle c_0, \ldots, c_{N-1} \rangle$ *is exhaustive and* 1*-competitive. Moreover, for* $1 \le k \le n$ *and when* $q^{k-1} \le i < q^k$*, the minimal distance of* $\{c_0, \ldots, c_i\}$ *is* $(n - k + 1)$.

PRACTICAL DISCUSSION: We notice that the procedure of computing c_i is extremely simple, practical and efficient (at most quadratic in n and $\log |F|$), and does not need to keep any state as i grows. In particular, it gives a practical solution to the "accounts problem" in the introduction.

4.2 First Look at Small Alphabets

In this section we discuss the Hamming space F^n over small alphabets (in particular, binary). As a first negative observation, we notice that the Hamming space is not incrementally perfect when $q < n$.

Lemma 2. *If* $q < n$*, then the Hamming space* $[q]^n$ *is not incrementally perfect, unless* $q = 2, n = 4$.

Proof. When $q < n$, we do not know the exact behavior of optimal codes. Luckily, we can establish the claim by looking only at the first $q + 1$ points of the code, when we still know what the optimum looks like. So let $q < n$ and assume that $[q]^n$ is incrementally perfect. Since the words $1^n \ldots q^n$ have pairwise distance n, we have that $\mathsf{opt}\text{-}d(q) = n$. Hence, if $[q]^n$ has a 1-competitive incremental code $\mathcal{C}$, it must be that $d(\mathcal{C}_q) = n$. Namely, any two of the first q words of $\mathcal{C}$ must differ in all the coordinates $1 \le i \le n$. In fact, we can assume w.l.o.g. that the words $0^n \ldots (q-1)^n$ are the first q words in $\mathcal{C}$.[1] This means, however, that the $(q + 1)$'st word of $\mathcal{C}$ must agree with one of the first q words in at least $\left\lceil \frac{n}{q} \right\rceil$ positions, implying that $d(\mathcal{C}_{q+1}) \le n - \left\lceil \frac{n}{q} \right\rceil$.

On the other hand, we now show that the optimal $(q+1)$-word code in $[q]^n$ has minimum distance at least $n - \left\lceil \frac{2n}{q(q+1)} \right\rceil$. Consider a $(q+1) \times n$ matrix, whose rows would correspond to $q + 1$ codewords. We describe how to fill this matrix so that every two rows would agree in at most $\lceil 2n/(q(q+1)) \rceil$ coordinates. Specifically, we fill the columns of this matrix in "chunks" of $\binom{q+1}{2}$ at a time, making sure that in each "chunk" every two rows agree in at most one coordinate. This is done as follows: the $\binom{q+1}{2}$ columns in each "chunk" correspond to all pairs of distinct indices $1 \le i < j \le q + 1$. Namely, the column $v(i, j)$ corresponding to the pair (i, j) has symbol q in positions i and j, and all the other symbols $1 \ldots (q-1)$ in the other $(q-1)$ positions of $v(i, j)$ (in arbitrary order). So within column $v(i, j)$, the only two positions that agree are positions i and j. It follows that within the current "chunk", any two rows i and j agree only in the column $v(i, j)$. And as there at most $\lceil n/\binom{q+1}{2} \rceil = \left\lceil \frac{2n}{q(q+1)} \right\rceil$ such "chunks", any two rows agree in at most $\left\lceil \frac{2n}{q(q+1)} \right\rceil$ coordinates.

[1] This is true since we can always permute the symbols in coordinate i of all the words of $\mathcal{C}$, without changing any of the distances of the code.

Since we assume that $\mathcal{C}$ is 1-competitive, we must have

$$n - \left\lceil \frac{2n}{q(q+1)} \right\rceil \le \text{opt-}d(q+1) = d(\mathcal{C}_{q+1}) \le n - \left\lceil \frac{n}{q} \right\rceil$$

i.e. $\left\lceil \frac{n}{q} \right\rceil \le \left\lceil \frac{2n}{q(q+1)} \right\rceil$. It is not hard to see that the only pair $n > q$ that satisfies this inequality is $n = 4$ and $q = 2$, i.e. no other space can be incrementally perfect. As for $\{0,1\}^4$, it is indeed incrementally perfect via the optimal code $\{0000, 1111, \text{ all words with two 1's, all the rest}\}$. □

Remark 4. Notice that a tighter analysis shows that for any q, if n is large enough, any code of size $(q+1)$ in $[q]^n$ has competitive ratio at least $1+1/\binom{q+2}{2}$. For example, for $q = 2$ we get the ratio of at least $7/6$. And indeed, the optimal incremental code of size 3 is $\{0^n, 0^{n/7}1^{6n/7}, 1^{4n/7}0^{3n/7}\}$. Indeed, opt-$d(2) = n$ and opt-$d(3) = 2n/3$ (via $\{0^{2n/3}1^{n/3}, 0^{n/3}1^{n/3}0^{n/3}, 1^{n/3}0^{2n/3}\}$), so both 2- and 3-prefixes have ratios $n/(6n/7) = 7/6$ and $(2n/3)/(4n/7) = 7/6$.

Constructions. We now turn to the question of efficient constructions of incremental codes over small alphabets. As a most trivial construction, consider a regular code with K codewords, minimum distance d and relative distance $\delta = d/n$. How well does it perform as an incremental code (under arbitrary ordering)? Without any additional knowledge about the code, the best competitive ratio we can get is $n/d = 1/\delta$. Still, if we take a family of *asymptotically good* codes[2], we get a family of incremental codes with constant rate and constant competitive ratio. Of course, this simplistic construction has several shortcomings. First, there is a pretty stringent tradeoff between the rate and the relative distance of the code, so we will either sacrifice the rate (make the code very sparse), or the competitive ratio. In particular, if the rate is close to 1, the completive ratio tends to ∞. Secondly, even on small prefixes our code can have the same minimal distance d, i.e. the distance does not necessarily "gradually decrease". Because of that, there is no point in using more sophisticated bounds than n on the optimal code's minimal distance. Thus, this approach does not address the essence of the problem at all.

Therefore, we use more sophisticated techniques that will give us better tradeoffs between the rate and the competitive ratio of incremental codes. Specifically, we examine three efficient constructions: (1) using algebraic-geometric codes (AG-codes) as a natural generalization of the RS-codes to small fields, (2) using concatenation theorem to reduce the alphabet size, and (3) using random linear codes. The latter construction will let us achieve our ultimate goal: have an absolute constant competitive ratio (close to 2), even when rate is $1 - o(1)$.

[2] A family of codes $\{\mathcal{C}^n\}_{n\in\mathcal{N}}$ is asymptotically good if both the relative distance and the rate of $\mathcal{C}^n$ is $\Omega(1)$.

4.3 Algebraic-Geometric Codes

Algebraic-Geometric Codes (AG-codes) are natural extensions of the RS-codes to small fields. Detailed treatment of AG-codes is beyond the scope of this paper, so we informally concentrate on the essentials only (see [S93] for more information). Rather than talking about polynomials of degree at most α which can be evaluated at n points in the field, AG-codes deal with algebraic functions with at most α "poles" at the "point of infinity" which can be evaluated at n "rational points" of the function field. In both cases, the valuation map returns n elements of F, and a given polynomial/algebraic function can have at most α zeros. We let $L(\alpha, \infty)$ denote the space of such functions, which turns out to be a linear space over F. The famous Riemann-Roch theorem says that the dimension of this space is at least $\alpha - g + 1$, where g is the "genus" of the algebraic field (for the RS-codes we can achieve $g = 0$, but for smaller fields g cannot be very small; see below). All together, AG-codes given by the space $L(\alpha, \infty)$ have the following parameters: the dimension $k \geq \alpha - g + 1$, the distance $d \geq n - \alpha$. Like with the RS-codes, we observe that $L(0; \infty) \subseteq L(1; \infty) \subseteq \ldots \subseteq L(n-1; \infty)$, which implies that AG-codes of increasing pole orders at infinity define an incremental code, the first q^k codewords of which have minimal distance at least $(n - k - g + 1)$. Since the singleton bound still says that the optimum distance is at most $(n - k + 1)$, we get

Theorem 3. *The competitive ratio of AG-codes of dimension k (listed in the order specified above) is at most $\frac{n-k+1}{n-k-g+1}$. In particular, setting $k = n - 2g + 1$ gives an incremental code with q^k codewords and competitive ratio at most* 2.

While the main advantage of the AG-codes is the fact that they are defined on small (e.g. constant size) alphabets, we briefly point out their limitations. In particular, the bound above is meaningful only when $k < n - g$, i.e. the rate can be at most $(1 - \frac{g}{n})$. It is known that $g \geq n/(\sqrt{q} - 1)$,[3] so the maximal rate we can hope to achieve is roughly $(1 - \frac{1}{\sqrt{q}-1})$.

4.4 Concatenation Theorem

We next address a general method of constructing an incremental code over small alphabet from the one over a large alphabet and a good regular error-correcting code over a small alphabet. This method is completely analogous to the one used when constructing regular codes over small alphabets, and is called the *concatenation* of codes.

Let $\mathcal{C} = \langle c_1 \ldots c_K \rangle$ be an incremental code in $[q]^n$, with competitive ratio A and rate $\tau = (\log_q K)/n$. Let $T = \{t(1) \ldots t(q)\}$ be a regular code in $[q_2]^{n_2}$ ($q_2 \ll q$), with distance d_2, relative distance $\delta_2 = d_2/n_2$, and rate $\tau_2 = (\log_{q_2} q)/n_2$. An incremental code $\mathcal{C}^* = \mathcal{C} * T = \langle c^*(1) \ldots c^*(K) \rangle \subseteq [q_2]^{n \cdot n_2}$, the *concatenation* of $\mathcal{C}$ and T, is defined as follows. If we write the i-th codeword of $\mathcal{C}$ as $c_i =$

[3] When n grows w.r.t. q, and this bound is tight for certain q's. The general bound is $g \geq (n - q - 1)/(2\sqrt{q})$.

$c_{i,1} \ldots c_{i,n} \in [q]^n$, and interpret each $c_{i,j}$ as an integer in $\{1 \ldots q\}$, then the i-th codeword of $\mathcal{C}^*$ is $c_i^* = t(c_{i,1}) \ldots t(c_{i,n}) \in [q_2]^{nn_2}$. The code $\mathcal{C}$ is called the "outer code", and T is called the "inner code".

Theorem 4. *$\mathcal{C}^*$ is an incremental code in $[q_2]^{nn_2}$ with K codewords, rate $\tau^* = (\log_{q_2} K)/(nn_2) = (\log_q K \cdot \log_{q_2} q)/(nn_2) = \tau\tau_2$, and competitive ratio A^* satisfying:*

$$A^* \leq \frac{A}{d_2} \cdot \max_{i \leq K} \frac{\mathsf{opt}\text{-}d(i;\ q_2, nn_2)}{\mathsf{opt}\text{-}d(i;\ q, n)} = \frac{A}{\delta_2} \cdot \max_{\rho \leq \tau} \frac{\mathsf{opt}\text{-}\delta(\rho\tau_2;\ q_2, nn_2)}{\mathsf{opt}\text{-}\delta(\rho;\ q, n)} \tag{3}$$

Proof. Take the i-prefix $\mathcal{C}_i^*$ of $\mathcal{C}^*$. We claim that $d(\mathcal{C}_i^*) \geq d(\mathcal{C}_i) \cdot d_2$, which is clear from the construction of $\mathcal{C}^*$. Using also A-competiveness of $\mathcal{C}$ (i.e. $\mathsf{opt}\text{-}d(\mathcal{C}_i) \leq A \cdot d(\mathcal{C}_i)$), we get

$$\begin{aligned} A^* &= \max_{i \leq K} \frac{\mathsf{opt}\text{-}d(i;\ q_2, nn_2)}{d(\mathcal{C}_i^*)} \leq \max_{i \leq K} \frac{\mathsf{opt}\text{-}d(i;\ q_2, nn_2)}{\mathsf{opt}\text{-}d(i;\ q, n)} \cdot \frac{\mathsf{opt}\text{-}d(i;\ q, n)}{d(\mathcal{C}_i) \cdot d_2} \\ &\leq \frac{A}{d_2} \cdot \max_{i \leq K} \frac{\mathsf{opt}\text{-}d(i;\ q_2, nn_2)}{\mathsf{opt}\text{-}d(i;\ q, n)} \end{aligned}$$

□

Clearly, we can try to substitute some known upper (resp. lower) bounds in place of $\mathsf{opt}\text{-}\delta(i;\ q_2, nn_2)$ (resp. $\mathsf{opt}\text{-}\delta(i;\ q, n)$), to get a more algebraic expression in the bound above. For example, in the asymptotic sense we can use the Hamming bound in the numerator, and the Gilbert-Varshamov bound in the denominator, and get

$$\max_{\rho \leq \tau} \frac{\mathsf{opt}\text{-}\delta(\rho\tau_2;\ q_2, nn_2)}{\mathsf{opt}\text{-}\delta(\rho;\ q, n)} \lesssim \max_{\rho \leq \tau} \frac{2 \cdot H_{q_2}^{-1}(1 - \rho\tau_2)}{H_q^{-1}(1 - \rho)} = \frac{2 \cdot H_{q_2}^{-1}(1 - \tau\tau_2)}{H_q^{-1}(1 - \tau)}$$

(recall, H_q is the q-ary entropy function). However, such generic bound are often not much easier to work with, and different such bounds could be more convenient for different constructions.

Below we illustrate the bound from Theorem 4 for the case where the outer code is the incremental RS-code constructed in Section 4.1, which is perhaps the most attractive case to consider. Recall, from Section 4.1 that these incremental codes are 1-competitive, can be used with any rate $\tau \leq 1$, and have the restriction that $q \geq n$.

Corollary 2. *Let $\mathcal{C}$ be the 1-competitive RS-code of rate τ, and T be as before. Then $\mathcal{C}^*$ has rate $\tau^* = \tau\tau_2$ and competitive ratio*

$$A^* \leq \frac{1 - \tau\tau_2}{\delta_2(1 - \tau)} \tag{4}$$

Proof. We notice that for the RS-code we have $A = 1$ and $\mathsf{opt}\text{-}\delta(\rho;\ q, n) = 1 - \rho + 1/n$. Now the most convenient bound to use for $\mathsf{opt}\text{-}\delta(\rho\tau_2;\ q_2, nn_2)$ seems to be the (very loose in general) singleton bound, which says that opt-$\delta(\rho\tau_2;\ q_2, nn_2) \leq 1 - \rho\tau_2 + 1/(nn_2)$. Using Equation (3) now, we get

$$\max_{\rho \leq \tau} \frac{1 - \rho\tau_2 + \frac{1}{nn_2}}{1 - \rho + \frac{1}{n}} \leq \max_{\rho \leq \tau} \frac{1 - \rho\tau_2}{1 - \rho} = \frac{1 - \tau\tau_2}{1 - \tau}$$

□

We notice the tradeoff that we obtain. In particular, $\frac{1-\tau\tau_2}{\delta_2} \geq \frac{1-\tau_2}{\delta_2} \geq 1$ (the latter part follows from the singleton bound applied to T). Thus, our guarantee on A^* cannot be better than $1/(1-\tau)$. This implies that if we want a constant competitive ratio from $\mathcal{C}^*$, we cannot make $\tau = 1 - o(1)$. In other words, even though we can extend the RS-code to all the rates up to 1, our analysis can no longer provide a constant guarantee on A^*. On a positive note, we can make the rate $\tau^* = \tau\tau_2$ of $\mathcal{C}^*$ arbitrarily close to 1 (at the expense of A^*). Finally, it is also interesting to compare the bound in Equation (4) with the trivial bound we get by simply viewing $\mathcal{C}^*$ as a code of minimal relative distance $\delta^* = \delta_2(1-\tau)$. We see that we would get the ratio $1/(\delta_2(1-\tau))$, which is a factor $(1-\tau\tau_2)$ worse than our bound.[4]

JUSTESON CODES. Justeson codes, which were the first known asymptotically good *binary codes*, are also constructed using the concatenation theorem: the outer code is again any RS-code, while the inner codes are special "good average distance" codes of rate 1/2. It is well known (see [MS81]) that Justeson codes can achieve any rate $\tau < 1/2$, while having relative distance at least $(1-2\tau)H_2^{-1}(1/2) \approx 0.1(1-2\tau)$. A straightforward extension of our concatenation theorem shows that such a code could be ordered into an incremental code of rate τ and competitive ratio at most $10(1-\tau)/(1-2\tau)$. Again, we save a factor $(1-\tau)$ over the trivial bound (order arbitrarily).

4.5 Random Linear Codes

We saw that the explicit (and efficient) constructions from the previous sections failed to achieve a constant competitive ratio for rates $(1-o(1))$. On the other hand, Theorem 1 shows the existence (and inefficient construction) of an exhaustive 2-competitive codes for any $[q]^n$. In this section we show that, with high probability, a *random linear code* will achieve a competitive rate $2(1+\epsilon)$ (for arbitrarily small ϵ, and possibly better if of understanding of optimal codes will improve), even for rate $(1-o(1))$ (specifically, dimension up to $n - \Theta(\log_q n)$). This gives an efficient (albeit randomized) procedure to generate competitive and almost exhaustive incremental codes.

Recall that a (standard) random linear code of dimension k in $[q]^n$, is the set of words which are spanned by the rows of a $k \times n$ random matrix G over $[q]$. As we are interested in *ordered* codes, we consider these words in a canonical

[4] In fact, a tighter gap implied by Theorem 4 is opt-$\delta(\tau\tau_2;\ q_2, nn_2)$, which is usually much less than $1 - \tau\tau_2$.

order of their coefficients. Namely, for a given matrix G, the order between two codewords $c = xG$ and $c' = x'G$ is determined by the lexicographic order between x and x'.[5] We note that with this ordering, all the codewords that are spanned by the first i rows of G, appear before all the codewords that depend also the $i+1$'st row. This means that for any $m \leq k$, the prefix $\mathcal{C}_{q^m}$ is itself a (random) linear code.

We first recall an easy lemma that bounds the minimum distance of a random linear code.

Lemma 3. *For $k \leq n$, let G be a random $k \times n$ q-ary matrix, and let $\mathcal{C} = \mathcal{C}(G) = \{c_1 \dots c_{q^k}\}$ be the ordered linear code that is spanned by G. Then for every $\epsilon < 1 - \frac{1}{q}$, we have*

$$\Pr\left[d(\mathcal{C}) < \epsilon n\right] < 2^{k-n(1-H_q(\epsilon))}$$

Proof. Let $S(\epsilon n)$ be the sphere of radius ϵn around the origin in Hamming space $[q]^n$. We know that $S(\epsilon n)$ contains $V_q(\epsilon n;\ n) \leq q^{nH_q(\epsilon)}$ words. For a fixed k-vector $x \neq \bar{0}$, we have $\Pr_G[xG \in S(\epsilon n)] \leq q^{nH(\epsilon)-n}$. Using the union bound, we get

$$\Pr\left[d(\mathcal{C}) < \epsilon n\right] \;=\; \Pr[\exists x \neq \bar{0},\ xG \in S(\epsilon n)] \;<\; q^k \cdot q^{-n(1-H(\epsilon))}$$

□

Corollary 3. *Let $k < n$, $\delta > 0$, and G be a random $k \times n$ q-ary matrix. With probability of at least $1-\delta$ (over the choice of G), the minimum distance of $\mathcal{C}(G)$ is at least $nH_q^{-1}\left(\frac{n-k-\log_q(1/\delta)}{n}\right)$.*

Below we prove that a random linear code has competitive ratio of at most $2(1+\epsilon)$ for rates up to $1 - \Theta(\frac{\log_q n}{n})$. For this proof, we first recall a somewhat weak variant of the Hamming bound, and one fact about the (inverse of the) q-ary entropy function H_q.

HAMMING BOUND. For any q and $k \leq n$, $\mathsf{opt}\text{-}d(q^k;\ q, n) \leq 2nH_q^{-1}\left(\frac{n-k-1+\log_q n}{n}\right)$.

Proposition 1. *For any constant $\epsilon > 0$, there exists a constant "threshold" $\rho = \rho(\epsilon) > 0$, so that for all $\delta \leq \rho$, all $z \geq 2\delta/\epsilon$ and all $q \geq 2$, it holds that $H_q^{-1}(z+\delta)/H_q^{-1}(z) < 1 + \epsilon$.*

Theorem 5. *For any q, any constant $\epsilon > 0$, any large enough n, and any $k \leq n - (2+\frac{6}{\epsilon})\log_q n$, a random linear code of dimension k in $[q]^n$ has competitive ratio $A \leq 2(1+\epsilon)$, with probability at least $1 - \frac{1}{n}$.*

[5] Notice, this is the same lexicographic order we used with the RS-codes in Section 4.1.

Proof. Let G be a random $k \times n$ q-ary matrix and let $\mathcal{C} = \mathcal{C}(G)$ be the ordered linear code spanned by G. For each integer $m \leq k$, Corollary 3 with $\delta = 1/n^2$ tells us that with probability at least $1 - 1/n^2$, the minimum distance of $\mathcal{C}_{q^m}$ (i.e., the code spanned by the first m rows of G) is at least

$$d_m = nH_q^{-1}\left(\frac{n - m - 2\log_q n}{n}\right)$$

Taking the union bound, we conclude that with probability at least $1 - k/n^2 > 1 - 1/n$, the above holds for all $1 \leq m \leq k$. This, in turn, implies that for any m and any $i \in \{q^{m-1} + 1, \ldots, q^m\}$, the minimum distance of $\mathcal{C}_i$ is at least d_m.

On the other hand, the Hamming bound tells us that the minimum distance of the optimal q^{m-1}-word code cannot be more than

$$d^*_{m-1} = 2nH_q^{-1}\left(\frac{n - (m-1) - 1 + \log_q n}{n}\right) = 2nH_q^{-1}\left(\frac{n - m + \log_q n}{n}\right)$$

This implies that also for every $i \in \{q^{m-1} + 1, \ldots, q^m\}$, the minimum distance of the optimal i-word code cannot be more than d^*_{m-1}. We conclude that with probability at least $1 - 1/n$, the competitive ratio of $\mathcal{C}$ is bounded below by $\max_m(d^*_{m-1}/d_m)$.

Fix any $m \leq k$, and denote $\delta = \frac{3\log_q n}{n}$ and $z = \frac{n-m-2\log_q n}{n}$. Since $m \leq k \leq n - (2 + \frac{6}{\epsilon})\log_q n$, it follows that $z \geq (\frac{6}{\epsilon}\log_q n)/n = 2\delta/\epsilon$. Also, for large enough n we have $\delta \leq \rho(\epsilon)$ (where ρ is the "threshold" function from Proposition 1), we can use Proposition 1 to conclude that

$$\frac{d^*_{m-1}}{d_m} = \frac{2nH_q^{-1}(z + \delta)}{nH_q^{-1}(z)} \leq 2(1 + \epsilon)$$

As the inequality above holds for any $m \leq k$, this completes the proof of the theorem. □

5 Conclusions

We have introduced the notion of incremental codes. We studied these codes for general as well as for specific metric spaces (including the most important Hamming space). We showed that many standard error-correcting codes can be made incremental by simply ordering them in a special way (in particular, *lexicographic* ordering seems to be great for linear codes). Thus, we advocated the practice of making codes *ordered*, since this is useful and can often be made at no extra cost. We believe incremental codes will prove useful in other application, and will be studied further. The most immediate open problems include: (1) both explicit and efficient construction of codes of constant competitive ratio and rate $1 - o(1)$ (maybe even exhaustive); and (2) trying to prove better (even matching) lower bounds on the competitive ratio for codes over small alphabets.

References

C00. B. Chazelle. The Discrepancy Method: Randomness and Complexity. *Cambridge University Press*, 2000.

DT97. M. Drmota, R. Tichy. Sequences, Discrepancies, and Applications *Lecture Notes in Mathematics 1651*, Springer, Berlin, 1997.

GJ79. M. Garay, D. Johnson. Computers and Intractability. *W.H. Freeman and Company*, New York, 1979.

G85. T. Gonzalez. Clustering to minimize the maximum inter-cluster distance. *Theoretical Computer Science*, (38)293–306, 1985.

MS81. F. MacWilliams, J. Sloane. Theory of Error-Correcting Codes, Amsterdam, 1981.

M99. J. Matousek. Geometric Discrepancy (An Illustrated Guide). *Springer-Verlag*, Berlin, 1999.

S93. H. Stichtenoth. Algebraic Function Fields and Codes. *Springer-Verlag*, Berlin, 1993.

A 3/2-Approximation Algorithm for Augmenting the Edge-Connectivity of a Graph from 1 to 2 Using a Subset of a Given Edge Set
(Extended Abstract)

Guy Even[1], Jon Feldman[2], Guy Kortsarz[3], and Zeev Nutov[3]

[1] Dept. of Electrical Engineering-Systems,
Tel-Aviv University, Tel-Aviv 69978, Israel
guy@eng.tau.ac.il
[2] Massachusetts Institute of Technology,
Laboratory for Computer Science, Cambridge, MA, 02139, USA
jonfeld@theory.lcs.mit.edu
[3] Open University of Israel,
16 Klauzner St, Tel-Aviv, Israel {guyk,nutov}@oumail.openu.ac.il

Abstract. We consider the following problem: given a connected graph $G = (V, \mathcal{E}E)$ and an additional edge set E, find a minimum size subset of edges $F \subseteq E$ such that $(V, \mathcal{E} \cup F)$ is 2-edge connected. This problem is NP-hard. For a long time, 2 was the best approximation ratio known. Recently, Nagamochi reported a $(1.875 + \varepsilon)$-approximation algorithm. We give a new algorithm with a better approximation ratio of 3/2 and a practical running time.

1 Introduction

The 2-Edge Connected Subgraph Problem Containing a Spanning Tree (2-ECST) is defined as follows. The input consists of an undirected graph $G(V, \mathcal{E})$ and a set of additional edges (called "links") $E \subseteq V \times V$. A subset $F \subseteq E$ is called an augmentation of G into a 2-edge-connected graph if $G(V, \mathcal{E} \cup F)$ is 2-edge connected. The goal is to find an augmentation F of G into a 2-edge-connected graph with fewest links. This problem is NP-Complete [2]. We present a 1.5-approximation algorithm for the 2-ECST problem.

The 2-edge-connected components of G form a tree. It follows that by contracting these components, one may assume that G is a tree. Hence, the 2-ECST Problem is equivalent to the Tree Augmentation Problem (TAP) defined as follows. The input consists of a tree $T(V, \mathcal{E})$ and a set of links $E \subseteq V \times V$. The goal is to find a smallest subset $F \subseteq E$ such that $G(V, \mathcal{E} \cup F)$ is 2-edge connected.

Previous Results. Frederickson & Jájá [7] presented a 2-approximation to the weighted version of TAP. Improving the approximation ratio below 2 was posed by Khuller [12] as one of the main open problems in graph augmentation. Nagamochi & Ibaraki [11] presented a 12/7-approximation algorithm for TAP. However, the proof of the approximation ratio in [11] contains an error. Recently,

M. Goemans et al. (Eds.): APPROX-RANDOM 2001, LNCS 2129, pp. 90–101, 2001.

Nagamochi [10] reported a $(1.875+\varepsilon)$-approximation algorithm. The time complexity of the algorithm in [10] depends exponentially on $1/\epsilon$ (alternatively, the approximation ratio can be increased to 1.9). Our paper uses a few techniques that appear in [10,11] including: leaf-close trees, shadows, stems, and the usage of maximum matchings that forbid certain links. Our lower bound (Claim 12) is a strengthening of [10, Lemma 4.2].

Related Results. The weighted version of 2-ECST (or equivalently TAP) is called Bridge-Connectivity Augmentation (BRA) in [7]. In the BRA problem, the input consists of a complete graph $G(V, E)$ and edge weights $w(e)$. The goal is to find a minimum weight subset of edges F such that $G'(V, F)$ is 2-edge connected. The 2-ECST is simply the case in which the edges have $\{0, 1, \infty\}$ weights; the edges of the connected graph have zero weight, links have unit weight, and all the rest have infinite weight.

There are a few 2-approximation algorithms for the BRA problem. The first algorithm, by Frederickson and Jájá [7] was simplified later by Khuller and Thurimella [13]. These algorithms are based on constructing a directed graph and computing a minimum weight arborescence. The primal-dual algorithm of Goemans & Williamson [8] and the iterative rounding algorithm of Jain [9] are LP-based 2-approximation algorithms. The approximation ratio of 2 for all these algorithms is tight. Fredrickson and Jájá [7] showed also that when the edge costs satisfy the triangle inequality, the Christofides heuristic leads to a 3/2-approximation algorithm.

Cheriyan et. al. [3] presented a 17/12-approximation algorithm for the special case of BRA in which edges have $\{1, \infty\}$ weights. A 4/3-approximation algorithm was presented for $\{1, \infty\}$ weights by Vempala & Vetta [14]. Eswaran & Tarjan [4] presented a linear time algorithm for the special case of BRA in which all edges have unit weights. Surveys for broad classes of augmentation problems can be found in [12] and [5].

Equivalent Problems. The following two problems are equivalent to TAP, in the sense that the corresponding reductions preserve approximation ratios (e.g., see [2]): **(1)** Given a laminar family $\mathcal{S}$ of proper subsets of a ground set U, and an edge set E on U, find a minimum subset $F^* \subseteq E$ that "covers" $\mathcal{S}$ (that is, for every $S \in \mathcal{S}$, there is an edge in F, with one endpoint in S and the other in $V - S$). **(2)** Given a k-connected graph $G = (U, E_0)$ with k odd, and an edge set E on U disjoint to E_0, find a min size edge set $F^* \subseteq E$ such that $G \cup F$ is $(k+1)$-edge connected. Due to space limitations, all proofs as well as details of the algorithm are omitted. See the full version for these parts.

2 Preliminaries

Let $T(V, \mathcal{E})$ denote a tree defined over a vertex set V. A *rooted tree* is a tree T with a node $r \in V$ designated as a root. The root induces a partial order on V as follows. For $u, v \in V$, we say that u is a *descendant* of v, and that v is an *ancestor* of u, if v lies on the path from r to u in T; if, in addition, (u, v) is an edge of T, then u is a *child* of v, and v is the *parent* of u. We denote the parent of u by $p(u)$. The *least common ancestor*, $\text{LCA}(u, v)$, of u and v is the common ancestor of u and v with the largest distance to the root. We refer to a rooted tree as a pair (T, r), and do not mention the root when it is clear which vertex is the root. In the TAP problem, we designate an arbitrary node r of T as the *root*.

The *leaves* of a rooted tree T are the nodes in $V - r$ having no descendants. We denote the leaf set of T by $L(T)$, or simply by L, when the context is clear.

Consider a rooted tree (T, r) and a vertex $v \in V$. The *rooted subtree* of T induced by the descendants of v (including v) is denoted by T_v. Note that v is the root of T_v. A subtree T' of T is called a *rooted subtree* if $T' = T_v$, for some vertex $v \in V$. We say that a rooted subtree T' of T is *minimal w.r.t. property* $\mathcal{P}$ or that T' is $\mathcal{P}$*-minimal* if T' satisfies property $\mathcal{P}$, but no proper rooted subtree of T' satisfies property $\mathcal{P}$.

To *contract* a subset X of V is to combine all nodes in X to make a single new node x. All edges and links with both endpoints in X are deleted. All edges and links with one endpoint in X now have x as their new endpoint, but retain their correspondence with the edge or link of the original graph. If $r \in X$, then x becomes the root of the new tree. If parallel links arise, we consider an arbitrary underlying set of links. For simplicity, we say that such x contains a node v if $v \in X$. For a set of links $F \subseteq E$ let T/F denote the tree obtained by contracting every 2-edge connected component of $T \cup F$ into a single node.

Let $p(u, v)$ denote the path in T between u and v. Consider a link $(u, v) \in E$. If we decide to add (u, v) to the solution F, then the vertices along the path $p(u, v)$ belong to the same 2-edge connected component of $T \cup F$. Hence, we would like to contract the vertices along $p(u, v)$. Instead of defining a new node into which the vertices are contracted, it is convenient to regard this contraction as a contraction of the vertices along $p(u, v)$ into the least common ancestor, $\text{LCA}(u, v)$. The advantage of this convention is that the contracted tree is a subtree of the original tree. Note that we may obtain T/F simply by contracting the paths corresponding to the links of F one by one. Moreover, the resulting tree does not depend on the order by which the links are contracted.

We say that a link (u, v) *covers* all the edges along the path $p(u, v)$. This terminology enables us to formulate TAP as a covering problem. Namely, find a minimum size set of links that covers the edges of T. We extend this terminology and refer to covering a an edge (u, v) where v is the parent of u as *covering the node* u.

We say that a link (u', v') is a *shadow* of a link (u, v) if $p(u', v') \subseteq p(u, v)$, and (u', v') is a *proper shadow* of (u, v) if the inclusion is proper, that is $p(u', v') \subset p(u, v)$. A link is *maximal* if it is not a proper shadow of any other link. We say that a cover F of T is *shadows-minimal* if for every link $(u, v) \in F$ replacing (u, v) by a proper shadow of (u, v) results in a link set that does not cover T.

An edge (u, v) of a tree T is *reducible* w.r.t. a link set E if there is an edge (u', v') such that every maximal link that covers (u', v') also covers (u, v).

Definition 1. *A tree T and an edge set E defined over the same vertex set are* proper *if (a) for every edge (u,v) of T, there are at least 2 maximal links covering it, and (b) no edge in T is reducible.*

Note that non-proper TAP instances can be reduced without effecting the size of the solution. Hence from this point we assume that every tree we wish to cover is proper. One convenient property of proper trees is that every parent of a leaf has at least two children. By induction, it follows that if (T,E) is a proper pair, then there cannot be a path (of length 2 or more) consisting of degree-2 vertices ending at a leaf.

The following assumption simplifies many of the arguments we make.

Assumption 2. *The set of links E is closed under shadows, that is, if $(u,v) \in E$ and $p(u',v') \subseteq p(u,v)$ then $(u',v') \in E$.*

Every instance can be made to be closed under shadows by adding shadows of existing links as new links (i.e. shadow completion). Shadow completion does effect the solution size, since every shadow can be replaced by a maximal link.

In what follows, the TAP instance consists of the input tree $T = (V, \mathcal{E})$ and the set of links E. We assume that E covers T, that is, the graph $(V, \mathcal{E} \cup E)$ is 2-connected; this can be tested beforehand by contracting all the links.

3 Motivation: A 2-Approximation Algorithm

We start the discussion with a 2-approximation algorithm. This rather simple algorithm introduces the notions of leaf-closed subtrees, up-links, and the disjointness condition. Our 3/2-approximation extends these ideas.

For a set of links or edges F and a vertex v, let $\deg_F(v)$ denote the degree of v in the graph (V,F). Consider an arbitrary cover $F \subseteq E$ of T. A simple well known lower bound follows from the degree-sum of F, namely, $\sum_v \deg_F(v)$, as follows. Every leaf of T has at least one link of F incident to it, so the degree-sum of F is at least $|L(T)|$, and hence $|F| \geq |L(T)|/2$.

The ratio between the value of an optimal solution and this lower bound can be arbitrarily large; if T is a path, and E consists of links parallel to the edges of T, then $|L(T)|/2 = 1$, but the value of an optimal solution is $|V| - 1$. An approximation algorithm based on this lower bound requires applying the lower bound to a subproblem of the original problem and that a local ratio argument be used. This means that we compare the cost invested in covering the subproblem with the cost the optimum invests in this subproblem. We recurse on the remaining subproblem. Since we compare the cost of our cover of the subproblem with the cost of an optimal solution of the subproblem, it is necessary that the an optimal solution covers each of the charged subproblems by disjoint sets of links. We refer to this requirement as the *disjointness condition*.

Suppose the subproblem we choose is to cover the leaves (namely, the edges in $\mathcal{E}$ incident to the leaves of T). This can be done using $|L(T)|$ links; we just choose a set of links F with one link per leaf. The cost of covering the leaves is $|L(T)|$, and the leaf lower bound is $|L(T)|/2$, so the local approximation ratio for solving the subproblem is 2. However, recursing on T/F and repeating the same approximation argument fails. The reason is that a link in an optimal solution can cover a leaf in $L(T)$ as well as a leaf in $L(T/F)$. Hence, the disjointness condition is not met. This motivates the following definition and lemma, which appeared in the paper of Nagamochi and Ibaraki [11] for $U = L(T)$.

Definition 3. *Let $U \subseteq V$. A rooted subtree T' of T is* U-closed *(w.r.t. E) if every link in E having one node in $U \cap V(T')$ has it other node in $V(T')$.*

For every $U \subseteq V$, the tree T is U-closed. Hence the set of rooted subtree that are U-closed is not empty. This implies that, for every $U \subseteq V$, there exists a rooted subtree that is U-closed minimal.

The highest link incident to a node u is denoted by $up(u)$. Namely, among the links emanating from u and whose other endpoint is along the path $p(r, u)$, $up(u)$ is the link that covers the largest subpath of $p(r, u)$. For $U \subseteq V$, let $up(U) = \bigcup\{up(u) : u \in U\}$.

Lemma 4. *Let $U \subseteq V$, and let T' be a U-closed minimal rooted subtree of T. Then* $\mathrm{up}(U \cap V(T'))$ *covers T'.*

Lemma 4 is interesting when $L(T') \subseteq U$ (otherwise T' consists of single leaf).

A rooted subtree T' of T is *leaf-closed* if it is $L(T')$-closed. Lemma 4 implies the following claim, that appears in [11].

Claim 5. *[11] A leaf-closed minimal subtree T' of T is covered by* $\mathrm{up}(L(T'))$.

We now present a simple 2-approximation algorithm. The algorithm is recursive and its parameter is a tree T and a link set F. The algorithm starts with the whole tree T and an empty link set $F = \emptyset$. It finds a minimally leaf-closed subtree T', adds $up(L(T'))$ to F, and recurses on $T/up(L(T'))$ and F, until T is contracted into a single node.

Initialization: $F \leftarrow \emptyset$.
Algorithm *cover*(T, F):
1. If $|V(T)| = 1$, then return F and stop.
2. Compute a leaf-closed minimal subtree T' of T.
3. $F \leftarrow F \cup up(L(T'))$, $T \leftarrow T/up(L(T'))$.
4. Recurse: *cover*(T, F).

Clearly, the algorithm returns a feasible cover. We can prove that the size of the computed cover is 2-optimal by induction on k, the number of iterations of the algorithm. The induction base $k = 1$ follows from Claim 5. For the induction step, consider a leaf-closed minimal subtree T' of the first iteration of the algorithm. Let F^* be an arbitrary optimal cover of T. Let F^*_L be the links in F^* incident to the leaves of T', and let F^*_{res} be the links in F^* that cover at least one link not in T'. Then $|F^*_L| \geq |L(T')|/2$, and since T' is leaf closed, all the links in F_L^* have their both endpoints in $V(T')$. Hence, F_L^* and F_{res}^* are disjoint. By the induction hypothesis, the recursive call for covering $T/up(L(T'))$ produces a solution F_{res} with at most $2|F_{res}^*|$ links. Hence, the total number of links in the solution produced by the algorithm is:

$$|up(L(T'))| + |F_{res}| = |L(T')| + |F_{res}| \leq 2|F_L^*| + 2|F_{res}^*| = 2|F^*|.$$

In the following section we present a lower bound that strengthens the leaf lower bound, as well as a generalization of the idea of leaf-closed minimal subtrees in order to improve the approximation ratio to 3/2.

4 Lower Bound and the Credit Scheme

In this section we present a new lower bound on the size of a minimum cover. This lower bound is based on a maximum matching consisting of a subset of links. A credit scheme is established based on the lower bound.

Notation and Preliminaries. A non-root node $s \in V - r$ is a *stem* of a tree T if it has exactly two children, both of them are leaves, such that there is a link joining them. Such two leaves are called a *twin pair*, and the link between them is called a *twin link*. The edge connecting a stem s and its parent is called a *stem-edge*. We denote the the set of stems of a tree T by $St(T)$.

Leaves and stems play a special role in our algorithm. For short, we say that node of T is *special* it is a leaf or a stem. We denote the set of special nodes in T by $Sp(T)$. We denote the set of non-special nodes in T by $\overline{Sp}(T)$. We often use L, St, Sp for short to denote the set of leaves, the set of stems, and the set of special nodes if the context is clear. A rooted subtree T' of T is *leaf-stem-closed* if it is $L(T) \cup St(T)$-closed.

For $X, Y \subseteq V$ and a set of links F, let $F_{X,Y}$ denote the be the set of links in F that have one endpoint in X and the other in Y. We denote the number of links in F that emanate from a node v by $\deg_F(v)$. For a subset $U \subseteq V$ of nodes we denote the degree-sum of nodes in U with respect to a set of links F by $\deg_F(U)$. Namely, $\deg_F(U) = \sum_{v \in U} \deg_F(v)$.

A link (u, v) is called a *backward link* if u is a descendant of v or vice-versa. A link that is not a backward link is called a *side link*. A link e with a non-special endpoint is called a *bad* link.

We can also take advantage of another property of shadow-minimality. Let F be a shadow minimal cover of T. Every two links in F that share an endpoint cover disjoint sets of edges. Therefore, there is exactly one link in F emanating from every leaf v of T, and $F_{L(T),L(T)}$ is a matching.

Consider a matching M that consists of leaf-to-leaf links. A leaf $\ell \in L$ is *unmatched* with respect to M if M lacks a link incident to ℓ. The set of unmatched leaves in T with respect to M is denoted by $UL_T(M)$. A stem $s \in S$ is *uncovered* with respect to M if the corresponding stem-edge is not covered by the links of M. The set of uncovered stems in T with respect to M is denoted by $US_T(M)$.

Let F^* denote an optimal shadows minimal cover of T. The optimality of F^* implies that it is not possible that both the links (s, ℓ_1) and (s, ℓ_2) are in F^*, for a stem s and its children ℓ_1 and ℓ_2. The reason is that these two links can be replaced by the link (ℓ_1, ℓ_2).

Assumption 6. *Without loss of generality, a cover F^* of T lacks links from a stem to one of its children.*

We use Assumption 6 to assume that a stem s is uncovered with respect to $F^*_{L,L}$ iff F^* contains the twin-link between the children of s. The following claim follows.

Claim 7. *Let s denote a stem. If F^* satisfies Assumption 6, then, $deg_{F^*}(s) \leq 1$. Moreover, $deg_{F^*}(s) = 1$ iff s is uncovered with respect to $F^*_{L,L}$.*

4.1 Two Special Small Trees

In this section we consider two special subtrees. Subtrees of the first kind are called H-structures and play an important role in the algorithm. Subtrees of the second kind can be reduced to H-structures. Part (A) of Figure 1 depicts a *twin-thorn subtree*. This is a rooted subtree with 3 leaves: twin leaves a_1, b_1, and a leaf b_2 called the "thorn". The leaves a_1 and b_1 are a twin-pair that are the children of a common stem s. The stem s is connected to the root-node v by a path, all the internal nodes along which are nodes of degree 2. The node v has two children, one is an ancestor of s, and the other is a leaf b_2.

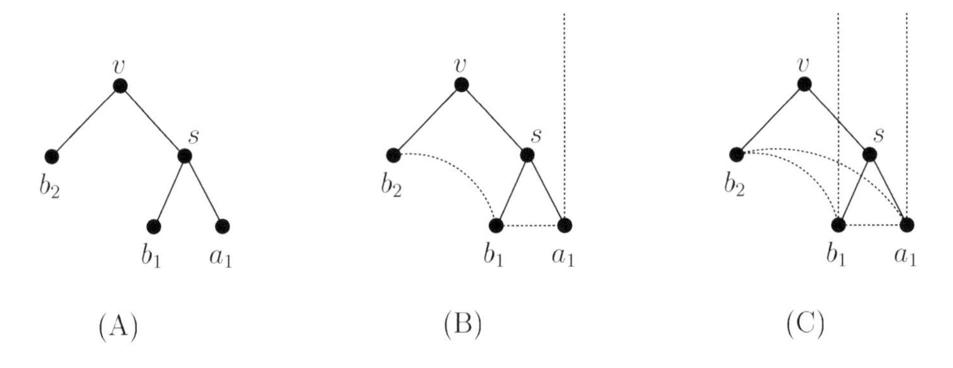

Fig. 1. (A) a twin-thorn subtree (B) an H-structure (C) A twin-thorn reducible to an H-structure.

Definition 8. *A subtree T_v is defined as an H-*structure *if: (i) T_v is isomorphic to a twin-thorn subtree, (ii) one of the twin-leaves is linked only to nodes that are its ancestors, and (iii) the other twin-leaf is linked to the thorn (b_1 may have other links incident to it). Part (B) of Figure 1 depicts an H-structure.*

The twin-leaf in an H-structure that is linked only to ancestors is called a *locked-leaf* (node a_1 in Fig. 1). The twin-link in an H-structure that has at least one side-link emanating from it is called a *locking-leaf* (node b_1 in Fig. 1). The link between the locking-leaf and the thorn is called the *locking-link* (link (b_1, b_2) in Fig. 1). A key observation is that if (a_1, b_1) does not belong to a feasible cover, then a_1 is covered by a bad link. Let F denote a set of links. An H-structure T_v is called *F-activated* if F contains the locking-link in T_v.

Reducible Twin-Thorn Subtrees. Consider the twin-thorn subtree T_v depicted in part (C) of Figure 1. Suppose that the links incident to nodes in T_v satisfy the following conditions: (a) (a_1, b_2) and (b_1, b_2) are links in E, and (b) the only side links emanating from a_1 and b_1 are (a_1, b_2) and (b_1, b_2), respectively.

Observe that if $(b_1, b_2) \in F^*$, then without loss of generality $up(a_1) \in F^*$. Similarly, if $(a_1, b_2) \in F^*$, then $up(b_1) \in F^*$. Assume that the path covered by the link $up(a_1)$ is not shorter than the path covered by the link $up(b_1)$. Informally, this means that $up(a_1)$ is "higher reaching" than $up(b_1)$. It follows that the links $up(a_1)$ and (b_1, b_2) cover all the edges that are covered by $up(b_1)$ and (a_1, b_2). Hence, we may discard the link (a_1, b_2) from E without increasing the size of an optimal solution. Note that this reduces T_v to an H-structure.

The above discussion justifies the first stage of the algorithm in which every twin-thorn subtree that satisfies the conditions described above is reduced to an H-structure. We summarize this reduction in the following assumption.

Assumption 9. *Let T_v denote a twin-thorn subtree. If the twin-leaves are linked only to the thorn or to their ancestors, then exactly one of the twin-leaves is linked to the thorn.*

4.2 Matchings

Our algorithm first computes maximum matching consisting of leaf-to-leaf links except for twin-links and locking-links. We denote this matching by M.

Now, consider an H-structure T_v. If both endpoints of a locking link (b_1, b_2) are unmatched leaves in $UL(M)$, we can add the locking-link (b_1, b_2) into the matching M. Let M^+ denote a maximal matching obtained by augmenting the matching M with all possible locking links. Let LL_{M^+} denote the set of locking-links in M^+, namely, $LL_{M^+} = M^+ - M$.

The following notation is now defined. Let M^* denote the set of leaf-to-leaf links in F^*, except for twin-links. Let *diff*(M, M^*) denote the set of links in M both endpoints of which are not matched by links in M^*. Observe that *diff*(M, M^*) is non-empty only if M^* is not a maximal matching. A link $e \in$ *diff*(M, M^*) is referred to as a *singleton*; the reason is that e shares no endpoints with other links in the symmetric difference of M^* and M. Let LL_{F^*} be the set of locking links in F^*. Note that $|LL_{F^*}|$ equals the number of F^*-activated H-structures.

Consider the following two terms: (a) $|M^*| - |M^+| +$ *diff*(M, M^*); the number of non-twin links in $F^*_{L,L}$ minus the number of links in M^+ plus the number of singletons, and (b) $|LL_{F^*}| - |LL_{M^+}|$; the number of F^*-active H-structures minus the number of M^+-active H-structures.

Claim 10. $|M^*| - |M^+| + |\text{diff}(M, M^*)| \leq |\text{LL}_{F^*}| - |\text{LL}_{M^+}|$

Observe that Claim 10 also holds if one considers a leaf-stem closed subtree T' and only considers the matching links and the H-structures within T'.

The Lower Bound. Consider a leaf-stem closed subtree T' of T. Recall that F^* denote a shadows-minimal optimal cover of T that satisfies Assumption 6. The following claim proves a lower bound on the number of links of F^* both endpoints of which are inside T' in terms of the number of leaves, uncovered stems, unmatched leaves and bad links in T'.

Claim 11. *Let T' denote a leaf-stem closed subtree, then*

$$|F^*_{V(T'),V(T')}| \geq \frac{1}{2} \cdot L + \frac{1}{3} \cdot |\text{US}_{T'}(F^*_{L,L})| + \frac{1}{6} \cdot |\text{UL}_{T'}(F^*_{L,L})|$$
$$+\frac{1}{3} \cdot deg_{F^*_{V(T'),V(T')}}(\overline{\text{Sp}}).$$

In the following lower bound claim, T' is a leaf-stem closed subtree, F^* is a shadows-minimal optimal cover of T, and M is a maximum matching consisting of leaf-to-leaf links without twin links or locking links. Let M^+ denote its augmentation with locking links. We denote by $M_{T'}$ and $M^+_{T'}$ the set of links in M and M^+ that incident to leaves in T'.

For every non-special node v let $\text{deg}'_{F^*_{V(T'),V(T')}}(v)$ denote the degree of v with respect to links of F^* both endpoints of which are in T' not including links connecting it to locked leaves in F^*-activated H-structures.

In the next claim M_S refers only to singletons internal to T'.

Claim 12. (The matching lower bound)

$$\frac{3}{2} \cdot |F^*_{V(T'),V(T')}| \geq \frac{3}{2} \cdot |M^+_{T'}| + \left(|L| - 2 \cdot |M^+_{T'}|\right) + \frac{1}{2}|\text{LL}_{M_{T'}}| + \frac{|M_S(F^*)|}{2}$$
$$+\frac{1}{2} \cdot deg'_{F^*_{V(T'),V(T')}}(\overline{\text{Sp}}).$$

5 A 3/2-Approximation Algorithm

In this section we present the techniques used in our 3/2-approximation algorithm. The algorithm is listed at the end of this section for reference. The algorithm is rather elaborate; more details and analysis will appear in the full version.

5.1 The Coupon Scheme

In order to use our lower bound, we need to apply a credit scheme. We say that a *coupon* can pay for a contraction of one link. We distribute coupons based on the matching lower bound from Claim 12. We assign 3/2 coupons to every pair of vertices that are matched by a link in M^+, and we assign 1 coupon to every leaf of T not covered by M^+. Furthermore, every M^+-activated H-structure gets half a coupon. By the matching lower bound, the number of coupons used is no more than $3|F^*|/2$.

In addition to coupons, we have tickets. Each ticket is worth half a coupon. We consider two types of tickets: (a) matching tickets - a matching ticket is given to each link in M_S. (b) golden tickets - a non-special node that has a link of F^* incident to it receives a golden ticket.

The difference between coupons and tickets is that the algorithm can assign coupons directly from the matching M^+ it computes. Tickets are harder to reveal and require a proof that they exist.

5.2 Algorithm Techniques

In the 2-approximation algorithm, we covered a leaf-closed subtree and recursed. In this algorithm, we are looking to cover a leaf-stem-closed subtree, and recurse. The main difficulty is to find enough coupons to pay for some leaf-stem-closed subtree. One important insight is that by using the coupon distribution described above, and maintaining the proper invariants, we can contract parts of the tree in order to find a leaf-stem-closed tree on which to recurse.

One invariant we maintain is a set A of *active* nodes, each of which always contain a coupon. We also build a set J of candidate links. As we add links to J, we contract the paths that the links of J cover. Hence we operate mostly on the tree T/J. The set A begins as the set of unmatched leaves.

The first step we take is to apply *greedy contractions*. An *α-greedy contraction* is defined as follows. Let T' be a subtree of T/J, (not necessarily rooted) coverable by α links (i.e. T' is a connected union of α paths, each covered by a link). If there are at least $\alpha+1$ coupons contained in T', then we can add these α links to J. The new node (created by contracting all the vertices of T' together) inherits the extra coupon, and becomes an active node. Thus we maintain the invariant that all active nodes contain a coupon.

After applying all possible 1-greedy and 2-greedy steps, we attempt to find a leaf-stem-closed subtree by finding an A-closed (or *active-closed*) subtree T' of $T/(J \cup M^+)$, the tree obtained by contracting all the links of J and those of the matching M^+. If this is leaf-stem-closed, we are done, and we can recurse, since we have enough coupons to pay for the links of J and M^+. The disjointness condition requires that the portions of F^* that are charged for solving each of the subproblems be disjoint. This way "double charging" is avoided. Note that the matching lower bound is applied to the links that are contained in a leaf-stem closed subtree T'. If the algorithm succeeds in contracting T', then the disjointness condition holds since F^* does not cover edges in $T - T'$ using links that are contained in T'. Hence the algorithm may recurse with $T - T'$.

If T' is not leaf-stem-closed, we need to be more careful. This requires some sophisticated case-analysis, and the details will appear in the full version of the paper.

Algorithm Tree-Cover(T)

1. If T contains a single node, then **Return**($\emptyset$).
2. Reduce the pair (T, E) to a proper pair.
3. Compute a maximum matching M consisting of leaf-to-leaf links that are not twin-links and not locking links. Augment it into M^+.
4. Define the set of active nodes A to be the set of unmatched leaves.
5. Initialize the set of candidate links: $J \leftarrow \emptyset$.
6. Apply 1-greedy contractions or 2-greedy contractions to T/J while possible. Update M^+, A, and J accordingly.
7. Find an active-closed minimal subtree $(T/(J \cup M^+))_v$ of $T/(J \cup M^+)$. Let T' denote the subtree $(T/J)_v$.
8. **While** T' is not leaf-stem closed, **do:**
   (a) If T' exhibits certain technical conditions, apply case analysis on the size and structure of T' (see full version for details).
   (b) Otherwise, cover T' by the basic cover. Formally,
      i. $J \leftarrow J \cup (M^+ \cap T') \cup up(active(T'))$,
      ii. $A \leftarrow A - active(T') + v$, and
      iii. $M^+ \leftarrow M^+ - T'$.
9. Cover T_v and recurse. Formally,
   (a) $J \leftarrow (J \cap T_v) \cup up(active((T/J)_v))) \cup (M^+ \cap (T/J)_v)$, and
   (b) **Return**($J \cup$ Tree-Cover(T/J)).

References

1. R. Bar-Yehuda, "One for the Price of Two: A Unified Approach for Approximating Covering Problems", Algorithmica 27(2), 2000, 131-144.
2. J. Cheriyan, T. Jordán, and R. Ravi, "On 2-coverings and 2-packing of laminar families", *Lecture Notes in Computer Science*, 1643, Springer Verlag, ESA'99, (1999), 510–520.
3. J. Cheriyan, A. Sebö, and Z. Szigeti, "An improved approximation algorithm for minimum size 2-edge connected spanning subgraphs", *Lecture Notes in Computer Science*, 1412, Springer Verlag, IPCO'98, (1998), 126–136.
4. K. P. Eswaran and R. E. Tarjan, "Augmentation Problems", *SIAM J. Computing*, 5 (1976), 653–665.
5. A. Frank, "Connectivity Augmentation Problems in Network Design", *Mathematical Programming*, State of the Art, Ed. J. R. Birge and K. G. Murty, 1994, 34–63.
6. A. Frank, "Augmenting Graphs to Meet Edge-Connectivity Requirements", *SIAM Journal on Discrete Mathematics*, 5 (1992), 25–53.
7. G. N. Frederickson and J. Jájá, "Approximation algorithms for several graph augmentation problems", *SIAM J. Computing*, 10 (1981), 270–283.
8. M. X. Goemans and D. P. Williamson, "A General Approximation Technique for Constrained Forest Problems", *SIAM J. on Computing*, 24, 1995, 296–317.
9. Kamal Jain, "Factor 2 Approximation Algorithm for the Generalized Steiner Network Problem", FOCS 1998, 448-457.

10. H. Nagamochi, "An approximation for finding a smallest 2-edge connected subgraph containing a specified spanning tree", TR #99019, (1999), Kyoto University, Kyoto, Japan.
`http://www.kuamp.kyoto-u.ac.jp/labs/or/members/naga/TC/99019.ps`
11. H. Nagamochi and T. Ibaraki, "An approximation for finding a smallest 2-edge-connected subgraph containing a specified spanning tree", *Lecture Notes In Computer Science*, vol. 1627, Springer-Verlag, 5th Annual International Computing and Combinatorics Conference, July 26-28, Tokyo, Japan, (1999) 31-40.
12. S. Khuller, Approximation algorithms for finding highly connected subgraphs, In *Approximation algorithms for NP-hard problems*, Ed. D. S. Hochbaum, PWS Publishing co., Boston, 1996.
13. S. Khuller and R. Thurimella, "Approximation algorithms for graph augmentation", *J. of Algorithms*, 14 (1993), 214–225.
14. S. Vempala and A. Vetta, "On the minimum 2-edge connected subgraph", Proc. of the 3rd Workshop on Approximation , Saarbrücken, 2000.

Approximation Algorithms for Budget-Constrained Auctions

Rahul Garg, Vijay Kumar*, and Vinayaka Pandit

IBM India Research Lab
IIT Delhi, New Delhi – 110016, India
{grahul,pvinayak}@in.ibm.com

Abstract. Recently there has been a surge of interest in auctions research triggered on the one hand by auctions of bandwidth and other public assets and on the other by the popularity of Internet auctions and the possibility of new auction formats enabled by e-commerce. Simultaneous auction of items is a popular auction format. We consider the problem of maximizing total revenue in the simultaneous auction of a set of items where the bidders have individual budget constraints. Each bidder is permitted to bid on all the items of his choice and specifies his budget constraint to the auctioneer, who must select bids to maximize the revenue while ensuring that no budget constraints are violated. We show that the problem of maximizing revenue is such a setting is NP-hard, and present a factor-1.62 approximation algorithm for it. We formulate the problem as an integer program and solve a linear relaxation to obtain a fractional optimal solution, which is then deterministically rounded to obtain an integer solution. We argue that the loss in revenue incurred by the rounding procedure is bounded by a factor of 1.62.

Keywords: auctions, winner determination, approximation algorithm, rounding.

1 Introduction

In recent years, there has been a great deal of interest in the research community in various formats and methodologies for auctions [2,7,8]. This has been driven in part by the widespread popularity of Internet auctions. Another major factor in the revival of interest in the theory of auctions has been the wave of privatization of public property in many parts of the world, which has involved the design and use of a variety of auction formats. Such auctions have ranged from the sale of spectrum rights in several countries [10] to the public assets in the former Soviet Union. Simultaneous auctions are also popular in electricity markets and equities trading [3,4]. Issues of interest include both the design of suitable auction formats as well as optimization problems related to winner determination.

In many scenarios, such as that of the auction of public assets, the simultaneous auction of several assets is likely to impose financial or liquidity constraints

* Vijay Kumar can be reached at amazon.com, 605 5th Avenue South, Seattle, WA 98104, USA. E-mail: vijayk@amazon.com

M. Goemans et al. (Eds.): APPROX-RANDOM 2001, LNCS 2129, pp. 102–113, 2001.

on the bidders. For instance, in the simultaneous auction of the expensive spectrum rights of several territories, bidders may be constrained in matching rivals' bids by the limitations of total resources available [1]. For efficient price discovery, it would be helpful to insulate the bidders against the risk of their winning bids exceeding their financial capacity.

One format that answers these concerns is a *budget-constrained auction*. In a budget-constrained auction, several items are simultaneously auctioned in sealed-bid auctions. Each bidder submits a collection of bids and informs the auctioneer of his aggregate financial capacity or *budget*. The auctioneer then allocates the objects to the bidders, charging each winner a price no higher than his bid for the item won, and making sure that the total charge for each bidder is within his specified budget. The auctioneer may for some item choose to charge a price lower than the corresponding bid in view of the winning bidder's budget constraint.

The problem of auction when bidders have budget constraints has been studied in several different contexts. Rothkopf [6] discusses how the computation of best responses is affected by budget constraints. J-P Benoit and Vijay Krishna [1] discuss ordering bids in sequential auctions when bidders have budget constraints. Palfrey [5] has studied the effects of budget constraints in a multiple-object setting with complete information. Both [1] and [5] are game theoretic in treatment, and try to prevent bidders from bringing down valuations in the face of complete information.

In this paper we consider the problem of winner determination in a budget-constrained auction, which we term the *Budget Constrained Auction Problem (BCAP)*. We show this problem to be NP-hard and present a polynomial-time 1.62-approximation for it. Our approach involves formulating the problem as an integer program, solving a linear relaxation, and deterministically rounding the resulting optimal fractional solution to derive a feasible integer solution. We argue that the feasible integer solution so obtained is within a factor of 1.62 of the optimal fractional solution.

More formally, *the budget constrained auction problem (BCAP)* with N bidders and M items can be formulated by the following integer program (IP1).

$$\max \sum_{i=1}^{N} \left(\sum_{j=1}^{M} b_{i,j} x_{i,j} - d_i \right) \tag{1}$$

subject to

$$\sum_{i=1}^{N} x_{i,j} \le Q_j \; j \in [1, M] \tag{2}$$

$$\sum_{j=1}^{M} b_{i,j} x_{i,j} - d_i \le D_i \; i \in [1, N] \tag{3}$$

$$x_{i,j} \quad \in \quad [0, q_{i,j}] \tag{4}$$

Q_j is an integer representing the available stock of item j, $b_{i,j}$ is the price per item bid by the bidder i on item j, $q_{i,j}$ is an integer representing the maximum quantity of item j that bidder i wants to buy, and D_i is the budget of bidder i. In any solution, the integer variables $x_{i,j}$ represent the number of items of type j assigned to bidder i. The variable d_i represents the *discount* given to bidder i – that is, the difference between the sum of the winning bids of bidder i and his budget (recall that the auctioneer may choose to charge a bidder less than the sum of his winning bids). The objective function represents the total revenue which is to be maximized. (2) represents the *stock constraint* of the items, (3) represents the *budget constraint* of the bidders, and (4) represents the *bid constraint*.

The structure of the integer program under consideration is similar to that of the *generalized assignment problem* [9]. Consider the variation of BCAP in which an item can be assigned to a bidder only at his bid value. In the special case where all the Q_j values are unity, this problem reduces directly to the generalized assignment problem, for which a factor-2 approximation is presented in [9]. It is possible to obtain such a reduction (and approximation) in the general case as well. Ostensibly, the flexibility provided by allowing the auctioneer to accept certain bids at below the quoted value permits an improved approximation factor of 1.62 (this may be an interesting fact from the point of view of its implications for the many scheduling problems [9] that have the structure of the generalized assignment problem).

1.1 Paper Outline

In Section 2 we present the basic framework of our approach. Section 3 contains the details of various procedures used to round a fractional solution into a integral one. The algorithm is analysed in Section 4. A proof of the NP-hardness of the problem under consideration is contained in Appendix A.

2 Our Algorithm

Our algorithm is based upon deterministic rounding of an optimal fractional solution. We begin by solving a linear program to obtain an optimal fractional solution. In this solution we make certain transformations or modificatons which, without reducing the value of the objective function, give us what we call a *simple* solution, which is a fractional solution with certain properties. Next we make a series of modifications to this simple solution, to obtain a succession of simple solutions each of which contains fewer fractional values than the previous. At the end of this process we are left with an integral solution. We show that the value of this integral solution is within a certain constant factor of the value of the initial simple solution.

Our linear program $LP1$ is obtained from $IP1$ by substituting the bid constraint (4) by the following *relaxed bid constraint*:

$$0 \leq x_{i,j} \leq q_{i,j} \tag{5}$$

In order to describe our algorithm, we need the following definitions. In the context of a solution $A = x^A_{i,j}$ of LP1, we define the *unsold stock* of item j as

$Q_j - \sum_{i=1}^{N} x^A_{i,j}$, the *sold quantity* of item j as $\sum_{i=1}^{N} x^A_{i,j}$, and the *amount spent* by bidder i as $\sum_{j=1}^{M} b_{i,j} x^A_{i,j}$. Item j is said to be *sold out* if its unsold stock is zero, *partially sold* if its unsold stock and sold quantity are both positive, and *unsold* if its sold quantity is zero. We say that bidder i is *unsatisfied* if the amount spent by him is less than the maximum permitted by his budget constraint and discount (that is, $\sum_{j=1}^{M} b_{i,j} x^A_{i,j} < D_i + d_i$ — incidentally, in our algorithm either this inequality is tight or $d_i = 0$). Define the *residual graph* $R(A)$ corresponding to a solution A as a bipartite graph with N vertices ($u_i, i \in [1, N]$) on the bidder side and M vertices ($v_j, i \in [1, M]$) on the item side. The edge (u_i, v_j) is present in $R(A)$ iff the value of $x^A_{i,j}$ is not an integer.

Given a path $P = (u_1, v_1, \ldots, u_k, v_k)$ in $R(A)$, let

$$
\begin{aligned}
e_1 &= 1 \\
e_{l+1} &= e_l \frac{b_{l+1,l}}{b_{l,l}} \text{ for } l \in [1, k-1] \\
b^P_{1,k} &= \frac{b_{k,k}}{e_k} \\
&= b_{k,k} \prod_{l=2}^{k} \left(\frac{b_{l-1,l-1}}{b_{l,l-1}}\right)
\end{aligned}
$$

We call $b^P_{1,k}$ the *effective bid* for item k by bidder 1 along the path P.

Let $P = (u_j, \ldots, v_k, u_l)$ be a path joining bidder vertices u_j and u_l. Define the *exchange ratio* between bidder j and l along the path P as $e(P) = \frac{b_{l,k}}{b^P_{j,k}}$.

We define a solution B of LP1 to be a *simple* solution if $R(B)$ contains

- no cycle,
- no path from a partially sold item to another partially sold item, and
- no path P from an unsatisfied bidder to another bidder such that $e(P) \leq 1$.

The last property implies that $R(B)$ can not contain a path P from an unsatisfied bidder to another unsatisfied bidder.

In the case of a simple solution A, the residual graph $R(A)$ has no cycles. Therefore the path joining any two bidder vertices is unique. Let $P = (u_j, \ldots, v_k, u_l)$ be the unique path joining bidder vertices u_j and u_l. Denote the exchange ratio between bidder j and l as $e(j, l) = e(P) = \frac{b_{l,k}}{b^Q_{j,k}}$, where $Q = (u_j, \ldots, v_k)$. Note that if u_i, u_j and u_k are connected bidder vertices, then $e(i, k) = e(i, j)e(j, k)$.

Theorem 1. *From a feasible solution A of LP1, a simple solution B of same or higher value can be constructed in polynomial time.*

In Section 3, we prove Theorem 1 by presenting a polynomial-time algorithm *simplify(A)* that, given a feasible solution A, constructs and returns a simple solution of equal or greater value.

Our algorithm can now be presented. It is described in Figure 1. The algorithm first obtains an optimal fractional solution of LP1, which is then converted into a simple solution. The algorithm then proceeds by selecting the bidder vertices by examining maximal paths in the residual graph and modifying x_{ij}'s

```
algorithm budget_constrained_auction
begin
      Let A be an optimal simple solution
s1:   while there are edges in R(A)
              Let P be a maximal path in R(A)
              if (P starts with an item vertex)
                  P = reverse(P)
              else if (both ends of P are bidder vertices) AND (e(P) > 1)
                  P = reverse(P)
              endif
              Let u_{i1} and v_{j1} be the first two vertices in P
              if (b_{i1 j1} x_{i1 j1} <= p D_{i1}), where p is a chosen constant
s2.1              mark (u_{i1})
s2.2              Obtain solution A' from A by setting x_{i1 j1} to 0
s2.3              A = simplify(A');
              else
s3.1              Modify A by setting d_{i1} to (1 - p) D_{i1}
s3.2              A = simplify(A);
s3.3              Modify A by setting d_{i1} to sum_{j=1}^{M} b_{i1,j} x^A_{i1,j} - D_{i1}
              end if
      end while
end
```

Fig. 1. Our Algorithm

corresponding to selected vertices. Each selected vertex is marked. The algorithm terminates when there are no edges left in the residual graph. The detail algorithm is presented in Figure 1, and its correctness and the performance are analysed is Section 4.

3 Obtaining Simple Solutions

Given a feasible solution, the procedure *simplify* converts it into a simple solution by repeated application of four simple transformations, which involve limited local redestribution of money and items among a set of bidders. These four transformations are described below.

3.1 Indirect Purchase

Consider a solution A that has a path P connecting an unsatisfied bidder to an unsold or partially sold item. WLOG, let the vertices of $R(A)$ be numbered so that $P = (u_1, v_1, \ldots, u_k, v_k)$.

Given such a path, an *indirect purchase* is the following transaction. Bidder 1, who has not yet spent all his budget, spends an additional amount δ to buy another small quantity of item 1. Item 1 may be sold out, in which case the required amount of item 1 is bought back from bidder 2. This leaves some spare money with bidder 2, which he spends to buy some more quantity of item 2, and so forth. This process terminates when we reach item k – an item which is not sold out. If δ is kept small enough, all the budget constraints, stock constraints and relaxed bid constraints are still satisfied and the solution so obtained is of higher value.

More formally, define

$$\begin{aligned} e_1 &= 1 \\ e_{l+1} &= e_l \frac{b_{l+1,l}}{b_{l,l}} \text{ for } l \in [1, k-1] \end{aligned} \tag{6}$$

From A, we obtain another solution B as follows.

$$x^B_{l,l} = x^A_{l,l} + \delta \frac{e_l}{b_{l,l}} \qquad 1 \le l \le k \tag{7}$$

$$x^B_{l+1,l} = x^A_{l+1,l} - \delta \frac{e_l}{b_{l,l}} \qquad 1 \le l \le k-1 \tag{8}$$

where

$$\delta = \min \begin{cases} \min_{l=1}^{k} (\lceil x^A_{l,l} \rceil - x^A_{l,l}) \frac{b_{l,l}}{e_l} \\ \min_{j=1}^{k-1} (x^A_{l+1,l} - \lfloor x^A_{l+1,l} \rfloor) \frac{b_{l,l}}{e_l} \\ D_1 - \sum_{j=1}^{M} b_{1,j} x^A_{1,j} \\ (Q_k - \sum_{i=1}^{N} x^A_{i,k}) \frac{b_{k,k}}{e_k} \end{cases} \tag{9}$$

For all other i, j,

$$x^B_{i,j} = x^A_{i,j}. \tag{10}$$

We refer to the perturbation of the solution A along a path P as given by (7 - 10) as an *indirect purchase* of item k by bidder 1 along the path P. Note that the value of δ is positive and non-zero, since the bidder 1 is unsatisfied, item k is not sold out and all the edges of the form $(u_l, v_l), l \in [1, k]$ and $(u_{l+1}, v_l), l \in [1, k-1]$ are present in $R(A)$.

It is easy to verify that

Lemma 1. *As a result of an indirect purchase along a path* $P = (u_1, v_1, \ldots, u_k, v_k)$*, bidder* 1 *spends more money and the sold quantity of item* k *increases, while the expenses of all other bidders and the sold quantities of all other items remain unchanged; and the resulting solution is a feasible solution.* □

In particular, bidder 1 spends δ more money and the sold quantity of item k increases by $\frac{\delta}{b^P_{1,k}}$.

3.2 Cycle Elimination

Consider a situation where the residual graph $R(A)$ corresponding to the current solution A has a cycle C. Let v_k, u_i and v_j be three consecutive vertices in C.

Let P be the longer path from u_i to v_k and Q the longer path from u_i to v_j along the cycle C. We observe that $b^P_{i,k} = \frac{b_{i,j} b_{i,k}}{b^Q i,j}$, which implies that either $b^P_{i,k} \geq b_{i,k}$ or $b^Q i, j > b_{i,j}$.

We call the following procedure a *cycle elimination*: if $b^Q_{i,j} \geq b_{i,j}$ then bidder i reduces by quantity ϵ his purchase of item j using his direct bid $b_{i,j}$ and uses all the money saved to indirectly purchase the same item along the path Q. Otherwise, bidder i reduces his direct purchase of item k by quantity ϵ and makes up for this reduction by an indirect purchase of item k along path P.

In either case, the expenses of all the bidders remain the same as in A and therefore the objective function value remains unchanged. The sold quantities of items j and k either remain the same or decrease. The sold quantity of all other items remain the same. By making ϵ large enough, we ensure that the transaction removes at least one edge from the residual graph.

3.3 Item Exchanges

Let $P = (v_i, u_j, \ldots, u_k, v_l)$ be a path from a partially sold item i to another partially sold item l. Let $Q = (u_j, \ldots, u_k, v_l)$ and $T = (u_k, \ldots, u_j, v_i)$. It is easily shown that $b^Q_{j,l} b^T_{k,i} = b_{j,i} b_{k,l}$, and it follows that either $b^Q_{j,l} \geq b_{j,i}$ or $b^T_{k,i} > b_{k,l}$. An *item exchange along the path* P is the following transaction: if $b^Q_{j,l} \geq b_{j,i}$ then bidder j reduces his purchase (using bid $b_{j,i}$) of item i and uses the saved money in an indirect purchase of item l along the path Q. Otherwise bidder k reduces his purchase of item l and instead indirectly purchases item i along the path T. The quantities involved can be chosen in such a way that either one edge in P is eliminated from the residual graph or one out of items i and l gets sold out. The amount of money spent by any bidder remains unchanged.

3.4 Bidder Exchanges

Let $P = (u_i, v_j, \ldots, v_k, u_l)$ be a path from unsatisfied bidder i to some bidder l such that $e(P) \leq 1$. Let $Q = (u_i, v_j, \ldots, v_k)$. We define a *bidder exchange along the path* P to be the following transaction: bidder l reduces by a small quantity ϵ his purchase of item k, and this quantity is bought by bidder i in an indirect purchase along the path Q. The quantities involved can be chosen in such a way that either one edge in P is eliminated from the residual graph or bidder i becomes satisfied. The sold quantity of every item remains unchanged. The total amount spent by all the bidders does not decrease, since $e(P) < 1 \Rightarrow b^Q_{i,k} \geq b_{l,k}$.

A proof of Theorem 1 is now straightforward: repeated application of cycle eliminations, item exchanges and bidder exchanges converts A into a simple solution B. As noted above, these procedures do not decrease the value of the objective function. □

Having thus obtained a simple solution, the procedure *simplify* proceeds to carry out as many indirect purchases as may be possible, thus possibly increasing the value of the objective function. Note that *simplify* terminates in polynomial time since each cycle elimination, item exchange, bidder exchange or indirect purchase reduces the number of edges in the residual graph.

4 Correctness and Performance

In the following, we compare the value of the integral solution obtained by our algorithm to that of the optimal solution of $LP1$. Consider the situation at any particular time during the execution of the algorithm. Let A be the current solution. We define the potential function $f(A)$ corresponding to solution A as:

$$f(A) = \frac{1}{1-p} \sum_{i \in S} \sum_{j=1}^{M} b_{ij} x_{ij} + \sum_{i \notin S} \left(\sum_{j=1}^{M} b_{ij} x_{ij} - d_i \right) \tag{11}$$

where $S = \{i \mid u_i \text{ is marked}\}$. The constant p is chosen such that $\frac{1}{1-p} = 2 - p$, and thus the selected value is $\frac{3-\sqrt{5}}{2}$.

Let OPT denote the value of the optimal solution of $LP1$. We will show that the algorithm maintains the following invariants at the entry of the outer loop.

Invariant 1 *At least one end point of every maximal path in $R(A)$ is a satisfied bidder vertex.*

Invariant 2 $f(A) \geq OPT$.

We begin by establishing some properties of the solutions we work with.

Lemma 2. *Let A be an optimal solution of LP1. There is no path in $R(A)$ joining an unsatisfied bidder to an unsold or partially sold item.*

Proof. If there were such a path P, then it follows from Lemma 1 that it is possible to obtain another feasible solution which increases the total revenue. □

Lemma 3. *Every sold-out item vertex in $R(A)$ either has no edge or at least two incident edges.*

Proof. Assume item j is sold-out, i.e. $\sum_{i=1}^{N} x_{ij} = Q_j$. Since Q_j is an integer, either none or at least two of the x_{ij} values must be fractional. Therefore either there is no incident edge, or there are two or more incident edges on v_j in $R(A)$. □

Lemma 4. *One of the endpoints of every maximal path in the residual graph of an optimal simple solution is a satisfied bidder vertex.*

Proof. Let P be any maximal path in $R(A)$ for which this is not true. Lemma 3 implies that neither endpoint of P is a sold-out item vertex. Thus P must connect two partially sold items, two unsatisfied bidders or a partially sold item to an unsatisfied bidder. The first two possibilities are ruled out since A is a simple solution, while the last is ruled out by Lemma 2. □

Theorem 2. *The algorithm maintains Invariants 1 and 2 at the time of every visit to step* **s1**.

Proof. The proof is by induction. We show that the invariants are true at the time time of the first visit to step **s1**, and we show that they are preserved from one visit to the next.

Lemma 4 implies that Invariant 1 is true when step **s1** is first visited. Since there are no marked vertices at this time, the Invariant 2 is also true.

Next, we show that the two invariants are preserved from one visit to step **s1** to the subsequent one. That is, the execution of the while loop beginning at step **s1** preserves the invariants.

The algorithm proceeds by selecting a maximal path P in the residual graph. From the first invariant, it follows that one end of P must be a satisfied bidder vertex (u_{i_1}). The algorithm orients P appropriately such that u_{i_1} is the first vertex of P. The rest of the body of the loop consists of a conditional statement in which either steps **2.1** – **2.3** or steps **3.1** – **3.3** are executed. We separately analyse the two possibilities resulting from the conditional statement.

Case 1: If the only edge from u_{i_1} contributes less than p times the amount spent by bidder i_1, then the edge is dropped, setting $x_{i_1 j_1}$ to zero, and u_{i_1} is marked. This could lead to a new maximal path ending at vertex v_{j_1}, from either a partially sold item vertex, or from a unsatisfied bidder vertex, either of which would violate Invariant 1. However, observe that paths of either kind are eliminated by the procedure *simplify*, which is executed at step **s2.3**. Invariant 1 is thus restored.

Since $b_{i_1 j_1} x_{i_1 j_1} \le p \sum_{j=1}^{M} b_{i_1 j_1} x_{i_1 j_1}$, the execution of steps **s2.1** and **s2.2** does not decrease the value of the potential function. Step **s2.3** does not decrease the value of the potential function since it only affects unmarked vertices (note that marked vertices are not present in the residual graph) and causes a non-negative change in the contribution of these vertices to the objective function (this contribution equals the second term of the potential function (11)). Thus Invariant 2 is maintained through the execution of the loop.

Case 2: If the only edge in $R(A)$ incident on u_{i_1} corresponds to an expense of more than pD_{i_1}, then at step **3.1** the discount d_{i_1} available to bidder i_1 is changed to $(1-p)D_{i_1}$. This makes the bidder vertex u_{i_1} unsatisfied and it is possible that Invariant 1 may not hold any more. As before, the invocation of *simplify* (at step **3.2**) removes any paths that may violate Invariant 1. At step **3.3** the discount d_{i_1} is reduced to the smallest quantity required to facilitate the current purchases of bidder i_1. Steps **3.1** and **3.3** do not affect the potential function while step **3.2** does not reduce it. Thus, Invariant 2 is maintained through the execution of steps **3.1** – **3.3**. □

The algorithm terminates in polynomial time because

Lemma 5. *Every iteration of the outer loop reduces at least one edge from the residual graph.*

Proof. In Case 1 above, the first edge of the path P is removed from $R(A)$. Consider Case 2. After step **3.1**, u_{i_1} is an unsatisfied vertex connected by path P to either a partially sold item or a satisfied bidder. Thus either an indirect purchase or a bidder exchange would be possible when the procedure *simplify* is invoked at step **3.2**, either of which would remove at least one edge from the residual graph. If that operation is not carried out, it must be that either vertex u_{i_1} has been converted into a satisfied vertex, or some edge of path P has been removed, during the execution of *simplify*. In either case, at least one edge has been removed from the residual graph. □

Theorem 3. *Our algorithm is a polynomial time factor-$\frac{1+\sqrt{5}}{2}$ approximation algorithm for the budget-constrained auction problem.*

Proof. Let A denote the integral solution X obtained by our algorithm and Obj the corresponding value of the objective function. Every bidder i contributes the amount $\sum_{j=1}^{M} b_{ij}x_{ij} - d_i$ to the objective function. Note that for $i \in S$, $d_i = 0$. Since $\frac{1}{1-p} = 2 - p$, we have

$$\frac{Obj}{1-p} = \frac{1}{1-p}\sum_{i \in S}\sum_{j=1}^{M} b_{ij}x_{ij} + (2-p)\sum_{i \notin S}\left(\sum_{j=1}^{M} b_{ij}x_{ij} - d_i\right)$$

$d_i > 0$ for some i only if $D_i < \sum_{j=1}^{M} b_{ij}x_{ij}$. Since $d_i \leq (1-p)D_i < \sum_{j=1}^{M} b_{ij}x_{ij}$,

$$\frac{Obj}{1-p} > \frac{1}{1-p}\sum_{i \in S}\sum_{j=1}^{M} b_{ij}x_{ij} + \sum_{i \notin S}\sum_{j=1}^{M} b_{ij}x_{ij} = f(A)$$

By Invariant 2, $f(A) \geq OPT$. Since $\frac{Obj}{1-p} > f(A)$, it follows that $Obj > (1-p)OPT$, and thus our algorithm is a $\frac{1}{1-p}$-approximation. Substituting $\frac{3-\sqrt{5}}{2}$ for p yields the desired approximation factor. □

The authors would like to thank David Shmoys, R. Ravi and Naveen Garg for helpful comments and suggestions.

A Hardness of BCAP

Using a reduction from the 3-SAT problem, we show that BCAP is NP-hard.

Given an instance I of a 3-SAT problem with n variables and m clauses, obtain an instance I' of BCAP as follows. Corresponding to each variable x_i in I, I' contains

- an item Y_i,
- two bidders, X_i and $\overline{X_i}$, both of which bid for Y_i with bid value 1.
- m items $\{X_{i1}, X_{i2}, \ldots, X_{im}\}$ called the *t-items* of X_i. X_i makes bids of value $\frac{1}{m}$ on each of the t-items of X_i.
- m items $\{\overline{X_{i1}}, \overline{X_{i2}}, \ldots, \overline{X_{im}}\}$ called the *f-items* of X_i. $\overline{X_i}$ makes bids of value $\frac{1}{m}$ on each of the f-items of X_i.

For each clause c_k, I' contains a bidder C_k which makes a bid of value $\frac{1}{m}$ for the object X_{ik} if variable x_i is present in clause c_k, and a bid of value $\frac{1}{m}$ for the object $\overline{X_{ik}}$ if $\overline{x_i}$ is present in clause c_k. Figure 2 shows the basic structure

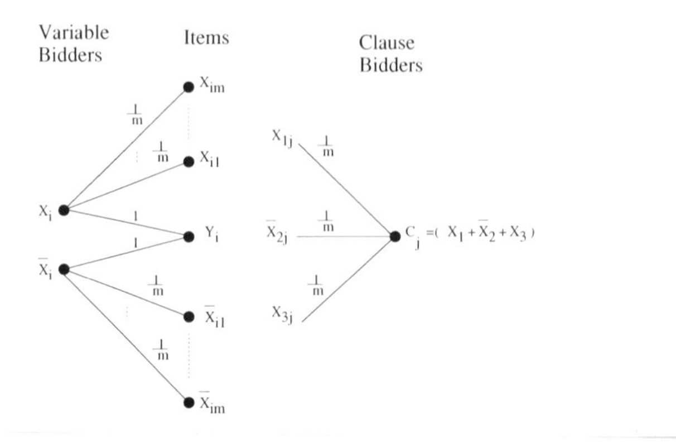

Fig. 2. The reduction

of the reduction for a variable x_i and a clause c_j. This structure is repeated for all variables and clauses of I, and appropriate bids are added. The budget constraint for each bidder of the type X_i or $\overline{X_i}$ is 1, and the budget constraint for each C_k is $\frac{1}{m}$. The maximum revenue possible in I' is $2n + 1$ where n is number of variables in I. Clearly I' can be obtained from I in polynomial time.

Lemma 6. *If I is satisfiable then there exists an assignment of items to bidders in I' such that the total revenue is $2n + 1$.*

Proof. Let A be asatifying assignment for I. If x_i is true in A then assign Y_i to X_i at price 1, and assign all f-items of X_i to $\overline{X_i}$ at price $\frac{1}{m}$ each. Otherwise assign Y_i to $\overline{X_i}$ at price 1, and assign all t-items of X_i to X_i at price $\frac{1}{m}$ each. For each clause c_j, there exists at least one true literal. Choose one such true literal x_i (or $\overline{x_i}$), and assign the j t-item (f-item in case of a negated literal) of X_i to C_j. It can be easily verified that we can assign exactly one item at price $\frac{1}{m}$ to each clause. □

Lemma 7. *If there exists an assignment of items to bidders in I' which results in revenue more than $(2n + 1 - \frac{1}{m} + \epsilon)$ for some $\epsilon > 0$, then I is satisfiable.*

Proof. Obtain a solution to I as follows: for all i, let x_i be true if Y_i has been assigned to X_i, and false otherwise.

As the revenue from all bidders of the type X_i or $\overline{X_i}$ can be no more than $2n$, the revenue from bidders of the type C_j is at least $(1 - \frac{1}{m} + \epsilon)$. Similarly, the revenue from each bidder of the type X_i or $\overline{X_i}$ must be at least $(1 - \frac{1}{m} + \epsilon)$. Clearly each C_j has been assigned an item at a price greater than zero. Consider any C_j and an item assigned to it. Let us suppose that the clause has been assigned a t-item (respectively f-item) of a variable X_i. Since the income from X_i (respectively $\overline{X_i}$) is at least $(1 - \frac{1}{m} + \epsilon)$, Y_i has been assigned to X_i (respectively $\overline{X_i}$). Thus x_i (respectively $\overline{x_i}$), one of the literals in clause c_j is true, and c_j is satisfied.

It follows from Lemmas 6 and 7 and the NP-completeness of 3-SAT that

Theorem 4. *The Budget Constrained Auction Problem is NP-hard and cannot be approximated to within a factor of* $1+\frac{1}{2n}$ *of the optimal, where* n *is the number of bidders.*

References

1. J-P Benoit and V. Krishna. Multiple-object auctions with budget constrained bidders. A working paper at http://econ.la.psu.edu/ vkrishna/multi10.ps, April 1998.
2. Y. Fujishima, K. Leyton-Brown, and Y. Shoham. Taming the computational complexity of combinatorial auctions: Optimal and approximate approaches. In *International Joint Conferences on Artificial Intelligence*, 1999.
3. R. P. McAfee and J. McMillan. Analyzing the airwaves auctions. *Journal of Economic Perspectives*, 10(1):159–175, 1996.
4. J. McMillan. Selling spectrum rights. *Journal of Economic Perspectives*, 8(3):145–162, 1994.
5. T. R. Palfrey. Multiple object, discriminatory auctions with bidding constraints: A game theoretic analysis. *Management Science*, 26:935–946, 1980.
6. M. Rothkopf. Bidding in simultaneous auctions with a constraint on exposure. *Operations Research*, 25:620–629, 1977.
7. M. H. Rothkopf, A. Pekec, and R. M. Harstad. Computationally manageable combinatorial auctions. *Management Science*, 44:1131–1147, August 1998.
8. T. Sandholm. An algorithm for optimal winner determination in combinatorial auction. Technical Report WUCS-99-01, Washington University, Department of Computer Science, January 1999.
9. D.B. Shmoys and E. Tardos. An approximation algorithm for the generalised assignment problem. *Mathematical Programming A*, 62:461–474, 1993.
10. Martin Spicer. International survey of spectrum assignment for cellular and pcs. Technical report, September 1996. Federal Telecommunications Commission, available at http://www.fcc.gov/wtb/auctions/papers/spicer.html.

Minimizing Average Completion of Dedicated Tasks and Interval Graphs

Magnús M. Halldórsson[1], Guy Kortsarz[2], and Hadas Shachnai[3]

[1] Dept. of Computer Science, University of Iceland, IS-107 Reykjavik, Iceland, and Iceland Genomics Corp., Reykjavik
mmh@hi.is

[2] Dept. of Computer Science, Open University, Ramat Aviv, Israel
guyk@shaked.openu.ac.il

[3] Dept. of Computer Science, Technion, Haifa, Israel
hadas@cs.technion.ac.il

Abstract. Scheduling dependent jobs on multiple machines is modeled as a graph *(multi)coloring* problem. The focus of this work is on the *sum of completion times* measure. This is known as the *sum (multi)coloring* of the conflict graph. We also initiate the study of the *waiting time* and the *robust throughput* of colorings. For uniform-length tasks we give an algorithm which simultaneously approximates these two measures, as well as sum coloring and the chromatic number, within constant factor, for any graph in which the k-colorable subgraph problem is polynomially solvable. In particular, this improves the best approximation ratio known for sum coloring interval graphs from 2 to 1.665.

We then consider the problem of scheduling non-preemptively tasks (of non-uniform lengths) that require exclusive use of dedicated processors. The objective is to minimize the sum of completion times. We obtain the first constant factor approximations for this problem, when each task uses a constant number of processors.

1 Introduction

We consider scheduling jobs which are dependent, since they utilize the same non-sharable resource. Coloring problems capture well this type of scheduling in multiple machine environment, as they inherently allow to select, at any time, a subset of non-conflicting jobs. Thus, scheduling with dependencies is modeled as graph *coloring*, when all jobs have the same (unit) execution times, and as graph *multicoloring* for arbitrary execution times; the vertices of the graph represent the jobs, and an edge in the graph between two vertices represents a dependency or conflict between the two corresponding jobs, which forbids scheduling these jobs at the same time. For simplicity, in formulating our results we do not refer to the number of machines in the system (i.e., it is implicitly assumed that the number of machines is "unbounded"); however, our study applies to a system with any given number of machines.

M. Goemans et al. (Eds.): APPROX-RANDOM 2001, LNCS 2129, pp. 114–126, 2001.

Objective Functions. It is often the case that no single measure of a quality of a schedule satisfies our needs. We seek to expand the set of measures that can be applied. The focus of our work is on the *sum of completion times* measure. For unit-length tasks, this is known as the *chromatic sum* or *sum coloring* of the conflict graph. We also initiate the study of two other measures of colorings.

The *waiting time* of a coloring ψ is the sum $\sum_v (\psi(v) - 1)$. Whereas the chromatic sum corresponds to the sum of completion times of the jobs represented by the vertices, the waiting time represents the total time jobs waited for execution. Given a coloring ψ of the conflict graph, G, the sum coloring of ψ is given by $\mathtt{SC}(G, \psi) = \sum_v \psi(v)$. The minimum chromatic sum of G is given by $\mathtt{SC}(G) = \min_\psi \mathtt{SC}(G, \psi)$. The waiting time approximation of ψ is then $\frac{\mathtt{SC}(G,\psi)-n}{\mathtt{SC}(G)-n}$. Thus, a ρ-approximation of waiting time immediately implies the same ρ-approximation for sum coloring.

The *throughput* of a partial schedule is the number of jobs that have been completed.[1] We can stop a coloring after any given time step, k, ask for the throughput at that point $Thr(\psi, k) = \sum_{i=1}^{k} |I_i|$, where I_i is the set of vertices colored with i, and compare that to the optimal throughput after k colors, $k\text{-}Col_S(G)$. The throughput of a coloring can be said to be robust, if it is good for each value of k simultaneously. The *robust throughput* measure of a coloring compares the throughput in each step to the best possible, and takes the maximum, that is,

$$Th(\psi) = \max_k \frac{k\text{-}Col_S(G)}{Thr(\psi, c)}.$$

Finally, recall that the *chromatic number* $\chi(G)$ of a graph G is the minimal number of colors required for (multi)coloring the vertices in G properly. In scheduling terms, this is the minimal total length (or *makespan*) of any legal schedule. This goal is important from system point of view, while for the users, the sum/average completion time is at least as important.

It is often hard to find a schedule that is optimal with respect to several measures. Indeed, a schedule that is optimal for one measure may perform poorly for other measures. A classical example is scheduling *independent* jobs on identical machines. The *shortest processing time (SPT)* rule, which minimizes the average completion time, may be far from the optimal for the total completion time measure. In this paper we consider schedules that minimize the average completion/waiting time of a job and the total completion time of the schedule, while maximizing the robust throughput of the system.

Sum Coloring Interval Graphs. Numerous applications of sum coloring (and sum multicoloring) interval graphs have been cited in the literature. One is a wire-minimization problem in VSLI design [NSS99]. Terminals lie on a single vertical line, and with unit spacings are vertical bus lanes. Pairs of terminals are to be connected via horizontal wires on each side to a vertical lane, with non-overlapping pair utilizing the same lane. With the vertical segments fixed, the

[1] In the weighted case, we take the sum of the weights of the jobs completed.

wire cost corresponds to the total length of horizontal segments. Numbering the lanes in increasing order of distance from the terminal line, lane assignment corresponds to coloring the interval by an integer. The wire-minimization problem then corresponds to sum coloring.

A certain warehouse storage allocation problem involves minimizing the total distance traveled by a robot [W97]. Goods are checked in and out at known times; thus, goods that are not in the warehouse at the same time can share the same location. Numbering the storage locations by their distance from the reception counter, the total distance corresponds to sum coloring the intervals formed by the goods.

Resource Constrained Scheduling. In resource-constrained scheduling in general, we are given a collection of n *jobs* of integral lengths and a collection of *resources*. Each job requires an exclusive access to a particular subset of the resources to execute.

A natural example of a limited resource is the set of processors in a multiprocessor system. In the biprocessor task scheduling problem we are given a set of jobs and a set of processors; each job requires the exclusive use of two dedicated processors. We are interested in finding a schedule, which minimizes the sum of completion times of the jobs. In scheduling terms, this problem is denoted by $P|\text{fix}_j|\sum C_j$ with $|\text{fix}_j| = 2$, where the second term indicates that each job requires a fixed subset of machines and the last term denotes the sum of completion times measure.

Another limited resource is the set of paths in communication network. Here we get *path coloring* problems, recently popularized in network design. Some additional applications, which involve sum multicoloring, include traffic intersection control [BH94] and compiler design [NSS99].

Our focus is on the case where each task uses up to k resources. Hence, the conflict graph is an intersection graph of a collection of sets of size at most k. Such graphs are called $k+1$-*claw free*, as they do not contain the star $K_{1,k+1}$ as an induced subgraph. In the special case of two resources per task, such as the biprocessor task problem, the conflict graph is a line graph. We focus on scheduling, or coloring such graphs, so as to minimize the sum of completion times of the jobs.

Our Results. Our first main result (in Section 2) is an algorithm, which approximates *simultaneously* sum coloring, waiting time, robust throughput and the chromatic number, to within constant factors from the optimal. The constant for sum coloring and waiting time is 1.6651, for robust throughput 1.4575, and for the chromatic number 3.237. This holds for graphs in which the maximum k-colorable subgraph problem is polynomially solvable, including comparability graphs and their complements, and chordal graphs with constant maximum clique size. Of particular interest is the class of interval graphs, for which we improve the best previous known approximation ratio of 2 for sum coloring. Note that Gonen [G01] has recently shown that SC is $\mathcal{APX}$-hard.

Table 1. Known results for sum (multi-)coloring problems

	SC		SMC	
	u.b.	*l.b.*	pSMC	npSMC
General graphs	·	$n^{1-\epsilon}$	$n/\log^2 n$	$n/\log n$
Perfect graphs	4	·	16	$O(\log n)$
Comparability	**1.6651** (4)	·	**6.66** (16)	·
Interval graphs	**1.6651** (2 [NSS99])	$1+\epsilon$ [G01]	·	·
Bipartite graphs	10/9 [BK98]	$1+\epsilon$ [BK98]	1.5	2.8
Line graphs	2	NPC	2	**12**
Partial k-trees	1 [J97]		PTAS [HK99]	FPTAS [HK99]
Planar graphs	·	NPC [HK99]	PTAS [HK99]	PTAS [HK99]
Trees	1 [K89]		PTAS [HK+99]	1 [HK+99]
$k+1$-claw free	$k+1$		$k+1$	$\mathbf{4k^2-2k}$

Our second main result (in Section 3) is a constant factor approximation algorithm for non-preemptive sum multicoloring on claw-free graphs. More specifically, for graph with no $k+1$-claw, we obtain a $2k(2k-1)$-approximation, and in particular a 12-approximation for line graphs. This encompasses dedicated processor scheduling, where the number $|\text{fix}_j|$ of processors involving a given job is at most k. The previously best ratio known for this problem was $\log n$ [BHK+99].

Related Work. The sum coloring (SC) problem was introduced in [K89] and the sum multicoloring (SMC) problems in [BHK+99]. Table 1 summarizes the known results for these problems in various classes of graphs. New bounds given in this paper are shown in boldface. The last two columns give known upper bounds for preemptive and non-preemptive SMC, denoted by pSMC and npSMC, respectively. Entries marked with · follow by inference, either using containment of graph classes (bipartite and interval graphs are perfect), or by SC being a special case of SMC. When omitted, [BBH+98] is the references for SC and [BHK+99] for SMC.

Resource-constrained scheduling has recently been investigated in the vast literature of scheduling algorithms (see, e.g., [BKR96,K96]). Kubale [K96] studied the complexity of scheduling biprocessor tasks. He also investigated special classes of graphs, and showed that npSMC of line graphs of trees is NP-hard in the weak sense. Recently, Afrati et al. [AB+00] have given a polynomial time approximation scheme for the problem that we consider, minimizing sum of completion times of dedicated tasks. However, their method applies only to the case where the total number of processors is a fixed constant.

2 Approximating Waiting Time and Throughput of Interval and Comparability Graphs

In this section, we consider algorithms for sum coloring and proceed to discuss the less studied measures of waiting time and robust throughput. We start by looking at two previous algorithms; then, we present a new method that obtains improved performance on several well-known classes of graphs.

The MaxIS Algorithm. The MaxIS coloring algorithm is the iterative greedy method, which colors in each round a maximum independent set among the yet-to-be colored vertices. It was first considered for SC in [BBH+98], where it was shown to give a 4-approximation. This was shown to be the tight bound for MaxIS in [BHK]. This ratio holds for graphs where the *maximum independent set (IS)* problem is polynomially solvable, including, e.g., perfect graphs. When IS is ρ-approximable, MaxIS yields a 4ρ-approximation for SC.

It is plausible that the performance of MaxIS for the waiting time is also 4, but the analysis is not straightforward. We can however make a simple argument to show that the performance ratio of MaxIS for waiting time is at most 6. For robust throughput, we can obtain a tight bound of $e/(e-1) \approx 1.58$ (Details are omitted).

The Algorithm of [NSS99] for Interval Graphs. An algorithm of Nicoloso et al. [NSS99] for sum coloring interval graphs starts by computing $G_1, G_2, \ldots, G_{\chi(G)}$, where G_i is a maximum i-colorable subgraph. They show that when G is an interval graph and the G_i's are computed by a left-to-right greedy algorithm, then (a) G_i contains G_{i-1}, and (b) the difference set $G_i - G_{i-1}$ is 2-colorable. Thus, the algorithm colors G by coloring G_1 with the color 1, and coloring $G_i - G_{i-1}$ with $2i-2$ and $2i-1$.

Nicoloso et al. showed that its performance for SC, and simultaneously for chromatic number, was 2, and that it was tight. From this description of the algorithm, it is easy to derive bounds on its performance for the other measures. Since for each k a maximum k-colorable subgraph is fully colored by color $2k-1$, a ratio of 2 follows for robust throughput.

Observation 1 *The algorithm of [NSS99] attains a simultaneous performance ratio of 2 for both the waiting time and robust throughput of interval graphs (as well as sum coloring and chromatic number).*

It should be fairly clear that these bounds on the performance of the algorithm cannot be improved. They also crucially depend on special properties of interval graphs. Also, it is not clear if these bounds hold when the algorithm is used on weighted graphs.

An Algorithm Based on Finding k-Colorable Subgraphs. We now give an algorithm for coloring, which simultaneously approximates the chromatic sum, waiting time, the robust throughput and the chromatic number.

A *k-colorable subgraph* in a graph G is a vertex subset $S \subset V$ that can be partitioned into k independent sets. In the *k-Col_S* problem, we need to find a k-colorable subgraph of maximum cardinality. The *k-Col_S* problem is solvable, for example, on interval graphs, by a greedy algorithm [YG87], on chordal graphs with constant maximum clique size (or k-trees), and on comparability graphs and their complements [F80]. This holds also for the vertex-weighted problem, where we seek a k-colorable subgraph of maximum total weight.

The algorithm Assign_Color_Sets (ACS) uses as a subroutine an algorithm **kIS**(G,k) for k-Col_S, where k can be as large as $\chi(G)$. It consists of repeatedly coloring k-colorable subgraphs, where k is taken from a geometrically increasing sequence, with a quotient q given as a parameter. The method is given in a randomized form, but can be derandomized with any desired precision by trying all sufficiently closely spaced values in the range $[0, 1]$.

ACS(G, q)
  $o \leftarrow$ uniformly random in $(0, 1]$
  Color a maximum independent set with color 0
  $i \leftarrow 0$; $\ell_i \leftarrow 0$
  while $(G \neq \emptyset)$ do
    $c_i \leftarrow \lfloor q^{i+o} \rfloor$
    $G_i \leftarrow$ kIS(G,c_i)
    Color G_i with colors $\ell_i + 1, \ell_i + 2, \ldots, \ell_i + c_i$
    $G \leftarrow G - G_i$
    $i \leftarrow i + 1$; $\ell_i \leftarrow \ell_{i-1} + c_{i-1}$
  end
end

The MaxIS algorithm of [BBH+98] corresponds to the choice of $q = 1$ and $o = 0$, i.e., in each step a single independent set is extracted from the graph.

Analysis. Note that $c_i = \lfloor q^{i+o} \rfloor$ is the number of colors used in iteration i of the algorithm, and ℓ_i is the number of colors used up to and including iteration $i - 1$. Let $A(c)$ denote the total number of vertices colored with the first c colors by ACS. Let $\mathcal{O}(c)$ denote the maximum number of vertices colorable with c colors, i.e., the cardinality of a maximum c-colorable subgraph. Much of the analysis hinges on the following useful property.

Observation 2 *$A(1 + \sum_{j=0}^{i} c_i) \geq \mathcal{O}(c_i)$, for $i = 0, 1, \ldots$.*

Proof: The claim follows from the fact that in each step i, $i \geq 0$, ACS uses a new set of colors, of size c_i, and finds a subgraph of maximum cardinality that can be colored with this set. Now, for any $i \geq 0$, consider a subgraph G_i that is c_i-colorable. Note that each vertex in G_i yet uncolored can be colored with the next c_i colors. Hence, by the end of step i (i.e., after ACS overall uses $\sum_{j=0}^{i} c_i$ colors, plus one for the initial independent set) at least $\mathcal{O}(c_i)$ vertices have been colored. □

This observation suggests that we can imagine a bijective mapping g of vertices as chosen by the algorithm to the vertices as colored in an optimal solution. The mapping has the property that if v is chosen in round i, then $g(v)$ is assigned a color greater than c_{i-1}.

Theorem 3. *For any* $1 < q \leq 5.828$, ACS *approximates* $\texttt{WT}(G)$ *(and thus also* $\texttt{SC}(G)$*) within a factor of* $\sqrt{q}(1/(q-1)+1/2)$.

Proof: We first bound the cost incurred by ACS for coloring an arbitrary vertex, v, for a fixed value of the random element o. Let ψ_{OPT} be a coloring of G optimizing the waiting time, and t be the color of $g(v)$ in that optimal solution, $t = \psi_{OPT}(g(v))$. If $t = 0$, then ACS also colored v in the initial independent set, paying the same zero amount. Otherwise, $q^{a-1+o} \leq t < q^{a+o}$, where a is some non-negative integer. We can write $t = q^{a-1+o+r}$, where $0 \leq r < 1$. Then, v is colored by ACS in round a.

The color used for v by ACS equals to the sum of two components: the number of colors already used, and the offset among the c_a colors used in round a. The former equals $\sum_{i=0}^{a-1} c_i$. The latter can be simplified by viewing all the vertices colored in round a together. These vertices are colored at an offset of at most $(c_a+1)/2$ on the average. Hence, the color of v is bounded by

$$\sum_{i=0}^{a-1} c_i + \frac{1+c_a}{2} = \sum_{i=0}^{a-1} \lfloor q^{i+o} \rfloor + \frac{\lfloor q^{a+o} \rfloor}{2} + \frac{1}{2}$$

$$\leq tq^{1-r}\left[\frac{1}{q-1} + \frac{1}{2}\right] + \frac{1}{2} - \frac{q^o}{q-1}.$$

Note that since o is selected randomly and uniformly in $(0,1]$, so is r. Hence, the expected value of the color assigned to v is $t\left(\sqrt{q}(1/(q-1)+1/2)\right) + 1/2 - \sqrt{q}/(q-1)$.

For $q \in (1, 5.828]$ we get that $\sqrt{q}/(q-1) \geq 1/2$. It follows that the waiting time of the coloring obtained is bounded by

$$\sum_{v \in G} \psi(v) \leq \sum_{v \in G} \sqrt{q}(1/(q-1)+1/2)\psi_{OPT}(g(v)) = \sqrt{q}(1/(q-1)+1/2)\texttt{WT}(G).$$

□

Since the analysis gave expected values for each vertex separately, these ratios hold also for the vertex-weighted variant problem. Intuitively, we can view each weighted vertex as a compatible collection of unweighted vertices

The best ratio is attained when q is about 4.12. This improves on the previously best known ratio of 2 by Nicoloso et al. [NSS99], stated for (unweighted) sum coloring.

Corollary 1. $\texttt{WT}(G)$ *and* $\texttt{SC}(G)$ *can be approximated within a factor of 1.6651 on interval graphs and comparability graphs and their complements, even in the weighted case.*

This yields a corollary for the *preemptive* sum multicoloring problem. In [BHK+99], a 4ρ-ratio approximation is obtained for graphs where SC can be approximated within a factor of ρ. We thus obtain an improvement, for graphs for which k-Col_S is polynomially solvable, from the previous 16-factor.

Corollary 2. pSMC *is approximable within* $4 \cdot 1.6651 \approx 6.66$ *on interval and comparability graphs and their complements.*

We proceed to analyze the throughput behavior of the algorithm.

Theorem 4. *Robust throughput can be approximated within a factor of 1.45751 on interval graphs and comparability graphs, even in the weighted case.*

Proof: For this, the random offset parameter is of no help; since it complicates the analysis, we shall fix its value as 1. Further, for simplicity, we look only at the case when $q = 2$, which we experimentally find the best one.

We compare the number of nodes $A(k)$ found by the algorithm in the first k color classes, to the $\mathcal{O}(k)$ nodes in the first k colors of the optimal solution. Let $m = \lfloor \lg k \rfloor$. Recall that $c_i = 2^i$ and $\ell_i = 2^i - 1$. The set of nodes found by ACS then consists of the $A(\ell_m)$ nodes found in the first $m-1$ rounds, and at least a $(k - \ell_m)/k$ fraction of the colors used in the m-th round. The analysis uses the following simple observation.

$$A(\ell_i) - A(\ell_{i-1}) \geq \frac{c_{i-1}}{k}(\mathcal{O}(k) - A(\ell_{i-1})).$$

Rewriting, we get $\mathcal{O}(k) - A(\ell_i) \leq (1 - c_{i-1}/k)(\mathcal{O}(k) - A(\ell_{i-1}))$. By induction, we have that

$$\mathcal{O}(k) - A(\ell_m) \leq \mathcal{O}(k) \prod_{i=0}^{m-1} (1 - 2^i/k).$$

In the last round, we count the final $k - \ell_m$ colors. The fraction still not colored after that round amounts to

$$\mathcal{O}(k) - A(k) \leq (1 - \frac{(k - (2^m - 1))}{k})(\mathcal{O}(k) - A(\ell_m)$$

$$= \mathcal{O}(k)\frac{2^m - 1}{k}\prod_{i=1}^{m}(1 - 2^i/k)$$

By computational analysis, we find that this value is maximized when $k/2^m \approx 1.3625$, converging to about $0.3139\ \mathcal{O}(k)$. This implies a performance ratio $\mathcal{O}(k)/A(k) \leq 1/(1 - 0.3139) \approx 1.4575$ for the throughput of the coloring, for a given value k. Since this bound is independent of k, we have derived a bound on the robust throughput measure. □

On graphs for which k-Col_S is not solvable but approximable within a ρ-factor, we easily obtain a 1.4575ρ ratio for robust throughput. Approximation-preserving arguments can also be made for waiting times, by repeating each round of the algorithm $\lceil \rho \rceil$ times.

Finally, we can argue a bound on the number of colors used.

Theorem 5. ACS *uses at most* $(\frac{q}{q-1} + q)\chi(G)$ *colors, or at most 4 times optimal when* $q = 2$. *The randomized algorithm uses an expected* $(\frac{q}{q-1} + \sqrt{q})\chi(G)$, *or about 3.237 times optimal when* $q = 3$.

We have seen that the same algorithm approximates all four measures: $\chi(G)$, $\mathtt{WT}(G)$, $\mathtt{SC}(G)$, and robust throughput. The parameters used to obtain optimum values were not the same; however with judicious choices of the parameters, one can obtain colorings that simultaneously approximate all the measures considered, as suggested in the following result.

Corollary 3. ACS, *with* $q = 2$, simultaneously *approximates robust throughput within a factor of 1.4575,* $\mathtt{SC}(G)$ *and* $\mathtt{WT}(G)$ *within a factor of 2.12, and* $\chi(G)$ *within a factor of 4.*

3 Scheduling Dedicated Tasks

In this section we give a constant factor approximation algorithm for npSMC on $k+1$ claw-free graphs. As noted earlier, this implies an $O(1)$ ratio approximation for the dedicated task scheduling problem $P|\text{fix}_j|\sum C_j$ with $|\text{fix}_j| = k$, for fixed k.

Sum Multicoloring. The problem of scheduling dependent tasks, to minimize their average completion time, can be modeled as a graph multicoloring problem. This is particularly useful when there is a structure to the task dependencies, that we can take advantage of. We consider here the situation where dependencies are caused by resource conflicts, and each job requires the exclusive use of up to k resources; thus, the resulting graph is $k+1$-claw free, i.e. the star graph $K_{1,k+1}$ is a forbidden subgraph.

An *instance* to multicoloring problems is a pair (G, x), where $G = (V, E)$ is a graph, and x is a vector of *color requirements* (or *lengths*) of the vertices. For a given instance, we denote by n the number of vertices, and by $p = \max_{v \in V} x(v)$ the maximum color requirement. A *multicoloring* of G is an assignment $\psi : V \rightarrow 2^N$, such that each vertex $v \in V$ is assigned a set of $x(v)$ distinct colors, and adjacent vertices receive non-intersecting sets of colors.

As schedules can be either non-preemptive or preemptive, so can the resulting colorings be either contiguous or arbitrary. A multicoloring, ψ, is called *non-preemptive* if the colors assigned to v are contiguous, i.e. if for any $v \in V$, $(\max_{i \in \psi(v)} i) - (\min_{i \in \psi(v)} i) + 1 = x(v)$.

Denote by $f_\psi(v) = \max_{i \in \psi(v)} i$ the largest color assigned to v by a multicoloring ψ. The *sum multicoloring* (SMC) of ψ on G is

$$\mathtt{SMC}(G, \psi) = \sum_{v \in V} f_\psi(v) \ .$$

The SMC problem is to find a multicoloring ψ, such that $\mathtt{SMC}(G, \psi)$ is minimized; the non-preemptive version is npSMC. When all the color requirements are equal to 1, the problem reduces to SC. Our focus here is on the *contiguous* version, npSMC.

Algorithm. Let G be a given $k+1$-claw-free graph, and let β be a parameter dependent on k, to be determined later. Informally, our strategy is the following. We allocate $x'(v) = (\beta+1)x(v)$ colors to each vertex v, or $\beta+1$ times more than required. We constrain this allocation so that the last $x(v)$ colors be contiguous, as they will form the actual set of colors assigned to v. The allocation is performed one color at a time, to a maximal independent set of vertices that have higher priority than others, either because they are shorter jobs, or because they have become *active*. Active vertices are those v that have received at least $\beta x(v)+1$ colors (but fewer than $(\beta+1)x(v)$), and thus must receive a contiguous set until fully colored. Namely, in each round, the algorithm colors a maximal independent set, I, satisfying two constraints: (i) I contains all currently active vertices, and (ii) vertices of small color requirements have priority over large ones. We assume that all color requirements are different, since ties can be broken in a fixed but arbitrary way.

The logic for allocating an additional $\beta x(v)$ colors is to build a buffer so that a long job does not accidentally become active and delay many short jobs for a long time. This way, all the neighbors of v have fair chance to be scheduled to completion before v becomes active. As they say, you have to pay your dues to become a player.

SG(G)
1. $j \leftarrow 1$.
2. **while** G is not fully colored **do**
   (a) $I_j \leftarrow$ all currently active vertices
   (b) Extend I_j to a maximal independent set, in such a way that each node outside the set has a neighbor in the set that is either active or of higher priority.
   (c) Assign color j to the vertices in set I_j, decrease their residue color requirement, update active vertices, and delete fully colored vertices.
   (d) $j \leftarrow j+1$.
3. The last $x(v)$ colors allocated to v form the coloring of v in the npSMC solution.

Note that step 2.(b) can be performed by greedily selecting compatible vertices of highest priority.

Analysis. We use the following notation and conventions in the analysis. Throughout this section, fix some optimum solution, OPT, for the npSMC instance. A vertex is *smaller* (*larger*) than its neighbor if its color-requirement is. The set of smaller (larger) neighbors of node v is denoted $N_s(v)$ ($N_l(v)$). Let O_v (respectively, O_v^s and O_v^l) denote the collection of neighbors of v (respectively, smaller and larger neighbors of v) scheduled before v by OPT. Similarly, let A_v, A_v^l and A_v^s be the nodes in $N(v)$, respectively, colored before v in SG, colored before v in SG and belonging to $N_l(v)$, and colored before v in SG and belonging to $N_s(v)$. Let $S_v = \sum_{u \in A_v} x(v)$, $S_v^l = \sum_{u \in A_v^l} x(v)$, and $S_v^s = \sum_{u \in A_v^s} x(v)$. Let $\mathcal{S}(G) = \sum_v x(v)$.

A vertex yet to be completed is either selected or delayed, in a given round. It is *selected* if placed in the current independent set, in which case it is either active or *chosen*. The vertex can either have a *good delay* in a round, if it has a smaller selected neighbor, or a *bad delay*, if it is delayed by a larger active neighbor.

We summarize this in the following fact. Let I be the current chosen independent set and I_a the set of active vertices, necessarily contained in I.

Fact 6 *In any round, exactly one of the following holds for a vertex v: (i)* Good delay: $I \cap N_s(v) \neq \emptyset$, (ii) Bad delay: $I_a \cap N_l(v) \neq \emptyset$, $I \cap N_s(v) = \emptyset$. (iii) Selected: *v is chosen or active.*

Let $d_g(v)$ ($d_b(v)$) denote the total good (bad) delay of v under SG. Fact 6 implies that the final color of v is given by $f_{SG}(v) = (\beta+1)x(v) + d_g(v) + d_b(v)$. We proceed to bound separately the good and bad delays. Define

$$Q(G) = \sum_v \sum_{w \in N_s(v)} x(w) = \sum_{vw \in E(G)} \min(x(v), x(w)).$$

The quantity $Q(G)$ provides an effective lower bound on the unrestricted multicolor sum, and thus also on the contiguous one.

Lemma 1. $Q(G) \leq k \cdot (\texttt{pSMC}(G) - \mathcal{S}(G))$.

Proof: Let D_v^i be the residual delay sum of v in OPT after round i. Namely, for any $u \in O_v$, if $x_i(u)$ time units of the job of u were already processed after round i by OPT, then $D_v^i = \sum_{u \in O_v}(x(u) - x_i(u))$.

We note that $D_v^{i+1} \geq D_v^i - k$, as in each round at most k of the neighbors of v are selected. Thus, the delay of v is at least D_v^0/k, and $\texttt{pSMC}(G) \geq \mathcal{S}(G) + \sum_v D_v^0/k$. Now, as every edge $e = (u, v)$ either contributes $x(u)$ to D_v^0 or $x(v)$ to D_u^0 we have: $\sum_v D_v^0 \geq Q(G)$. The required lemma follows. □

Lemma 2. *For any graph G, $\sum_v d_g(v) \leq (\beta+1) \cdot Q(G)$.*

Proof: A smaller neighbor u of v is selected for at most $(\beta+1) \cdot x(u)$ time units, and can delay v by at most that much. Thus, $d_g(v) \leq (\beta+1) \sum_{w \in N_s(v)} x(w)$. □

Lemma 3. *For any vertex v, $d_b(v) \leq (k-1)/\beta - k + 1) \cdot d_g(v)$.*

Proof: Each $u \in A_v^l$ was chosen $\beta \cdot x(u)$ times, because it became active. During rounds when a node in A_v^l was chosen, v was not selected, leaving us with cases (i) and (ii) of Fact 6. Some of these events may occur in parallel, i.e., several nodes in A_v^l can be chosen in the same round. At most k neighbors of v can be selected in a given round, since the graph is $k+1$-claw free. In any round when v is delayed, there is either a smaller selected neighbor or a larger active neighbor, and therefore at most $k-1$ larger chosen (i.e. selected but not active) neighbors.

Recall that $S_v^l = \sum_{u \in A_v^l} x(u)$. The total number of times that a node in A_v^l was chosen is exactly $\beta \cdot S_v^l$. Thus, the total delay of v, $d_g(v) + d_b(v)$ is at least $\beta \cdot S_v^l/(k-1)$, the total count of nodes in A_v^l chosen divided by the number of nodes simultaneously chosen. By the definition of A_v^l, $d_b(v)$ is at most S_v^l. Then, we have that $d_g(v) + d_b(v) \geq \beta d_b(v)/(k-1)$, and the lemma follows. □

Theorem 7. *SG approximates* npSMC(G) *on* $k+1$*-claw free graphs by a factor of* $2k(2k-1)$*. In particular, it achieves a factor 12 on line graphs and proper interval graphs.*

Proof: Let $\beta = 2(k-1)$. Let $SG(G)$ denote the multicolor sum of our algorithm on G. Combining Lemmas 2, 3 and 1, we have that $SG(G) \leq (\beta+1)\mathcal{S}(G) + (\beta+1)(1+\frac{k-1}{\beta-k+1})Q(G) \leq 2k(2k-1)\texttt{pSMC}(G) - (2k-1)^2\mathcal{S}(G)$. □

The above result can be applied to any claw-free graph. For example, it holds for the intersection graph of families of radius one circles (which are 7−claw free).

The above proof implies that the optimum preemptive and non-preemptive multicolor sums are within a constant factor on these graphs. Also, it gives a simultaneous approximation of the makespan within the same constant factor. The proofs can be extended to release times and to weights on jobs. We omit these extensions for lack of space.

References

AB⁺00. F. Afrati, E. Bampis, A. Fishkin, K. Jansen, and C. Kenyon. Scheduling to minimize the average completion time of dedicated tasks. In *FSTTCS 2000*, LNCS, Delhi.

BK98. A. Bar-Noy and G. Kortsarz. The minimum color-sum of bipartite graphs. *Journal of Algorithms*, 28:339–365, 1998.

BBH⁺98. A. Bar-Noy, M. Bellare, M. M. Halldórsson, H. Shachnai, and T. Tamir. On chromatic sums and distributed resource allocation. *Information and Computation*, 140:183–202, 1998.

BH94. D. Bullock and C. Hendrickson. Roadway traffic control software. *IEEE Transactions on Control Systems Technology*, 2:255–264, 1994.

BHK. A. Bar-Noy, M. M. Halldórsson, G. Kortsarz. Tight Bound for the Sum of a Greedy Coloring. *Information Processing Letters*, 1999.

BHK⁺99. A. Bar-Noy, M. M. Halldórsson, G. Kortsarz, H. Shachnai, and R. Salman. Sum Multicoloring of Graphs. *Journal of Algorithms*, 37(2):422–450, November 2000.

BKR96. P. Brucker and A. Krämer. Polynomial algorithms for resource-constrained and multiprocessor task scheduling problems. *European Journal of Operational Research*, 90:214–226, 1996.

F80. A. Frank. On Chain and Antichain Families of a Partially Ordered Set. *J. Combinatorial Theory*, Series B, 29: 176–184, 1980.

G01. M. Gonen, Coloring Problems on Interval Graphs and Trees. M.Sc. Thesis. School of Computer Science, The Open Univ., Israel, 2001.

HK99. M. M. Halldórsson and G. Kortsarz. Multicoloring Planar Graphs and Partial k-trees. In *Proceedings of the Second International Workshop on Approximation algorithms* (APPROX '99). Lecture Notes in Computer Science Vol. 1671, Springer-Verlag, August 1999.

HK$^+$99. M. M. Halldórsson, G. Kortsarz, A. Proskurowski, R. Salman, H. Shachnai, and J. A. Telle. Multi-Coloring Trees. In *Proceedings of the Fifth International Computing and Combinatorics Conference (COCOON)*, Tokyo, Japan, Lecture Notes in Computer Science Vol. 1627, Springer-Verlag, July 1999.

J97. K. Jansen. The Optimum Cost Chromatic Partition Problem. *Proc. of the Third Italian Conference on Algorithms and Complexity (CIAC '97)*. LNCS 1203, 1997.

K89. E. Kubicka. The Chromatic Sum of a Graph. PhD thesis, Western Michigan University, 1989.

K96. M. Kubale. Preemptive versus nonpreemptive scheduling of biprocessor tasks on dedicated processors. *European Journal of Operational Research* 94:242–251, 1996.

NSS99. S. Nicoloso, M. Sarrafzadeh and X. Song. On the Sum Coloring Problem on Interval Graphs. *Algorithmica*, 23:109–126,1999.

W97. G. Woeginger. Private communication, 1997.

YG87. M. Yannakakis and F. Gavril. The maximum k-colorable subgraph problem for chordal graphs. *Inform. Proc. Letters*, 24:133–137, 1987.

A Greedy Facility Location Algorithm Analyzed Using Dual Fitting

Mohammad Mahdian[1], Evangelos Markakis[2], Amin Saberi[2], and Vijay Vazirani[2]

[1] Department of Mathematics, MIT, MA 02139, USA
mahdian@mit.edu
[2] College of Computing, Georgia Institute of Technology, GA 30332, USA
{vangelis,saberi,vazirani}@cc.gatech.edu

Abstract. We present a natural greedy algorithm for the metric uncapacitated facility location problem and use the method of dual fitting to analyze its approximation ratio, which turns out to be 1.861. The running time of our algorithm is $O(m \log m)$, where m is the total number of edges in the underlying complete bipartite graph between cities and facilities. We use our algorithm to improve recent results for some variants of the problem, such as the fault tolerant and outlier versions. In addition, we introduce a new variant which can be seen as a special case of the concave cost version of this problem.

1 Introduction

A large fraction of the theory of approximation algorithms, as we know it today, is built around the theory of linear programming, which offers the two fundamental algorithm design techniques of rounding and the primal–dual schema (see [18]). It also offers the method of dual fitting for analyzing combinatorial algorithms. The latter has been used on perhaps the most central problem of this theory, the set cover problem [13,4]. Although this method appears to be quite basic, to our knowledge, it does not seem to have found use outside of the set cover problem and its generalizations [15]. Perhaps the most important contribution of this paper is to apply this method to the fundamental metric uncapacitated facility location problem.

The method can be described as follows, assuming a minimization problem: The basic algorithm is combinatorial – in the case of set cover it is in fact a simple greedy algorithm. Using the linear programming relaxation of the problem and its dual, one shows that the primal integral solution found by the algorithm is fully paid for by the dual computed; however, the dual is infeasible. The main step in the analysis consists of dividing the dual by a suitable factor and showing that the shrunk dual is feasible, i.e., it fits into the given instance. The shrunk dual is then a lower bound on OPT, and the factor is the approximation guarantee of the algorithm.

Our combinatorial algorithm for the metric uncapacitated facility location problem is a simple greedy algorithm. It is a small modification of Hochbaum's

M. Goemans et al. (Eds.): APPROX-RANDOM 2001, LNCS 2129, pp. 127–137, 2001.

greedy algorithm for this problem [7]. The latter was in fact the first approximation for this problem, with an approximation guarantee of $O(\log n)$. In contrast, our greedy algorithm achieves an approximation ratio of 1.861 and has a running time of $O(m \log m)$, where m is the number of edges of the underlying complete bipartite graph between cities and facilities, i.e., $m = n_c \times n_f$, where n_c is the number of cities and n_f is the number of facilities. Although this approximation factor is not the best known for this problem, our algorithm is natural and simple, and achieves the best approximation ratio within the same running time. For a metric defined by a sparse graph, Thorup [17] has obtained a $(3 + o(1))$-approximation algorithm with running time $\tilde{O}(|E|)$.

The first constant factor approximation algorithm for this problem was given by Shmoys, Tardos, and Aardal [16]. Later, the factor was improved by Chudak and Shmoys [3] to $1 + 2/e$. This was the best known algorithm until the recent work of Charikar and Guha [1], who slightly improved the factor to 1.728. The above mentioned algorithms are based on LP-rounding, and therefore have high running times. Jain and Vazirani [9] gave a primal–dual algorithm, achieving a factor of 3, and having the same running time as ours (we will refer to this as the JV algorithm). Their algorithm was adapted for solving several related problems such as the fault-tolerant and outlier versions, and the k-median problem [9,10,2]. Mettu and Plaxton [14] used a restatement of the JV algorithm for the on-line median problem.

Strategies based on local search and greedy improvement for facility location problem have also been studied. The work of Korupolu et. al. [11] shows that a simple local search heuristic proposed by Kuehn and Hamburger [12] yields a constant factor approximation for the facility location problem. Guha and Khuller [5] showed that greedy improvement can be used as a post-processing step to improve the approximation guarantee of certain facility location algorithms. The best approximation ratio for facility location [1] was obtained by combining the JV algorithm, greedy augmentation, and the best LP-based algorithm known. They also combined greedy improvement and cost scaling to improve the factor of the JV algorithm. They proposed two algorithms with approximation factors of $2.41+\epsilon$ and 1.853 and running times of $\tilde{O}(n^2/\epsilon)$ and $\tilde{O}(n^3)$ respectively, where n is the total number of vertices of the underlying graph. Regarding hardness results, Guha and Khuller [5] showed that the best approximation factor that we can get for this problem is 1.463, assuming $NP \not\subseteq DTIME[n^{O(\log \log n)}]$.

Our greedy algorithm is quite similar to the greedy set cover algorithm: iteratively pick the most cost-effective choice at each step, where cost-effectiveness is measured as the ratio of the cost incurred to the number of new cities served. In order to use LP-duality to analyze this algorithm, we give an alternative description which can be seen as a modification of the JV algorithm. This algorithm constructs a primal and dual solution of equal cost. Let us denote the dual by $\boldsymbol{\alpha}$. In general, $\boldsymbol{\alpha}$ may be infeasible. We write a linear program that captures the worst case ratio, R, by which $\boldsymbol{\alpha}$ may be infeasible. We then find a feasible solution to the dual of this latter LP having objective function value 1.861, thereby showing that $R \leq 1.861$. Therefore, $\frac{\boldsymbol{\alpha}}{1.861}$ is a dual feasible solution to the facility location LP, and hence a lower bound on the cost of an optimal solution to the facility location problem. As a consequence, the approximation guarantee of our algorithm is 1.861.

We have run our algorithm on randomly generated instances to obtain experimental results. The cost of the integral solution found is compared against the solution of the LP-relaxation of the problem. The results are good: The error is at most 7.1%.

We also use our algorithm to improve some recent results for some variants of the problem. In the facility location problem with outliers we are not required to connect all cities to some open facilities. In the robust version of this variant we are asked to choose l cities and connect the rest of them to some open facilities. In facility location with penalties we can either connect a city to a facility, or pay a specified penalty. Both versions were motivated by commercial applications, and were proposed by Charikar et al. [2]. In this paper we will modify our algorithm to obtain a factor 2 approximation algorithm for these versions, improving the best known result of factor 3.

In the fault tolerant variant, each city has a specified number of facilities it should be connected to. This problem was proposed in [10] and the best factor known is 2.47 [6]. We can show that we can achieve a factor 1.861 algorithm, when all cities have the same connectivity requirement. In addition, we introduce a new variant which can be seen as a special case of the concave cost version of this problem: the cost of opening a facility at a location is specified and it can serve exactly one city. In addition, a *setup cost* is charged the very first time a facility is opened at a given location.

Recently, Jain, Mahdian, and Saberi [8] have shown that a small modification of our greedy algorithm, analyzed using dual fitting, achieves an approximation factor of 1.61. This becomes the current best factor for the metric uncapacitated facility location problem. The running time of their algorithm is higher than ours, and is $O(n^3)$.

2 The Algorithm

Before stating the algorithm, we give a formal definition of the problem.

Metric Uncapacitated Facility Location: Let G be a bipartite graph with bipartition (F, C), where F is the set of facilities and C is the set of cities. Suppose also that $|C| = n_c$ and $|F| = n_f$. Thus, the total number of vertices in the graph is $n = n_c + n_f$ and the total number of edges is $m = n_c \times n_f$. Let f_i be the cost of opening facility i, and c_{ij} be the cost of connecting city j to facility i. The connection costs satisfy the triangle inequality. We want to find a subset $I \subseteq F$ of facilities that should be opened and a function $\phi : C \to I$ assigning cities to open facilities, that minimizes the total cost of opening facilities and connecting cities to them.

In the following algorithm we use a notion of cost-effectiveness. For each pair (i, C'), where i is a facility and $C' \subseteq C$ is a subset of cities, we define its cost-effectiveness to be $\left(f_i + \sum_{j \in C'} c_{ij}\right) / |C'|$.

Algorithm 1

1. In the beginning all cities are unconnected and all facilities are closed.
2. While $C \neq \emptyset$:
   - Among all pairs of facilities and subsets of C, find the most cost effective one, (i, C'), open facility i, if it is not already open, and connect all cities in C' to i.
   - Set $f_i := 0$, $C := C \setminus C'$.

Note that a facility can be chosen again after being opened, but its opening cost is counted only once since we set f_i to zero after the first time the facility is picked by the algorithm. As far as cities are concerned, every city j is removed from C, when connected to an open facility, and is not taken into consideration again. Also, notice that although the number of pairs of facilities and subsets of cities is exponentially large, in each iteration the most cost-effective pair can be found in polynomial time. For each facility i, we can sort the cities in increasing order of their connection cost to i. It can be easily seen that the most cost-effective pair will consist of a facility and a set, containing the first k cities in this order, for some k.

The idea of cost-effectiveness essentially stems from a similar notion in the greedy algorithm for the set cover problem. In that algorithm, the cost effectiveness of a set S is defined to be the cost of S over the number of uncovered elements in S. In each iteration, the algorithm picks the most cost-effective set until all elements are covered. The most cost-effective set can be found either by using direct computation, or by using the dual program of the linear programming formulation for the problem. The dual program can also be used to prove the approximation factor of the algorithm. Similarly, we will use the LP-formulation of facility location to analyze our algorithm. As we will see, the dual formulation of the problem helps us to understand the nature of the problem and the greedy algorithm.

Consider the following integer program for this problem. In this program y_i is an indicator variable denoting whether facility i is open, and x_{ij} is an indicator variable denoting whether city j is connected to facility i. The first constraint ensures that each city is connected to at least one facility and the second that this facility should be open.

$$\begin{array}{lll} \text{minimize} & \sum_{i \in F, j \in C} c_{ij} x_{ij} + \sum_{i \in F} f_i y_i & \\ \text{subject to} & \sum_{i \in F} x_{ij} \geq 1 & \forall j \in C \\ & y_i - x_{ij} \geq 0 & \forall i \in F, j \in C \\ & x_{ij}, y_i \in \{0, 1\} & \forall i \in F, j \in C \end{array} \tag{1}$$

The LP-relaxation of this program can be obtained if we allow x_{ij} and y_i to be non-negative real numbers. The dual program of the LP-relaxation will then be

$$\begin{array}{lll} \text{maximize} & \sum_{j \in C} \alpha_j & \\ \text{subject to} & \alpha_j - \beta_{ij} \leq c_{ij} & \forall i \in F, j \in C \\ & \sum_{j \in C} \beta_{ij} \leq f_i & \forall i \in F \\ & \alpha_j, \beta_{ij} \geq 0 & \forall i \in F, j \in C \end{array} \tag{2}$$

There is an intuitive way of interpreting the dual variables. We can think of α_j as the contribution of city j. This contribution goes towards opening some facility and connecting city j to it. Using the inequalities of the dual program, we will have $\sum_{j \in C} \max(0, \alpha_j - c_{ij}) \leq f_i$. We can now see how the dual variables can help us find the most cost-effective pair in each iteration of the greedy algorithm: if we start raising the dual variables of all unconnected cities simultaneously, the most cost-effective pair (i, C') will be the first pair for which $\sum_{j \in C'} \max(0, \alpha_j - c_{ij}) = f_i$. Hence we can restate Algorithm 1 based on the above observation. This is in complete analogy to the greedy algorithm and its restatement using LP-formulation for set-cover.

Algorithm 2

1. We introduce a notion of time, so that each event can be associated with the time at which it happened. The algorithm starts at time 0. Initially, each city is defined to be unconnected, all facilities are closed, and α_j is set to 0 for every j.
2. While $C \neq \emptyset$, for every city $j \in C$, increase the parameter α_j simultaneously, until one of the following events occurs (if two events occur at the same time, we process them in arbitrary order).
   (a) For some unconnected city j, and some open facility i, $\alpha_j = c_{ij}$. In this case, connect city j to facility i and remove j from C.
   (b) For some closed facility i, we have $\sum_{j \in C} \max(0, \alpha_j - c_{ij}) = f_i$. This means that the total contribution of the cities is sufficient to open facility i. In this case, open this facility, and for every unconnected city j with $\alpha_j \geq c_{ij}$, connect j to i, and remove it from C.

In each iteration of algorithm 1 the process of opening a facility and connecting some cities to it can be thought of as an event. It is easy to prove the following theorem by induction.

Theorem 1. *The events executed by algorithms 1 and 2 are identical.*

Algorithm 2 can also be seen as a modification of JV algorithm [9]. The only difference is that in JV algorithm cities, when connected to an open facility, are not excluded from C, hence they might contribute towards opening several facilities. Due to this fact they have a second cleanup phase, in which some of the already open facilities will be closed down.

3 Analysis of the Algorithm

In this section we will give an LP-based analysis of the algorithm. As stated before, the contribution of each city goes towards opening at most one facility and connecting the city to an open facility. Therefore the total cost of the solution produced by our algorithm will be equal to the sum $\sum_j \alpha_j$ of the contributions. However, $(\boldsymbol{\alpha}, \boldsymbol{\beta})$, where $\beta_{ij} = max(\alpha_j - c_{ij}, 0)$, is no longer a dual feasible solution as it was in the JV algorithm. The reason is that $\sum_j max(\alpha_j - c_{ij}, 0)$ can be greater than f_i and hence the second constraint of the dual program is violated.

However, if we show that for some $R > 1$, we can define $\boldsymbol{\beta}$ such that $(\boldsymbol{\alpha}/R, \boldsymbol{\beta}/R)$ is a feasible dual solution, then by the weak duality theorem, $(\sum_j \alpha_j)/R$ is a lower bound for the optimum solution, and therefore the approximation ratio of the algorithm is R.

Theorem 2. *Let α_j $(j = 1, \ldots, n_c)$ denote the contribution of city j when algorithm 2 terminates. If for every facility i, every set of k cities, and a fixed number R, we have $\sum_{j=1}^{k} \alpha_j \leq R(f_i + \sum_{j=1}^{k} c_{ij})$, then the approximation ratio of the algorithm is at most R.*

Proof. Let $\beta_{ij} = max(\alpha_j - Rc_{ij}, 0)$. We will show that $(\boldsymbol{\alpha}/R, \boldsymbol{\beta}/R)$ is a feasible dual solution. To see that the first condition of the dual program is satisfied, we need to show that $\alpha_j - max(\alpha_j - Rc_{ij}, 0) \leq Rc_{ij}$. We can verify that this holds by considering the two possible cases $(\alpha_j > Rc_{ij})$ and $(\alpha_j \leq Rc_{ij})$. As far as the second constraint of the dual program is concerned, we need to show that $\sum_{j=1}^{n_c} max(\alpha_j - Rc_{ij}, 0) \leq Rf_i$. Let S be the set of cities for which $\alpha_j - Rc_{ij} > 0$. Then $\sum_{j=1}^{n_c} max(\alpha_j - Rc_{ij}, 0) = \sum_{j \in S}(\alpha_j - Rc_{ij})$. Thus the constraint becomes equivalent to the condition $\sum_{j \in S} \alpha_j \leq R(f_i + \sum_{j \in S} c_{ij})$, which is true due to the assumptions of the theorem. Hence by the weak duality theorem, $(\sum_j \alpha_j)/R$ is a lower bound for the optimal solution. We also know that the cost of the solution produced by our algorithm is $\sum_j \alpha_j$. This completes the proof.

From now on, we assume without loss of generality that $\alpha_1 \leq \alpha_2 \leq \ldots \leq \alpha_{n_c}$. For the rest of the analysis, we need the following lemmas.

Lemma 1. *For every two cities j, j' and facility i, $\alpha_j \leq \alpha_{j'} + c_{ij'} + c_{ij}$.*

Proof. If $\alpha_{j'} \geq \alpha_j$, the inequality obviously holds. Assume $\alpha_j > \alpha_{j'}$. Let i' be the facility that city j' is connected to by our algorithm. Thus, facility i' is open at time $\alpha_{j'}$. The contribution α_j cannot be greater than $c_{i'j}$ because in that case city j could be connected to facility i' at some time $t < \alpha_j$. Hence $\alpha_j \leq c_{i'j}$. Furthermore, by triangle inequality, $c_{i'j} \leq c_{i'j'} + c_{ij'} + c_{ij} \leq \alpha_{j'} + c_{ij'} + c_{ij}$.

Lemma 2. *For every city j and facility i, $\sum_{k=j}^{n_c} \max(\alpha_j - c_{ik}, 0) \leq f_i$.*

Proof. Assume, for the sake of contradiction, that for some j and some i the inequality does not hold, i.e., $\sum_{k=j}^{n_c} \max(\alpha_j - c_{ik}, 0) > f_i$. By the ordering on cities, for $k \geq j$, $\alpha_k \geq \alpha_j$. Let time $t = \alpha_j$. By the assumption, facility i is fully paid for before time t. For any city k, $j \leq k \leq n_c$ for which $\alpha_j - c_{ik} > 0$ the edge (i, k) must be tight before time t. Moreover, there must be at least one such city. For this city, $\alpha_k < \alpha_j$, since the algorithm will stop growing α_k as soon as k has a tight edge to a fully paid for facility. The contradiction establishes the lemma.

Subject to the constraints introduced by these lemmas, we want to find a factor R such that for every facility i and every set of k cities, $\sum_{j=1}^{k} \alpha_j \leq R(f_i + \sum_{j=1}^{k} c_{ij})$.

This suggests considering the following program.

$$
\begin{array}{llll}
z_k = \text{maximize} & \dfrac{\sum_{j=1}^{k} \alpha_j}{f + \sum_{j=1}^{k} d_j} & & \\
\text{subject to} & \alpha_j \le \alpha_{j+1} & \forall j \in \{1, \ldots, k-1\} & \\
& \alpha_j \le \alpha_l + d_j + d_l & \forall j, l \in \{1, \ldots, k\} & (3) \\
& \sum_{l=j}^{k} \max(\alpha_j - d_l, 0) \le f & \forall j \in \{1, \ldots, k\} & \\
& \alpha_j, d_j, f \ge 0 & \forall j \in \{1, \ldots, k\} &
\end{array}
$$

For a facility i and a set of k cities, S, the variables f and d_j's of this maximization program will correspond to the opening cost of i, and the costs of connecting each city $j \in S$ to i. Note that it is easy to write the above program as an LP.

In the following theorem, we prove, by demonstrating an infinite family of instances, that the approximation ratio of Algorithm 2 is not better than $\sup_{k \ge 1}\{z_k\}$. The proof is not difficult and is omitted.

Theorem 3. *For every k, there is an instance of the facility location problem for which Algorithm 2 outputs a solution of cost at least z_k times the optimum solution.*

The following theorem combined with theorems 2 and 3 shows that the factor of our algorithm is exactly equal to $\sup_{k \ge 1}\{z_k\}$. The proof is easy using Lemmas 1 and 2 and is omitted.

Theorem 4. *For every facility i and every set of k cities, $1 \le k \le n_c$, we have $\sum_{j=1}^{k} \alpha_j \le z_k(f_i + \sum_{j=1}^{k} c_{ij})$.*

Hence, the following theorem shows that the approximation ratio of our algorithm is 1.861.

Theorem 5. *For every $k \ge 1$, $z_k \le 1.861$.*

Proof. omitted.

Numerical computations using the software package AMPL show that $z_{300} \approx 1.81$. Thus, the approximation factor of our algorithm is between 1.81 and 1.861. We still do not know the exact approximation ratio of our algorithm. The example in Fig. 1 shows that the approximation factor of the algorithm is at least 1.5. The cost of the missing edges in this figure are equal to the cost of the shortest path in the above graph.

3.1 Running Time Analysis

In order to implement algorithm 2, for each facility, we can use a heap data structure to maintain a list of events. There are three types of event: (a) City j starts contributing towards opening facility i; (b) Facility i is being paid for and hence opened; and (c) City j connects to an open facility i. It is easy to see that there are at most m events of type (a), n_f events of type (b), and n_c events of type (c), and these events can be processed in times $O(\log m)$, $O(n_c)$ and $O(n_f \log m)$, respectively. Hence, the total running time of the algorithm is $O(m \log m)$.

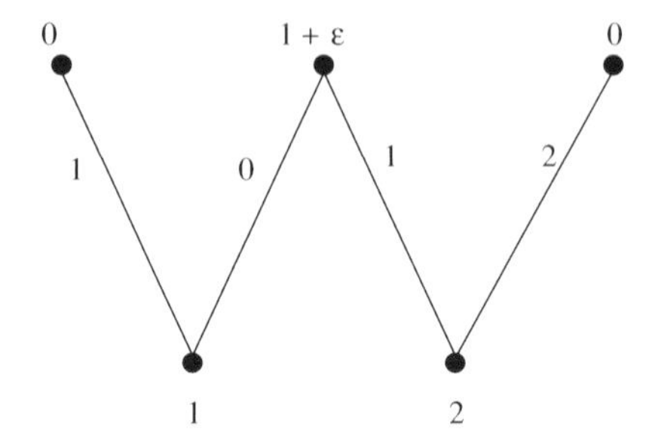

Fig. 1. The approximation ratio of the algorithm is at least 1.5

Table 1. Experimental results

n_c	n_f	# of instances	average ratio	worst ratio
50	20	20	1.033	1.070
100	20	20	1.025	1.071
100	50	20	1.026	1.059
200	50	20	1.032	1.059
200	100	20	1.027	1.064
300	50	20	1.034	1.070
300	80	20	1.030	1.057
300	100	20	1.033	1.053
300	150	20	1.029	1.048
400	100	20	1.030	1.060
400	150	20	1.030	1.050

4 Experimental Results

We implemented our algorithm in C to see how it behaves in practice. The test bed of our experiments consisted of randomly generated instances. In each instance, cities and facilities were points, drawn uniformly from a 10000×10000 grid, the connection costs were according to the Euclidean distance, and facility opening costs were random integers between 0 and 9999. We used the optimal solution of the LP-relaxation, computed using the package AMPL, as a lower bound for the optimal solution of each instance. The results of our experiments are summarized in Table 1.

5 Variants

In this section, we show that our algorithm can also be applied to several variants of the metric facility location problem.

Arbitrary Demands: In this version, for each city j, a non-negative integer demand d_j, is specified. An open facility i can serve this demand at the cost of $c_{ij}d_j$. The best way to look at this modification is to reduce it to unit demand case by making d_j copies of city j. This reduction suggests that we need to change Algorithm 2, so that each city j raises its contribution α_j at rate d_j. Note that the modified algorithm still runs in $O(m \log m)$ in more general cases, where d_j is fractional or exponentially large, and achieves an approximation ratio of 1.861.

Fault Tolerant Facility Location with Uniform Connectivity Requirements: We are given a connectivity requirement r_j for each city j, which specifies the number of open facilities that city j should be connected to. We can see that this problem is closely related to the set multi-cover problem, in the case at which every set can be picked at most once [15]. The greedy algorithm for set-cover can be adapted for this variant of the multi-cover problem achieving the same approximation factor. We can use the same approach to deal with the fault tolerant facility location: The mechanism of raising dual variables and opening facilities is the same as in our initial algorithm. The only difference is that city j stops raising its dual variable and withdraws its contribution from other facilities, when it is connected to r_j open facilities. We can show that when all r_j's are equal, this algorithm has an approximation ratio of 1.861.

Facility Location with Penalties: In this version we are not required to connect every citiy to an open facility; however, for each city j, there is a specified penalty, p_j, which we have to pay, if it is not connected to any open facility. We can modify Algorithm 2 for this problem as follows: If α_j reaches p_j before j is connected to any open facility, the city j stops raising its dual variable and keeps its contribution equal to its penalty until it is either connected to an open facility or all remaining cities stop raising their dual variables. At this point, the algorithm terminates and unconnected cities remain unconnected. Using the same proof as the one we used for Algorithm 2, we can show that the approximation ratio of this algorithm is 2, and its running time is $O(m \log m)$.

Robust Facility Location: In this variant, we are given a number l and we are only required to connect $n_c - l$ cities to open facilities. This problem can be reduced to the previous one via Lagrangian relaxation. Very recently, Charikar et al. [2] proposed a primal-dual algorithm, based on JV algorithm, which achieves an approximation ratio of 3. As they showed, the linear programming formulation of this variant has an unbounded integrality gap. In order to fix this problem, they use the technique of parametric pruning, in which they guess the most expensive facility in the optimal solution. After that, they run JV algorithm on the pruned instance, where the only allowable facilities are those that are not more expensive than the guessed facility. Here we can use the same idea, using Algorithm 1 rather than the JV algorithm. Using similar methods, we can prove that this algorithm solves the robust facility location problem with an approximation factor of 2.

Dealing with Capacities: In real applications, it's not usually the case that the cost of opening a facility is independent of the number of cities it will serve. But we can assume that we have *economy of scales*, i.e., the cost of serving each city decreases when the number of cities increases. In order to capture this property, we define the following variant of the capacitated metric facility location problem. For each facility i, there is an initial opening cost f_i. After facility i is opened, it will cost s_i to serve each city. This variant can be solved using metric uncapacitated facility location problem: We just have to change the metric such that for each city j and facility i, $c'_{ij} = c_{ij} + s_i$. Clearly, c' is also a metric and the solution of the metric uncapacitated version to this problem can be interpreted as a solution to the original problem with the same cost.

We can reduce the variant of the capacitated facility location problem in which each facility can be opened many times [9] to this problem by defining $s_i = f_i/u_i$. If in the solution to this problem k cities are connected to facility i, we open this facility $\lceil k/u_i \rceil$ times. The cost of the solution will be at most two times the original cost so any α-approximation for the uncapacitated facility location problem can be turned into a 2α-approximation for this variant of the capacitated version.

Acknowledgments

The first and third authors would like to thank Dr. Mohammad Ghodsi, Computer Engineering Department, Sharif University of Technology, for proposing the problem. We would also like to thank Nisheet K. Vishnoi for valuable discussions.

References

1. M. Charikar and S. Guha. *Improved combinatorial algorithms for facility location and k-median problems.* Proceedings of the 40th Annual IEEE Symposium on Foundations of Computer Science, pp. 378-388, October 1999.
2. M. Charikar, S. Khuller, D. Mount, and G. Narasimhan. *Algorithms for Facility Location Problems with Outliers.* Proceedings of the 12th ACM-SIAM Symposium on Discrete Algorithms, 2001.
3. F. Chudak and D. Shmoys. *Improved approximation algorithms for the capacitated facility location problem.* Proceedings of the 10th Annual ACM-SIAM Symposium on Discrete Algorithms, pp. 875-876, 1999.
4. V. Chvatal. *A greedy heuristic for the set covering problem.* Math. Oper. Res. 4, pp. 233-235, 1979.
5. S. Guha and S. Khuller. *Greedy strikes back: Improved facility location algorithms.* Journal of Algorithms 31, pp. 228-248, 1999.
6. S. Guha, A. Meyerson, and K. Munagala. *Improved Approximation Algorithms for Fault-tolerant Facility Location.* Proceedings of the 12th ACM-SIAM Symposium on Discrete Algorithms, 2001.
7. D. S. Hochbaum. *Heuristics for the fixed cost median problem.* Math. Programming 22, 148-162, 1982.
8. K. Jain, M. Mahdian, and A. Saberi. *A new greedy approach for facility location problems*, manuscript.
9. K. Jain and V. V. Vazirani. *Primal-dual approximation algorithms for metric facility location and k-median problems.* Proceedings of the 40th Annual IEEE Symposium on Foundations of Computer Science, pp. 2-13, October 1999.

10. K. JAIN and V. VAZIRANI. *An approximation algorithm for the fault tolerant metric facility location problem.* Proceedings of APPROX 2000, pp. 177-183, 2000.
11. M. R. KORUPOLU, C. G. PLAXTON, and R. RAJARAMAN. *Analysis of a local search heuristic for facility location problems.* Proceedings of the 9th Annual ACM-SIAM Symposium on Discrete Algorithms, pp. 1-10, January 1998.
12. A. A. KUEHN and M. J. HAMBURGER. *A heuristic program for locating warehouses.* Management Science 9, pp. 643-666, 1963.
13. L. LOVASZ. *On the ratio of Optimal Integral and Fractional Covers.* Discrete Math. 13, pp. 383-390, 1975.
14. R. METTU and G. PLAXTON. *The online median problem.* Proceedings of 41st IEEE FOCS, 2000.
15. S. RAJAGOPALAN and V. V. VAZIRANI. *Primal-dual RNC approximation of covering integer programs.* SIAM J. Comput. 28, pp. 526-541, 1999.
16. D. B. SHMOYS, E. TARDOS, and K. AARDAL. *Approximation algorithms for facility location problems.* Proceedings of the 29th Annual ACM Symposium on Theory of Computing, pp. 265-274, May 1997.
17. M. THORUP. *Quick k-median, k-center, and facility location for sparse graphs.* To appear in ICALP 2001.
18. V. V. VAZIRANI. *Approximation Algorithms*, Springer-Verlag, Berlin, 2001.

0.863-Approximation Algorithm for MAX DICUT

Shiro Matuura[1] and Tomomi Matsui[2]

[1] University of Tokyo, Tokyo, Japan
shiro@misojiro.t.u-tokyo.ac.jp
[2] University of Tokyo, Tokyo, Japan
tomomi@misojiro.t.u-tokyo.ac.jp

Abstract. In this paper, we propose 0.863-approximation algorithm for MAX DICUT. The approximation ratio is better than the previously known result by Zwick, which is equal to 0.8596434254.
The algorithm solves the SDP relaxation problem proposed by Goemans and Williamson for the first time. We do not use the 'rotation' technique proposed by Feige and Goemans. We improve the approximation ratio by using hyperplane separation technique with skewed distribution function on the sphere. We introduce a class of skewed distribution functions defined on the 2-dimensional sphere satisfying that for any function in the class, we can design a skewed distribution functions on any dimensional sphere without decreasing the approximation ratio. We also searched and found a good distribution function defined on the 2-dimensional sphere numerically.

1 Introduction

In this paper we propose an approximation algorithm for the optimization problem called MAX DICUT. Let $D = (V, A)$ be a complete directed graph with vertex set $V = \{1, 2, \dots, n\}$ and arc set $A = \{(i, j) \in V \times V \mid i \neq j\}$. For each arc $(i, j) \in A$, we associate the non-negative arc weight w_{ij}. An arc subset $A' \subseteq A$ is called a *dicut* if and only if there exists a vertex subset $U \subseteq V$ satisfying that $A' = \{(i, j) \in A \mid i \in U \text{ and } j \in V \setminus U\}$. The weight of an arc-subset A' is the sum total of the weights of arcs in A'. The MAX DICUT is the problem for finding a dicut which maximizes its weight. The MAX DICUT is formulated as follows;

$$\text{(DI) maximize} \sum_{i \in U,\ j \in V \setminus U} w_{ij} \quad \text{subject to } U \subseteq V.$$

This problem is NP-Hard and so there are some algorithms for finding an approximate solution. As known well, Goemans and Williamson [5] proposed a randomized polynomial time algorithm for MAX CUT, MAX 2SAT and MAX DICUT. Their algorithm is based on Semi-Definite Programming (SDP) relaxation and random hyperplane separation technique. The approximation ratio of their algorithm for MAX DICUT is 0.79607. More precisely, their algorithm finds a dicut whose weight is at least 0.79607 times the optimal value.

M. Goemans et al. (Eds.): APPROX-RANDOM 2001, LNCS 2129, pp. 138–146, 2001.

In the paper [4], Feige and Goemans proposed an approximation algorithm for MAX DICUT which achieves 0.859387 of approximation ratio. Their algorithm based on two ideas. First, they added some constraints introduced by Feige and Lovász in [3] to SDP relaxation problem. Next, they proposed the 'rotation' technique which modifies the solution obtained by SDP relaxation. They calculated the approximation ratio of their algorithm numerically.

Recently, Zwick refined the rotation technique and proposed an algorithm whose approximation ratio is 0.8596434254 in [8]. He also showed that the approximation ratio of his algorithm almost completely matches upper bounds that we can obtain by any rotation technique.

As a related work, in the APPROX 2000, Ageev, Hassin and Sviridenko proposed an approximation algorithm for MAX DICUT with given sizes of parts [1].

In this paper, we propose an approximation algorithm without rotation technique whose approximation ratio is 0.863. Our algorithm solves the SDP relaxation problem by Goemans and Williamson with the constraints used in Feige and Goemans' algorithm. We improve the approximation ratio by using hyperplane separation technique with skewed distribution function on the sphere. Although, the use of hyperplane separation technique with skewed distribution is suggested by Feige and Goemans in the paper [4], there is a non-trivial problem to design a good distribution function. More precisely, the performance of skewed distribution functions depends on the dimension of the corresponding sphere. First, we show a non-trivial relation between the skewed distribution functions on the 2-dimensional sphere and the n-dimensional sphere. We introduce a class of skewed distribution functions defined on the 2-dimensional sphere satisfying that for any distribution function in the class, we can design a skewed distribution function defined on any dimensional sphere without decreasing the approximation ratio. Second, we searched and found a good distribution function on the 2-dimensional sphere numerically. By using the above results, we can design a good skewed distribution function on any dimensional sphere. It means that the distribution function of our algorithm changes with respect to the dimension of the corresponding sphere.

In Section 2, we review the SDP relaxation and hyperplane separation technique briefly. In Section 3, we describe the outline of our algorithm. In Section 4, we discuss some relations between the skewed distribution functions on the 2-dimensional sphere and the n-dimensional sphere. In Section 5, we describe a numerical method used for finding a good distribution function defined on the 2-dimensional sphere.

2 Semi-definite Programming Relaxation

Here we describe a SDP relaxation of MAX DICUT and review the hyperplane separation technique. First, we formulate the MAX DICUT problem as an integer programming problem as follows;

$$\text{(DI')} \text{ maximize } (1/4) \sum_{(i,j)\in A} w_{ij}(1 + v_0 v_i - v_0 v_j - v_i v_j),$$

$$\text{subject to } v_0 = 1, \quad v_i \in \{-1, 1\} \quad (\forall i \in V).$$

The above problem is equivalent to the original problem (DI). In the paper [5], Goemans and Williamson relaxed this problem by replacing each variable $v_i \in \{-1, 1\}$ with a vector on the n-dimensional unit sphere $\boldsymbol{v}_i \in \boldsymbol{S}_n$ where $\boldsymbol{S}_n \stackrel{\text{def.}}{=} \{\boldsymbol{v} \in \boldsymbol{R}^{n+1} \mid ||\boldsymbol{v}|| = 1\}$. This relaxation is proposed by Lovász [6] originally. And We add some valid constraints used in papers [3,4], and obtain the following relaxation problem;

$$\begin{aligned}
(\overline{\text{DI}}) \text{ maximize } & (1/4) \sum_{(i,j)\in A} w_{ij}(1 + \boldsymbol{v}_0 \cdot \boldsymbol{v}_i - \boldsymbol{v}_0 \cdot \boldsymbol{v}_j - \boldsymbol{v}_i \cdot \boldsymbol{v}_j), \\
\text{subject to } & \boldsymbol{v}_0 = (1, 0, \ldots, 0)^\top, \quad \boldsymbol{v}_i \in \boldsymbol{S}_n \quad (\forall i \in V) \\
& \boldsymbol{v}_0 \cdot \boldsymbol{v}_i + \boldsymbol{v}_0 \cdot \boldsymbol{v}_j + \boldsymbol{v}_i \cdot \boldsymbol{v}_j \geq -1 \quad (\forall (i,j) \in A), \\
& -\boldsymbol{v}_0 \cdot \boldsymbol{v}_i - \boldsymbol{v}_0 \cdot \boldsymbol{v}_j + \boldsymbol{v}_i \cdot \boldsymbol{v}_j \geq -1 \quad (\forall (i,j) \in A), \\
& -\boldsymbol{v}_0 \cdot \boldsymbol{v}_i + \boldsymbol{v}_0 \cdot \boldsymbol{v}_j - \boldsymbol{v}_i \cdot \boldsymbol{v}_j \geq -1 \quad (\forall (i,j) \in A).
\end{aligned}$$

It is well-known that we can transform the above problem to semidefinite programming problem [5] and so we can solve the problem in polynomial time by using an interior point method [2,7].

Next, we describe the hyperplane separation technique proposed by Goemans and Williamson. Let $(\overline{\boldsymbol{v}}_1, \overline{\boldsymbol{v}}_2, \ldots, \overline{\boldsymbol{v}}_n)$ be an optimal solution of $\overline{\text{DI}}$. We generate a vector $\boldsymbol{r} \in \boldsymbol{S}_n$ uniformly and construct the vertex-subset $\overline{U} = \{i \in V \mid \text{sign}(\boldsymbol{r} \cdot \boldsymbol{v}_0) = \text{sign}(\boldsymbol{r} \cdot \overline{\boldsymbol{v}}_i)\}$ and the corresponding dicut $\overline{A} = \{(i,j) \in A \mid i \in U \text{ and } j \notin U\}$. We denote the expected weight of the dicut $\overline{A}$ by $\text{E}(\overline{U})$. Then the linearity of the expectation implies that;

$$\text{E}(\overline{U}) = \sum_{(i,j)\in A} w_{ij} \frac{\arccos(\overline{\boldsymbol{v}}_i \cdot \overline{\boldsymbol{v}}_j) + \arccos(\boldsymbol{v}_0 \cdot \overline{\boldsymbol{v}}_j) - \arccos(\boldsymbol{v}_0 \cdot \overline{\boldsymbol{v}}_i)}{2\pi}.$$

Then we can estimate the approximation ratio of the algorithm by calculating α defined by;

$$\alpha \stackrel{\text{def.}}{=} \min_{(\boldsymbol{v}_i, \boldsymbol{v}_j) \in \Omega} \frac{(1/2\pi)(\arccos(\boldsymbol{v}_i \cdot \boldsymbol{v}_j) + \arccos(\boldsymbol{v}_0 \cdot \boldsymbol{v}_j) - \arccos(\boldsymbol{v}_0 \cdot \boldsymbol{v}_i))}{(1/4)(1 + \boldsymbol{v}_0 \cdot \boldsymbol{v}_i - \boldsymbol{v}_0 \cdot \boldsymbol{v}_j - \boldsymbol{v}_i \cdot \boldsymbol{v}_j)},$$

where

$$\Omega \stackrel{\text{def.}}{=} \left\{ (\boldsymbol{v}_i, \boldsymbol{v}_j) \in \boldsymbol{S}_2 \times \boldsymbol{S}_2 \;\middle|\; \begin{array}{r} \boldsymbol{v}_0 \cdot \boldsymbol{v}_i + \boldsymbol{v}_0 \cdot \boldsymbol{v}_j + \boldsymbol{v}_i \cdot \boldsymbol{v}_j \geq -1, \\ -\boldsymbol{v}_0 \cdot \boldsymbol{v}_i - \boldsymbol{v}_0 \cdot \boldsymbol{v}_j + \boldsymbol{v}_i \cdot \boldsymbol{v}_j \geq -1, \\ -\boldsymbol{v}_0 \cdot \boldsymbol{v}_i + \boldsymbol{v}_0 \cdot \boldsymbol{v}_j - \boldsymbol{v}_i \cdot \boldsymbol{v}_j \geq -1, \\ \boldsymbol{v}_0 \cdot \boldsymbol{v}_i - \boldsymbol{v}_0 \cdot \boldsymbol{v}_j - \boldsymbol{v}_i \cdot \boldsymbol{v}_j \geq -1 \end{array} \right\},$$

and $\boldsymbol{v}_0 = (1, 0, 0)^\top$. Clearly, the following inequalities hold;

$$\text{E}(\overline{U}) \geq \alpha(\text{optimal value of } (\overline{\text{DI}})) \geq \alpha(\text{optimal value of (DI)}).$$

So, the expected weight $\text{E}(\overline{U})$ is greater than or equal to α times the optimal value of (DI). It is known that $\alpha > 0.79607$ [5].

Feige and Goemans' algorithm solves the problem ($\overline{\text{DI}}$) and modifies the obtained optimal solution by using rotation technique. The approximation ratio is 0.859387. Zwick refined the rotation technique and proposed an algorithm whose approximation ratio is equal to 0.8596434254. Our algorithm does not use the rotation technique and so we will not describe the technique here.

3 Hyperplane Separation by Skewed Distribution on Sphere

Goemans and Williamson's algorithm generates a separating hyperplane at random. Our algorithm generates a separating hyperplane with respect to a distribution function defined on $\boldsymbol{S}_n$ which is skewed towards $\boldsymbol{v}_0$ but is uniform in any direction orthogonal to $\boldsymbol{v}_0$. Given the n-dimensional sphere $\boldsymbol{S}_n$, we define the class of skewed distribution function $\mathcal{F}_n$ by;

$$\mathcal{F}_n \stackrel{\text{def.}}{=} \left\{ f : \boldsymbol{S}_n \to \boldsymbol{R}_+ \;\middle|\; \begin{array}{l} \int_{\boldsymbol{S}_n} f(\boldsymbol{v})\, \mathrm{d}s = 1, \;\; f(\boldsymbol{v}) = f(-\boldsymbol{v}) \; (\forall \boldsymbol{v} \in \boldsymbol{S}_n), \\ {[\boldsymbol{v}_0 \cdot \boldsymbol{v} = \boldsymbol{v}_0 \cdot \boldsymbol{v}' \to f(\boldsymbol{v}) = f(\boldsymbol{v}')]} \; (\forall \boldsymbol{v}, \forall \boldsymbol{v}' \in \boldsymbol{S}_n) \end{array} \right\}.$$

Let $f \in \mathcal{F}_n$ be a skewed distribution function defined on $\boldsymbol{S}_n$. Now consider the probability that arc (i, j) are contained in a dicut obtained by hyperplane separation technique based on f. For any pair $(\boldsymbol{v}_i, \boldsymbol{v}_j) \in \boldsymbol{S}_n$, we define

$$p(\boldsymbol{v}_i, \boldsymbol{v}_j \mid f) \stackrel{\text{def.}}{=} \Pr \begin{bmatrix} \text{sign}(\boldsymbol{r} \cdot \boldsymbol{v}_0) = \text{sign}(\boldsymbol{r} \cdot \boldsymbol{v}_i) \text{ and} \\ \text{sign}(\boldsymbol{r} \cdot \boldsymbol{v}_0) \neq \text{sign}(\boldsymbol{r} \cdot \boldsymbol{v}_j) \end{bmatrix}.$$

Then the expectation of the weight of the dicut with respect to a feasible solution $(\overline{\boldsymbol{v}}_1, \overline{\boldsymbol{v}}_2, \ldots, \overline{\boldsymbol{v}}_n)$ of $\overline{\text{DI}}$ based on the distribution function f is $\sum_{(i,j)\in A} w_{ij}\, p(\overline{\boldsymbol{v}_i}, \overline{\boldsymbol{v}_j} \mid f)$.

When we use a skewed distribution function $f \in \mathcal{F}_n$ defined on $\boldsymbol{S}_n$, the approximation ratio can be estimated by the distribution function $\widehat{f}$ defined by projection of a vector on $\boldsymbol{S}_n$ to the linear subspace spanned by $\{\boldsymbol{v}_0, \boldsymbol{v}_i, \boldsymbol{v}_j\}$. We define $\widehat{f}$ more precisely. Let H be the 3-dimensional linear subspace including $\{\boldsymbol{v}_0, \boldsymbol{v}_i, \boldsymbol{v}_j\}$. The distribution function $\widehat{f} \in \mathcal{F}_2$ is defined as follows;

$$\widehat{f}(\boldsymbol{v}') \stackrel{\text{def.}}{=} \int_{T(\boldsymbol{v}')} f(\boldsymbol{v})\, \mathrm{d}s,$$

where

$$T(\boldsymbol{v}') \stackrel{\text{def.}}{=} \{\boldsymbol{v} \in \boldsymbol{S}_n \mid \text{ the projection of } \boldsymbol{v} \text{ to } H \text{ is parallel to } \boldsymbol{v}'\}.$$

Here we note that the distribution function $\widehat{f}$ is invariant with respect to the 3-dimensional subspace H including $\boldsymbol{v}_0$, since f is uniform in any directions orthogonal to $\boldsymbol{v}_0$. For any distribution function $f' \in \mathcal{F}_2$ we define

$$\alpha_{f'} \stackrel{\text{def.}}{=} \min_{(\boldsymbol{v}_i, \boldsymbol{v}_j) \in \Omega} \frac{p(\boldsymbol{v}_i, \boldsymbol{v}_j \mid f')}{(1/4)(1 + \boldsymbol{v}_0 \cdot \boldsymbol{v}_i - \boldsymbol{v}_0 \cdot \boldsymbol{v}_j - \boldsymbol{v}_i \cdot \boldsymbol{v}_j)},$$

here we note that $p(\boldsymbol{v}_i, \boldsymbol{v}_j \mid f')$ is defined on $S_2 = H \cap S_n$. Then the approximation ratio of the algorithm using skewed distribution function $f \in \mathcal{F}_n$ is bounded by $\alpha_{\widehat{f}}$ from below.

For constructing a good skewed distribution function, we need to find a function $f' \in \mathcal{F}_2$ such that the value $\alpha_{f'}$ is large. In Section 5, we describe a numerical method for finding a good skewed distribution function in $\mathcal{F}_2$.

Even if we have a good distribution function in $\mathcal{F}_2$, a non-trivial problem still remains. For applying hyperplane separation technique, we need a skewed distribution function on the n-dimensional sphere. However, when $n > 2$, not every distribution function $f' \in \mathcal{F}_2$ has a distribution function $f \in \mathcal{F}_n$ satisfying $\widehat{f} = f'$. For example, it is easy to show that there does not exists any distribution function $f \in \mathcal{F}_3$ satisfying the conditions that

$$\widehat{f}(\boldsymbol{v}) = \begin{cases} 1/(2\sqrt{2}\pi) & (-0.5 \leq \boldsymbol{v}_0 \cdot \boldsymbol{v} \leq 0.5), \\ 0 & (\text{otherwise}). \end{cases}$$

In Section 4, we propose a class of functions in $\mathcal{F}_2$ such that a corresponding skewed distribution function exists for any sphere $\boldsymbol{S}_n$ with $n \geq 3$.

4 Main Theorem

For any function $f \in \mathcal{F}_n$, we can characterize f by the function $P_f : [0, \pi/2] \to \boldsymbol{R}_+$ defined by

$$P_f(\theta) \stackrel{\text{def.}}{=} f(\boldsymbol{v})|_{\cos\theta = |\boldsymbol{v}_0 \cdot \boldsymbol{v}|}.$$

The following theorem gives a class of permitted skewed distribution function in $\mathcal{F}_n$.

Theorem 1 *Let $f \in \mathcal{F}_n$ be a skewed distribution function with $n \geq 2$ satisfying*

$$P_f(\theta) = \frac{1}{a} \sum_{k=0}^{\infty} a_k \cos^k \theta.$$

Then the function $P_{\widehat{f}}(\phi)$ can be described as

$$P_{\widehat{f}}(\phi) = \frac{1}{a} \sum_{k=0}^{\infty} \frac{S^{(k+n)}(1)}{S^{(k+2)}(1)} a_k \cos^k \phi,$$

where a is a coefficient used for normalizing the total probability to 1 and $S^{(n)}(r)$ is the area of the n dimensional sphere whose radius is equal to r.

Proof. First, we notate some well-known formulae;

$$\Gamma(0) = 1, \ \ \Gamma(\frac{1}{2}) = \sqrt{\pi}, \ \ \Gamma(x+1) = x\Gamma(x),$$
$$\int_0^{\frac{\pi}{2}} \sin^p x \cos^q x \, \mathrm{d}x = \frac{\Gamma(\frac{p+1}{2})\Gamma(\frac{q+1}{2})}{2\Gamma(\frac{p+q+2}{2})}, \ \ S^n(r) = \frac{2\pi^{\frac{n+1}{2}}}{\Gamma(\frac{n+1}{2})}.$$

When we fix ϕ and $\mathrm{d}\phi$, we have the following;

$$\begin{aligned}&2\pi \sin\phi P_{\widehat{f}}(\phi)\mathrm{d}\phi\\&= \int_0^1 P_f(\arccos(r\cos\phi))(2\pi r\sin\phi)\left(S^{(n-3)}\left(\sqrt{1-r^2}\right)\right)\\&\left(\frac{\sqrt{1-r^2\cos^2\phi}}{\sqrt{1-r^2}}r\frac{\mathrm{d}\phi}{\cos\phi}\right)\left(\frac{\cos\phi}{\sqrt{1-r^2\cos^2\phi}}\right)\mathrm{d}r.\end{aligned}$$

Thus we have

$$P_{\widehat{f}}(\phi) = \int_0^1 P_f(\arccos(r\cos\phi))S^{(n-3)}\left(\sqrt{1-r^2}\right)r^2\frac{\mathrm{d}r}{\sqrt{1-r^2}}.$$

When we replace r by $\sin\alpha$ and $P_f(\theta)$ by $(1/a)\sum_{k=0}^{\infty} a_k \cos^k\theta$, we can describe $P_{\widehat{f}}(\phi)$ as;

$$\begin{aligned}P_{\widehat{f}}(\phi) &= \int_0^{\frac{\pi}{2}}\left(\frac{1}{a}\sum_{k=0}^{\infty} a_k \sin^k\alpha\cos^k\phi\right)\frac{2\pi^{\frac{n-2}{2}}}{\Gamma(\frac{n-2}{2})}\cos^{n-3}\alpha\ \sin^2\alpha\ \mathrm{d}\alpha\\&= \frac{1}{a}\sum_{k=0}^{\infty}\frac{2\pi^{\frac{n-2}{2}}}{\Gamma(\frac{n-2}{2})}a_k\cos^k\phi\int_0^{\frac{\pi}{2}}\sin^{k+2}\alpha\ \cos^{n-3}\alpha\ \mathrm{d}\alpha\\&= \frac{1}{a}\sum_{k=0}^{\infty}\frac{2\pi^{\frac{n-2}{2}}}{\Gamma(\frac{n-2}{2})}a_k\cos^k\phi\frac{\Gamma(\frac{k+3}{2})\Gamma(\frac{n-2}{2})}{2\Gamma(\frac{n+k+1}{2})}\\&= \frac{1}{a}\sum_{k=0}^{\infty}\frac{\Gamma(\frac{k+3}{2})}{2\pi^{\frac{k+3}{2}}}\frac{2\pi^{\frac{n+k+1}{2}}}{\Gamma(\frac{n+k+1}{2})}a_k\cos^k\phi = \frac{1}{a}\sum_{k=0}^{\infty}\frac{S^{(k+n)}(1)}{S^{(k+2)}(1)}a_k\cos^k\phi.\end{aligned}$$

And so we have done. □

The above theorem directly implies the following.

Corollary 1 *Let $f' \in \mathcal{F}_2$ be a distribution function satisfying*

$$P_{f'} = \frac{1}{b}\sum_{k=0}^{\infty} b_k\cos^k\phi$$

with the condition that $b_k \geq 0$. Then, for any $n \geq 2$, there exists a distribution function $f \in \mathcal{F}_n$ satisfying $\widehat{f} = f'$ and

$$P_f = \frac{1}{b}\sum_{k=0}^{\infty}\frac{S^{(k+2)}(1)}{S^{(k+n)}(1)}b_k\cos^k\theta,$$

where b is a coefficient used for normalizing the total probability to 1.

The following theorem extends the class of tractable distribution functions.

Theorem 2 *Let $f \in \mathcal{F}_n$ be a distribution function satisfying that* $P_f(\theta) = (1/a) \sum_{p \in X} a_p \cos^p \theta$ *and X is a finite set of non-negative real numbers. Then the distribution function $\widehat{f}$ satisfies* $P_{\widehat{f}}(\phi) = (1/a) \sum_{p \in X} c_p a_p \cos^p \phi$, *where a is a normalization coefficient and*

$$c_p = \frac{2\pi^{\frac{n-2}{2}}}{\Gamma(\frac{n-2}{2})} \int_0^{\frac{\pi}{2}} \sin^{p+2} \alpha \cos^{n-3} \alpha \, d\alpha \text{ for each } p \in X.$$

Proof. We can prove in a similar way with the proof of Theorem 1 and so proof is omitted. □

This theorem implies the following.

Corollary 2 *Let $f' \in \mathcal{F}_2$ be a distribution function satisfying* $P_{f'}(\phi) = (1/b) \sum_{p \in X} b_p \cos^p \phi$ *and* $b_p \geq 0$ $(\forall p \in X)$ *where X is a finite set of positive real numbers. Then there exists a distribution function $f \in \mathcal{F}$ satisfying* $\widehat{f} = f'$ *and* $P_f(\theta) = (1/b) \sum_{p \in X} d_p b_p \cos^p \theta$ *where b is a normalization coefficient and*

$$d_p = \left(\frac{2\pi^{\frac{n-2}{2}}}{\Gamma(\frac{n-2}{2})} \int_0^{\frac{\pi}{2}} \sin^{p+2} \alpha \cos^{n-3} \alpha \, d\alpha \right)^{-1} \text{ for each } p \in X.$$

The above corollaries imply that if we have a good distribution function $f' \in \mathcal{F}_2$ satisfying that $P_{f'}(\phi)$ is a finite sum of non-negative power of $\cos \phi$, then we can construct an approximation algorithm for MAX DICUT whose approximation ratio is greater than $\alpha_{f'}$.

5 Numerical Method for Designing Algorithm

First, we designed a distribution function $f' \in \mathcal{F}_2$ satisfying $P_{f'}(\phi) = (1/a)(a_0 + a_1 \cos \phi + a_2 \cos^2 \phi)$, $(a_0, a_1, a_2) \geq \mathbf{0}$ and $a_0 + a_1 + a_2 = 3$. We tried every triplet

$$(a_0, a_1, a_2) \in \{0.01(x_0, x_1, x_2) \mid (x_0, x_1, x_2) \in \mathrm{Z}_+^3, \ 0.01(x_0 + x_1 + x_2) = 3\}.$$

We choose a triplet (a_0^*, a_1^*, a_2^*) which maximizes the approximation ratio $\alpha_{f'}$. Then we decrease the grid size around (a_0^*, a_1^*, a_2^*) and tried all the triplets in the set

$$\left\{ 0.0001(x_0, x_1, x_2) \middle| \begin{array}{l} (x_0, x_1, x_2) \in \mathrm{Z}_+^3, \ 0.0001(x + y + z) = 3, \\ |0.0001 x_i - a_i^*| < 0.01 \ (\forall i \in \{0, 1, 2\}) \end{array} \right\}.$$

As a result, we found that the following function

$$P_{f'}(\phi) = (1/a)(1.1404 + 0.7754 \cos \phi + 1.0842 \cos^2 \phi)$$

satisfies that the approximation ratio is greater than 0.859.

Next, we designed a distribution function $f' \in \mathcal{F}_2$ satisfying $P_{f'}(\phi) = (1/b) \cos^{1/\beta} \phi$, for each $\beta \in \{1.5, 1.6, \ldots, 2.5\}$. As a result, we found that the following function

$$P_{f'}(\phi) = \cos^{(1/1.8)} \phi$$

satisfies that the corresponding approximation ratio is greater than 0.863.

For each function $P_{f'}(\phi)$, we calculate the approximation ratio $\alpha_{f'}$, as follows. We discretize the 2-dimensional sphere $\boldsymbol{S}_2$ and choose every pair of points $(\boldsymbol{v}_i, \boldsymbol{v}_j)$ from the set

$$\left\{ (x, y, z)^\top \in \boldsymbol{S}_2 \,\middle|\, \begin{array}{l} \exists \eta, \exists \xi \in \{-32\pi/64, -31\pi/64, \ldots, 32\pi/64\} \\ x = \cos \eta,\ y = \sin \eta \cos \xi,\ z = \sin \eta \sin \xi, \end{array} \right\},$$

and calculate the value

$$\frac{p(\boldsymbol{v}_i, \boldsymbol{v}_j \mid f')}{(1/4)(1 + \boldsymbol{v}_0 \cdot \boldsymbol{v}_i - \boldsymbol{v}_0 \cdot \boldsymbol{v}_j - \boldsymbol{v}_i \cdot \boldsymbol{v}_j)}.$$

Next, we choose minimum, 2nd minimum and 3rd minimum pairs of points. For each pair $(\boldsymbol{v}_i^*, \boldsymbol{v}_j^*)$ of chosen three pairs, we decrease the grid size and checked every pair of points $(\boldsymbol{v}_i, \boldsymbol{v}_j)$ satisfying that

$$\boldsymbol{v}_i \in \left\{ (x, y, z) \in \boldsymbol{S}_2 \,\middle|\, \begin{array}{l} \exists \eta, \exists \xi \in \{-64\pi/4096, -63\pi/4096, \ldots, 64\pi/4096\} \\ x = \cos(\eta_i^* + \eta),\ y = \sin(\eta_i^* + \eta) \cos(\xi_i^* + \xi), \\ z = \sin(\eta_i^* + \eta) \sin(\xi_i^* + \xi) \end{array} \right\},$$

and

$$\boldsymbol{v}_j \in \left\{ (x, y, z) \in \boldsymbol{S}_2 \,\middle|\, \begin{array}{l} \exists \eta, \exists \xi \in \{-64\pi/4096, -63\pi/4096, \ldots, 64\pi/4096\} \\ x = \cos(\eta_j^* + \eta),\ y = \sin(\eta_j^* + \eta) \cos(\xi_j^* + \xi), \\ z = \sin(\eta_j^* + \eta) \sin(\xi_j^* + \xi) \end{array} \right\},$$

where $\boldsymbol{v}_i^* = (\cos \eta_i^*, \sin \eta_i^* \cos \xi_i^*, \sin \eta_i^* \sin \xi_i^*)^\top$ and $\boldsymbol{v}_j^* = (\cos \eta_j^*, \sin \eta_j^* \cos \xi_j^*, \sin \eta_j^* \sin \xi_j^*)^\top$. For each pair of points $(\boldsymbol{v}_i, \boldsymbol{v}_j)$ we calculated the value $p(\boldsymbol{v}_i, \boldsymbol{v}_j | f)$ by numerical integration.

6 Conclusion

In this paper, we proposed an approximation algorithm for MAX DICUT problem whose approximation ratio is 0.863. Our algorithm solves the SDP relaxation problem proposed by Goemans and Williamson with additional valid constraints introduced in [3,4]. We generate a dicut by using hyperplane separating technique based on skewed distribution function $f \in \mathcal{F}_n$ satisfying that $P_{\widehat{f}}(\theta) = \cos^{(1/1.8)} \phi$.

References

1. A. Ageev, R. Hassin and M. Sviridenko, "An approximation algorithm for MAX DICUT with given sizes of parts", *Proc. of APPROX 2000*, LNCS 1913(2000), 34–41.

2. F. Alizadeh, "Interior point methods in semidefinite programming with applications to combinatorial optimization", *SIAM Journal on Optimization*, 5(1995), 13–51.
3. U. Feige and L. Lovász, "Two-prover one-round proof systems: Their power and their problems", *Proc. of the 24th Annual ACM Symposium on the Theory of Computing*, 1992, 733–744.
4. U. Feige and M. X. Goemans, "Approximating the value of two prover proof systems, with applications to MAX 2SAT and MAX DICUT", *Proc. of 3rd Israel Symposium on the Theory of Computing and Systems*, 182–189, 1995.
5. M. X. Goemans and D. P. Williamson, "Improved approximation algorithms for maximum cut and satisfiability problems Using Semidefinite Programming", *Journal of the ACM*, 42(1995), 1115-1145.
6. L. Lovász, "On the Shannon capacity of a graph", *IEEE Transactions on Information Theory*, 25(1979), 1–7.
7. Y. Nesterov and A. Nemirovskii *Interior point polynomial methods in convex programming*, SIAM Publications, SIAM, Philadelphia, USA, 1994.
8. U. Zwick, "Analyzing the MAX 2-SAT and MAX DI-CUT approximation algorithms of Feige and Goemans", currently available from http://www.math.tau.ac.il/~zwick/my-online-papers.html.

The Maximum Acyclic Subgraph Problem and Degree-3 Graphs

Alantha Newman

Laboratory for Computer Science
Massachusetts Institute of Technology
Cambridge, MA 02139
alantha@theory.lcs.mit.edu

Abstract. We study the problem of finding a maximum acyclic subgraph of a given directed graph in which the maximum total degree (in plus out) is 3. For these graphs, we present: (i) a simple combinatorial algorithm that achieves an 11/12-approximation (the previous best factor was 2/3 [1]), (ii) a lower bound of 125/126 on approximability, and (iii) an approximation-preserving reduction from the general case: if for any $\epsilon > 0$, there exists a $(17/18 + \epsilon)$-approximation algorithm for the maximum acyclic subgraph problem in graphs with maximum degree 3, then there is a $(1/2 + \delta)$-approximation algorithm for general graphs for some $\delta > 0$. The problem of finding a better-than-half approximation for general graphs is open.

1 Introduction

Given a directed graph $G = (V, E)$, the maximum acyclic subgraph problem is to find a maximum cardinality subset E' of E such that $G' = (V, E')$ is acyclic. The problem is NP-hard [2] and the best-known polynomial-time computable approximation factor for general graphs is $\frac{1}{2}$.

In this paper, we focus on graphs in which every vertex has total degree (in-degree plus out-degree) at most 3. Throughout this paper, we refer to these graphs as *degree-3* graphs. The problem remains NP-hard for these graphs [2]. In Section 2, we present an algorithm that finds an $\frac{11}{12}$-approximation. This improves on the previous best guarantees of $\frac{2}{3}$ for graphs with maximum degree 3 and $\frac{13}{18}$ for 3-regular graphs [1]. The algorithm is purely combinatorial and relies heavily on exploiting the structure of degree-3 graphs. As a simple corollary of a Theorem in [4,5], we obtain an approximation lower bound of $\frac{125}{126}$ in Section 3. Finally, in Section 4, we show that the problem for this special class of graphs has a close connection to the general problem. Specifically, we show that if for any $\epsilon > 0$, there exists a $(\frac{17}{18} + \epsilon)$-approximation algorithm for the maximum acyclic subgraph in degree-3 graphs, then there is a $(\frac{1}{2} + \delta)$-approximation algorithm for general graphs, for some $\delta > 0$. Finding a $(\frac{1}{2} + \delta)$-approximation algorithm for general graphs is an open problem. Our methods open up the possibility of finding such an approximation via an algorithm for the special case of degree-3 graphs.

M. Goemans et al. (Eds.): APPROX-RANDOM 2001, LNCS 2129, pp. 147–158, 2001.

Fig. 1.

2 Combinatorial Approximation Algorithms

In [1], Berger and Shor present an algorithm that returns an acyclic subgraph of size at least $\frac{2|E|}{3}$ for degree-3 graphs that do not contain 2-cycles. For 3-regular graphs (note that the set of 3-regular graphs is a proper subset of the set of degree-3 graphs) with no 2-cycles, an algorithm that returns an acyclic subgraph of size $\frac{13|E|}{18}$ is given in [1]. In this section, we show that the problem in degree-3 graphs (with or without 2-cycles) can be approximated to within $\frac{11}{12}$ of optimal using simple combinatorial methods. First we give an $\frac{8}{9}$-approximation algorithm to illustrate some basic arguments. Then we extend these arguments to give an $\frac{11}{12}$-approximation algorithm.

2.1 An $\frac{8}{9}$-Approximation

Given a degree-3 graph $G = (V, E)$ for which we want to find an acyclic subgraph $S \subseteq E$, we can make the following assumptions.

(i) All vertices in G have in-degree and out-degree at least 1 and total degree exactly 3.
(ii) G contains no directed or undirected 2- or 3-cycles.

The explanation for assumption (i) is as follows. If G contains any vertices with in- or out-degree 0, we can immediately add all edges adjacent to these vertices to the acyclic subgraph S, since these edges are contained in any maximal acyclic subgraph. Additionally, we can contract all vertices in G that have in-degree 1 and out-degree 1. For example, say that vertex j in G has in-degree 1 and out-degree 1 and G contains edges (i, j) and (j, k). Then at least one of these two edges will be included in any maximal acyclic subgraph of G. Thus, contracting vertex j is equivalent to contracting edge (i, j) and adding it to the acyclic subgraph S.

Now we explain assumption (ii). We can contract multi-edges without adding cycles to the graph, thus removing any undirected 2-cycles. This is shown in Figure 1A. The edges in the undirected 2-cycle are added to S since they are included in any maximal acyclic subgraph. In Figure 1, the dotted edges are added to S. Similarly, we can remove any undirected 3-cycle by contracting it and adding its edges to S. This results in a degree-3 vertex as shown in Figure 1B. Contracting an undirected 3-cycle will not introduce any new cycles into the graph since each of the vertices in the 3-cycle has in-degree and out-degree at least 1 by (i).

In the case of directed 2- and 3-cycles, we can remove the minimum number of edges from the graph while breaking all such cycles. For directed 2-cycles, consider the two adjacent non-cycle edges of a 2-cycle. If they are both in edges, or both out edges, as in Figure 2A, then we can break the 2-cycle by removing an arbitrary edge. If one is out and the other is in, as in Figure 2B, then only one of the edges in the 2-cycle is consistent with the direction of a possible cycle containing both of edges that are not in the 2-cycle. For example, in Figure 2B, we would remove edge (i, j). For directed 3-cycles, consider Figure 2C. In this case, or in the analogous case where three edges point towards the 3-cycle, we can remove any edge from the 3-cycle. In the other case, we remove an edge from the 3-cycle, so that the path from the single in edge or to the single out edge is broken. For example, in Figure 2D, we would remove edge (j, k).

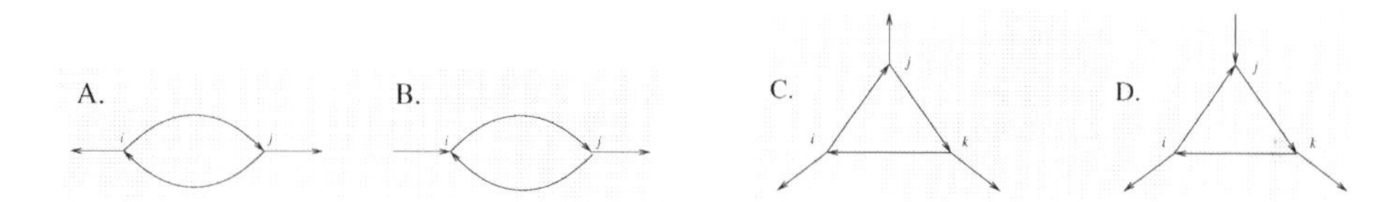

Fig. 2.

We now consider two subclasses of degree-3 graphs. We will use the following definition.

Definition 1. *An* α-edge *is an edge (i, j) such that vertex i has in-degree 2 and out-degree 1 and vertex j has in-degree 1 and out-degree 2.*

For example, edge (j, k) in Figure 2D is an α-edge. First, we consider the case where G contains no α-edges. If there are no α-edges, then we can find the maximum acyclic subgraph in polynomial time. We will use the following lemma.

Lemma 1. *If G is a 3-regular graph and contains no α-edges, then all cycles in G are edge disjoint.*

Proof. Assume that there are two cycles in G that have an edge (or a path) in common. First case: assume that these two cycles have a single edge (i, j) in common, i.e. edge (i, j) belongs to both cycles, but edges (a, i) and (j, b) each belong to only one of these cycles. Then vertex i must have in-degree 2 and vertex j must have out-degree 2. Thus, edge (i, j) is a α-edge, which is a contradiction. Second case: assume these two cycles have a path $\{i, \dots, j\}$ and that this path is maximal, i.e. edge (a, i) and (j, b) each belong to only one of these cycles. Vertex i must have in-degree 2 and vertex j must have out-degree 2. Therefore, at least one of the edges on the path must be an α-edge, which is a contradiction. □

Since all the cycles in a graph with no α-edges are edge disjoint, we can find the maximum acyclic subgraph of such a graph in polynomial time. Given a graph G containing no α-edges, we simply find a cycle in G, throw away any edge from this cycle, and add edges to the acyclic subgraph S by contracting appropriate edges in G or removing appropriate edges from G until G satisfies properties (i) and (ii). We repeat until there are no more cycles in G.

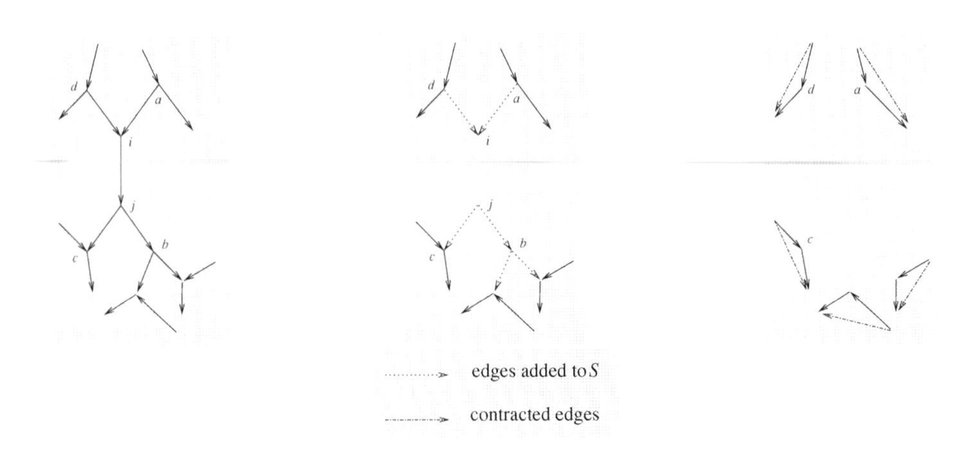

Fig. 3. An illustration of step 4.

If G contains α-edges, then the problem is NP-hard. For this case, we give the following $\frac{8}{9}$-approximation algorithm. Define $C(e)$ as the connected component containing edge e. Define $E(e)$ as the set of edges adjacent to edge e, i.e. the edges that share an endpoint with e. For example, if e is edge (i, j) in the first picture in Figure 3, then $E(e)$ contains edges $(d, i), (a, i), (j, c)$, and (j, b). S is the solution set. The first part of the algorithm is the following procedure. An illustration of step 4 is shown in Figure 3.

While G contains α-edges, do the following:

1. Make sure G is 3-regular and remove all 2- and 3-cycles from G (see explanation of assumptions (i) and (ii)).
2. Find an α-edge e in G.
3. If $|C(e)| = 9$, solve this component exactly.
4. Else remove e from G. Add $E(e)$ and any other edges with in- or out-degree 0 to S. Contract any vertices with in-degree and out-degree 1.

When there are no more α-edges in G, then we can solve for the maximum acyclic subgraph in polynomial time as discussed previously. Then, we uncontract every edge in S that corresponds to a path contracted in some execution of step 1 or step 4. For every edge *not* in S that corresponds to some contracted path, we throw away one edge from the path, and add the remaining edges to S. Thus, every time we contract a vertex, we guarantee that at least one more edge will be added to S.

Theorem 1. *The algorithm is an $\frac{8}{9}$-approximation for the maximum acyclic subgraph problem in degree-3 graphs.*

The proof of this theorem is based on the idea that for each α-edge we remove, we can add at least eight edges to S. For example, in Figure 3, we immediately add six edges to S and make five contractions. This results in a total of eleven edges added to S. The full proof of Theorem 1 is simple, but is omitted here.

2.2 An $\frac{11}{12}$-Approximation

We now show how to extend the previous algorithm to obtain an $\frac{11}{12}$-approximation algorithm. In our $\frac{8}{9}$-approximation algorithm, we arbitrarily choose α-edges to remove. There are degree-3 graphs such that if we arbitrarily choose α-edges to remove, then we may obtain an acyclic subgraph with size only $\frac{8}{9}$ of optimal. We will show that if we choose the α-edges to remove carefully, then we can always ensure that the resulting graph contains certain α-edges whose removal allows us to add eleven edges (rather than eight) to the solution set.

In order to analyze the steps of the algorithm more easily, we consider a further modification of a given degree-3 graph. We contract any pair of adjacent vertices in which each vertex has in-degree 1 or each vertex has out-degree 1. An example of such a pair of adjacent vertices is shown in Figure 4. Here, j, k is a pair of vertices both with in-degree 1 and f, i is a pair of vertices both with out-degree 1, so we contract edges (f, i) and (j, k). In order to account for the contracted edges, if a vertex has d out edges or d in edges after an edge was contracted, then the *value* of these edges is $2d - 2$, since this is the number of edges they represent in the original graph. For example, in Figure 4, there are now three incoming edges to vertex i. These three edges represent four edges in the original degree-3 graph, so they have value 4. In other words, if the three edges coming into vertex i are added to the acyclic subgraph for the modified graph, then this is equivalent to adding all four edges to the acyclic subgraph for the original graph. After contracting the relevant edges, the resulting graph will no longer be a degree-3 graph, but will correspond to a degree-3 graph. However, every edge still has in- or out-degree 1 and total degree at least 3. Hence, we can still handle undirected and directed 2- and 3-cycles as described in Section 2.1 and thus property (ii) holds. We now have the additional assumption about the given graph G for which we want to find an acyclic subgraph.

(iii) G contains no adjacent vertices such that both vertices have in-degree 1 or both vertices have out-degree 1.

When we remove an edge e from a graph G, the graph $G - e$ will represent the graph that is obtained by removing edge e from G, removing all edges adjacent to a vertex with in- or out-degree 0 in the resulting graph, contracting all resulting vertices that have in-degree 1 and out-degree 1 and all edges (i, j) such that both i and j have in-degree or out-degree 1. We will use the following definitions.

Definition 2. *Edge (i, j) is a* profitable α-edge *if either i or j has in-degree or out-degree at least 3.*

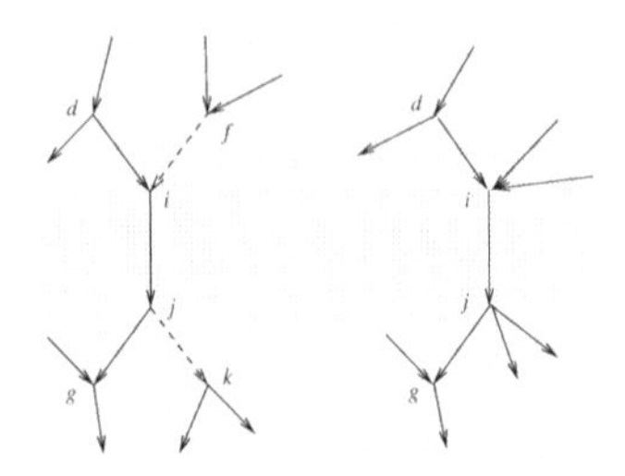

Fig. 4. Edges (f, i) and (j, k) will be contracted.

Definition 3. *A* super-profitable graph *is a graph that contains either a 4-cycle or an α-edge (i, j) in which the in-degree of i plus the out-degree of j is at least 6.*

Our algorithm will use the following lemmas. The proofs of these lemmas are omitted here.

Lemma 2. *If e is a profitable α-edge, then removing e from G allows us to add 11 edges to the solution set S.*

Lemma 3. *If G is not super-profitable and G does not contain any profitable α-edges, then G contains an edge e such that the graph $G - e$ contains a profitable α-edge.*

Lemma 4. *If G is not a super-profitable graph and G contains a profitable α-edge, then there is some set $\{e_1, \ldots e_k\}$ of edges for some $k \in \{1, 2, 3\}$ such that $G - \{e_1, \ldots e_k\}$ contains a profitable α-edge and removing these k edges from G allows us to add at least $11k$ edges to the solution set S.*

Lemma 5. *If G is a super-profitable graph, then G contains some α-edge whose removal allows us to add at least 14 edges to the solution set S.*

We have now stated all the lemmas that we will use to show that we can approximate our problem to within $\frac{11}{12}$. The algorithm is similar to the previous algorithm, except that the while loop is more complex. We will give a high-level description of the steps that make up this loop. A precise statement of the algorithm and the proof of its approximation guarantee (Theorem 2) are omitted here.

The main idea is that during each iteration of the while loop, we want to remove a profitable α-edge from the graph and simultaneously ensure that the resulting graph also contains a profitable α-edge or is super-profitable. We can assume that the given degree-3 graph G contains a profitable α-edge. If it does not, we can use Lemma 3 to obtain a graph that does. We will only discard one edge in the process and since G contains at least one cycle (otherwise it is already acyclic), the number of edges in the new graph is no less than the maximum acyclic subgraph of the original graph. Then we have two cases. In the first case, if this graph is super-profitable, by Lemma 5, we can remove an α-edge and add 14 edges to S. If after removing this edge, we are not left with a graph that is super-profitable or contains a profitable α-edge, then we can use Lemma 3 again to obtain a graph that contains a profitable α-edge. Thus we will discard two edges and add at least 22 edges to S. In the second case, if the graph is not super-profitable, we can use Lemma 4 remove a set of $k \in \{1, 2, 3\}$ edges and add a set of $11k$ edges to S so that the resulting graph contains a profitable α-edge.

Theorem 2. *The algorithm is an* $\frac{11}{12}$*-approximation algorithm for the maximum acyclic subgraph problem in degree-3 graphs.*

3 A Lower Bound

We can make a straightforward modification of the gadgets in [4,5] to obtain the following lower bound for degree-3 graphs. Specifically, we can add edges to the gadgets so that the graphs obtained in the reduction are degree-3 graphs.

Theorem 3. *It is NP-hard to approximate the maximum acyclic subgraph of a 3-regular graph to within* $\frac{125}{126} + \epsilon$ *for any* $\epsilon > 0$.

4 Bounded-Degree Graphs

In this section, we investigate the maximum acyclic subgraph problem restricted to Eulerian graphs and to the maximum acyclic subgraph problem restricted to degree-3 graphs. Somewhat surprisingly, the general problem can be reduced to these special cases. First, we present an approximation-preserving reduction from the problem in general graphs to the problem restricted to Eulerian graphs. We then give an approximation-preserving reduction from Eulerian graphs to degree-3 graphs.

4.1 Reduction to Eulerian Graphs

This reduction is due to Fang Chen and László Lovász [3].

Theorem 4. *If for any* $\delta > 0$*, there exists a* $(\frac{1}{2} + \delta)$*-approximation algorithm for the maximum acyclic subgraph problem in Eulerian graphs, then there exists a* $(\frac{1}{2} + \frac{\delta}{8})$*-approximation algorithm for the maximum acyclic subgraph problem in general graphs.*

Proof. Let $d^+(v)$ be the in-degree of a vertex v, let $d^-(v)$ be the out-degree of a vertex v, and let $d(v) = d^+(v) + d^-(v)$. We say that a graph $G = (V, E)$ is ϵ-far from Eulerian if there is a $v \in V$ such that $|d^+(v) - d^-(v)| > 2\epsilon d(v)$. We call such a vertex v an ϵ-far vertex. Assume that for some $\delta > 0$, there exists a

$(\frac{1}{2}+\delta)$-approximation algorithm for the maximum acyclic subgraph problem in Eulerian graphs. Then we will consider two cases.

(i) If $\epsilon \geq \frac{\delta}{8}$, then we can make at least an ϵ (or $\frac{\delta}{8}$) gain at each ϵ-far vertex by placing $\max\{d^+(v), d^-(v)\}$ edges in the solution set and throwing $\min\{d^+(v), d^-(v)\}$ edges away. When there are no ϵ-far vertices remaining, the resulting graph is less than $\frac{\delta}{8}$-far from Eulerian.

(ii) If $\epsilon < \frac{\delta}{8}$, then G is less than $\frac{\delta}{8}$-far from Eulerian. In this case, we add a new vertex v^* to G to obtain a new graph $G+v^*$. To the vertex v^*, we attach in edges and out edges from and to each vertex in G for which $|d^+(v)-d^-(v)| > 0$, thus making $G+v^*$ Eulerian. Let $OPT(G)$ denote the size of the maximum acyclic subgraph of G. So we have:

$$d(v^*) = \sum_{v\in G} |d^+(v)-d^-(v)| \leq \sum_{v\in G} 2\epsilon d(v) \leq 2\epsilon \sum_{v\in G} d(v) \leq 4\epsilon|E| \qquad (1)$$

Since $\frac{|E|}{2} \leq OPT(G)$, we have, $|E| \leq 2OPT(G)$. Thus:

$$4\epsilon|E| \leq 8\epsilon OPT(G) \qquad (2)$$

Combining (1) and (2), we have:

$$d(v^*) \leq 8\epsilon OPT(G) \qquad (3)$$

Let $OPT(G+v^*)$ denote the size of the maximum acyclic subgraph of the Eulerian graph $G+v^*$. Then the following is true of $G+v^*$:

$$OPT(G+v^*) \geq OPT(G) + \frac{1}{2}d(v^*) \qquad (4)$$

Say $0 \leq \delta \leq \frac{1}{2}$. Then from (4), we have:

$$\begin{aligned}
&(\frac{1}{2}+\delta)OPT(G+v^*) - d(v^*) \geq \\
&(\frac{1}{2}+\delta)OPT(G) + (\frac{1}{2}+\delta)\frac{1}{2}d(v^*) - d(v^*) \geq \\
&(\frac{1}{2}+\delta)OPT(G) - \frac{3}{4}d(v^*) \qquad (5)
\end{aligned}$$

Combining (3) and (5), we have:

$$(\frac{1}{2}+\delta)OPT(G) - 6\epsilon OPT(G) \geq (\frac{1}{2}+\delta-6\epsilon)OPT(G) \qquad (6)$$

Since $\epsilon < \frac{\delta}{8}$:

$$(\frac{1}{2}+\delta-6\epsilon) > (\frac{1}{2}+\delta-\frac{6}{8}\delta) = (\frac{1}{2}+\frac{\delta}{4}) \qquad (7)$$

Combining (5), (6), and (7), we have:

$$(\frac{1}{2}+\delta)OPT(G+v^*) - d(v^*) \geq (\frac{1}{2}+\frac{\delta}{4})OPT(G) \qquad (8)$$

Fig. 5. Steps 0 and 1.

Thus, if there is algorithm with an approximation guarantee of $(\frac{1}{2} + \delta)$ for the Eulerian graph $G + v^*$, then we can use this algorithm to obtain a $(\frac{1}{2} + \frac{\delta}{4})$-approximation for a graph that is less than $\frac{\delta}{8}$-far from Eulerian. Therefore, in both cases (i) and (ii), we have an algorithm with an approximation guarantee of at least $(\frac{1}{2} + \frac{\delta}{8})$. □

4.2 Reduction to 3-Regular Graphs

In this section, we give an approximation-preserving reduction from the maximum acyclic subgraph problem restricted to Eulerian graphs to degree-3 graphs. Applying Theorem 4 yields an approximation-preserving reduction from the problem in general graphs to the problem restricted to degree-3 graphs.

Theorem 5. *If for any $\epsilon > 0$, there exists a $(\frac{17}{18} + \epsilon)$-approximation algorithm for the maximum acyclic subgraph problem in degree-3 graphs, then there exists some $\delta > 0$ such that there is a $(\frac{1}{2}+\delta)$-approximation algorithm for the maximum acyclic subgraph problem in general graphs.*

To prove this theorem we first introduce the following lemmas.

Lemma 6. *Given an Eulerian graph $G = (E, V)$, we can construct a 3-regular graph $G' = (E', V')$ with $|E'| = 9|E| - 9|V|$ such that the size of the minimum feedback arc set in G is the same size as the minimum feedback arc set in G'.*

Proof. Since G is Eulerian, we use $d(v)$ to denote both the in-degree and the out-degree of vertex v for the duration of this proof. We will construct G' as follows. Step 0 (Figure 5) depicts a vertex v in G. We first place a vertex in the middle of each edge of G as shown in step 1 (Figure 5). The number of vertices in G' is now $|V| + |E|$. Then for each vertex $v \in V$, we match each incoming edge with a distinct outgoing edge by adding an edge from the vertex placed on the incoming edge to the vertex placed on the outgoing edge. (Since G is Eulerian, we can always find such a matching.) This is shown in step 2 (Figure 6).

All of the $|E|$ new vertices now have in- and out-degree 2. We want to replace the vertices from V with a new set of vertices in which each vertex has in- and out-degree 2 so that the entire graph is 4-regular and Eulerian. Consider vertex $v \in V$. We build a binary tree from v to the $d(v)$ new vertices placed on the outgoing edges as shown in step 3 (Figure 6).

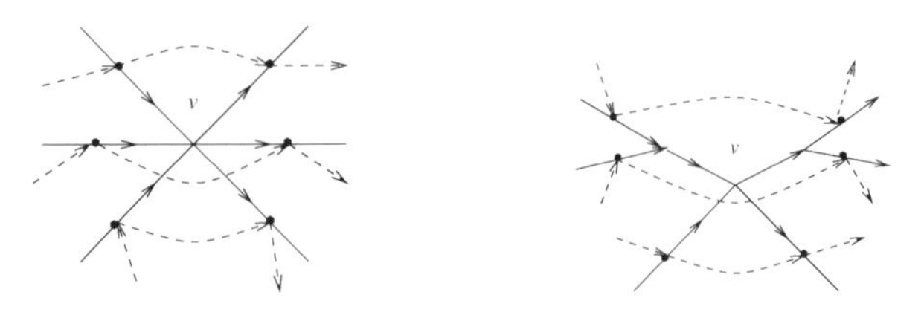

Fig. 6. Steps 2 and 3.

Fig. 7. Steps 4 and 5.

This requires $d(v) - 2$ new vertices. We can see this by the following reasoning: let p be the highest power of 2 not greater than $d(v)$. For a binary tree that connects p vertices to v, we need $\frac{p}{2} + \frac{p}{4} + \cdots + 2 = p - 2$ new vertices or vertices that are internal nodes on the binary tree. For each of the remaining $d(v) - p$ vertices, we connect two vertices to one of the p leaves thus adding just one internal vertex for each of these vertices. Therefore, we have a total of $d(v) - p + p - 2 = d(v) - 2$ internal or new vertices on the binary tree.

We also build a binary tree from v to the $d(v)$ new vertices placed on the incoming edges. Then we match each vertex from the incoming binary tree with a distinct vertex from the outgoing binary tree by adding an edge from the former to the latter as shown in step 4 (Figure 7). We build these two binary trees and match their vertices as we just described for all vertices $v \in V$. The total number of new vertices needed to build these two binary trees for vertex v is $2d(v) - 4$. The total number of new vertices needed to build two binary trees for all $v \in V$ is therefore $\sum_{v \in V} 2d(v) - 4 = 2|E| - 4|V|$.

We now have a 4-regular Eulerian graph with $(|E| + |V|) + (2|E| - 4|V|) = 3|E| - 3|V|$ vertices. We are not yet finished constructing G' since we want G' to be a 3-regular graph and we currently have a 4-regular graph. We will refer to this 4-regular graph as G'' further on in this proof. The last step in the construction of G' is to "stretch" each vertex in the current graph into an edge so that G' is 3-regular. This is shown in step 5 (Figure 7). After this final step in the construction of G', $|E'| = 3(3|E| - 3|V|) = 9|E| - 9|V|$. Recall the definition (Definition 1) of an α-edge from Section 2.1. Note that the edges in G' that correspond to "stretched" vertices are α-edges.

We now show that this reduction preserves the size of minimum feedback arc set, i.e. the size of the minimum feedback arc set of G' is equal to the size of the minimum feedback arc set of G. Specifically, we show that given a feedback arc set in G, we can construct a feedback arc set in G' of the same size. Conversely, given a feedback arc set in G', we can construct a feedback arc set in G of size at most the size of the feedback arc set in G'.

(i) Suppose F is a feedback arc set of G. We construct F', a feedback arc set of G'. For each edge $e_{ij} \in F$, we add to F' the α-edge from E' that corresponds to the vertex used to subdivide edge e_{ij} in the first stage of the construction of G'. Then $|F'| = |F|$ and $E' - F'$ is acyclic by the following proof.

Assume $E' - F'$ is not acyclic. Note that every α-edge corresponds to an original edge, to an original vertex, or to a vertex on a binary tree in G''. If we take a walk on the graph G' starting from a α-edge corresponding to an edge e_{ij} in G (i.e. to a vertex used to subdivide an edge in the first step of the construction) we either follow a path on the binary tree leading to a α-edge corresponding to edge e_{jk} or we walk directly to edge e_{jk} for some k. Therefore, if we find a cycle in G', the set of edges in that cycle that correspond to edges in E are still present in G implying that there is a cycle in G, which is a contradiction to the fact that F is a feedback arc set.

(ii) Suppose F' is a feedback arc set in G'. Assume all edges in F' are α-edges. If they are not, we can replace them with α-edges adjacent to the non-α-edges and obtain a feedback arc set of equal or smaller size. Then for every edge in F' that corresponds to an edge in E, add the corresponding edge in E to F. Then $E - F$ is acyclic by the following proof.

Assume $E - F$ is not acyclic. Then consider some cycle in $E - F$. The α-edges corresponding to each edge e_{ij} in the cycle are still in $E' - F'$. Since $E' - F'$ is acyclic, at least one of the edges used to connect vertices that we placed in the middle of the edges in E must have been removed. But this is a contradiction, because these edges are not α-edges, and we converted F' to a feedback arc set that contained only α-edges. □

Corollary 1. *A maximum acyclic subgraph in G of size S corresponds to a maximum acyclic subgraph in G' of size $S + 8|E| - 9|V|$.*

Let $MAS(G)$ denote the value of a maximum acyclic subgraph of G.

Lemma 7. *We can convert an acyclic subgraph of G' of size at least $(\frac{17}{18} + \epsilon)MAS(G')$ to an acyclic subgraph of G of size at least $(\frac{1}{2} + \delta)MAS(G)$ for some constants $\epsilon, \delta > 0$.*

Proof. We assume that $MAS(G)$ is at least βE for some fixed $\beta < 1$. We will discuss the exact value of β later on in the proof. If $MAS(G)$ is less than βE, then we can find an acyclic subgraph in G with at least half the edges of G thereby obtaining a $(\frac{1}{2} + \epsilon)$-approximation for some $\epsilon > 0$.

Say we are given an α-approximation algorithm for 3-regular graphs. We can take an Eulerian graph G and convert it a 3-regular graph G' using the construction described previously. Then we can find an acyclic subgraph S' for G' that is of size at least $\alpha MAS(G')$. By Corollary 1 we have:

$$\alpha MAS(G') = \alpha(MAS(G) + 8E - 9V)$$

To find an acyclic subgraph S of G given S', we remove all edges from S' that do not correspond to α-edges representing original edges of G. There are at most $8E - 9V$ such edges, since E of the edges in G' correspond to edges in G. So when we remove these edges from S', we are left with a set S of size at least:

$$\alpha(MAS(G) + 8E - 9V) - (8E - 9V) = \alpha MAS(G) + (8\alpha - 8)E + (9 - 9\alpha)V$$

Since $E \leq \frac{MAS(G)}{\beta}$, $8\alpha - 8 < 0$, and $9 - 9\alpha > 0$, we have:

$$\alpha MAS(G) + (8\alpha - 8)E + (9 - 9\alpha)V \geq (\alpha + \frac{8\alpha - 8}{\beta})MAS(G)$$

We can set β so that the following is true:

$$\alpha + \frac{8\alpha}{\beta} - \frac{8}{\beta} > \frac{1}{2} \Rightarrow \alpha + \frac{8\alpha}{\beta} > \frac{1}{2} + \frac{8}{\beta}$$

Since $\beta < 1$, we have:

$$\alpha > \frac{(\beta + 16)}{(\beta + 8)} \frac{1}{2} > \frac{17}{18} \tag{9}$$

Therefore, if we found a $(\frac{17}{18} + \delta)$-approximation for 3-regular graphs, then we could find some β such that equation (9) is true. If $MAS(G) \geq \beta E$, then we could use the reduction and the $(\frac{17}{18} + \delta)$-approximation algorithm for degree-3 graphs to find a $(\frac{1}{2} + \epsilon)$-approximation for Eulerian graphs, which would lead to a $(\frac{1}{2} + \frac{\epsilon}{8})$-approximation for general graphs by Theorem 4. □

Note that $\frac{17}{18} = \frac{119}{126}$ and Theorem 3 states that is NP-hard to approximate the maximum acyclic subgraph of 3-regular graphs to within $\frac{125}{126}$.

Acknowledgements

I thank Santosh Vempala for many discussions on the maximum acyclic subgraph problem.

References

1. Bonnie Berger and Peter W. Shor. Tight Bounds on the Maximum Acyclic Subgraph Problem, *Journal of Algorithms*, vol. 25, pages 1–18, 1997.
2. Richard M Karp. Reducibility Among Combinatorial Problems, *Complexity of Computer Computations*, Plenum Press, 1972.
3. Fang Chen and László Lovász. Personal communication via Santosh Vempala.
4. Alantha Newman. Approximating the Maximum Acyclic Subgraph, M.S. Thesis, MIT, June 2000.
5. Alantha Newman and Santosh Vempala. Fences Are Futile: On Relaxations for the Linear Ordering Problem, *Proceedings of IPCO 2001*, Springer-Verlag.

Some Approximation Results for the Maximum Agreement Forest Problem*

Estela Maris Rodrigues[1,**], Marie-France Sagot[2], and Yoshiko Wakabayashi[1,***]

[1] Universidade de São Paulo, Brazil
{estela,yw}@ime.usp.br
[2] Institut Pasteur and Université de Marne-la-Vallée, France
sagot@pasteur.fr

Abstract. There are various techniques for reconstructing phylogenetic trees from data, and in this context the problem of determining how distant two such trees are from each other arises naturally. Various metrics (NNI, SPR, TBR) for measuring the distance between two phylogenies have been defined. Another way of comparing two trees $\mathcal{T}$ and $\mathcal{U}$ is to compute the so called *maximum agreement forest* of these trees. Informally, the number of components of an agreement forest tells how many edges need to be cut from each of $\mathcal{T}$ and $\mathcal{U}$ so that the resulting forests agree, after performing some forced edge contractions. This problem is known to be $\mathcal{NP}$-hard. It was introduced by Hein et al. [3], who presented an approximation algorithm for it, claimed to have approximation ratio 3. We present here a 3-approximation algorithm for this problem and show that the performance ratio of Hein's algorithm is 4.

1 Introduction

Phylogenetic trees or *phylogenies* are a standard model for representing evolutionary processes, mostly involving biological entities such as species or genes. By a phylogenetic tree we mean a rooted unordered tree whose leaves are uniquely labeled with elements of some set S, and whose internal nodes are unlabeled and have exactly two children. The elements of S stand for the contemporary taxa whose evolutionary relationships one intends to model. These taxa correspond to the leaves of the tree, whereas the ancestral taxa correspond to its internal nodes, so that for each ancestral taxon all of its nearest derived taxa are depicted as its children in the tree.

There are various techniques for reconstructing phylogenetic trees from data, and in this setting the problem of determining how distant two such trees are

* This research is part of CAPES-COFECUB Project 272/99-II.

** Supported by CNPq Grant Proc. 142307/97-1. Also supported by CAPES Grant BEX 0650-99/4 during her visit to Institut Pasteur, where part of this research was done.

*** Partially supported by CNPq (Procs. 304527/89-0 and 464114/00-4) and by ProNEx Project 107/97 (Proc. CNPq 664107/97-4).

M. Goemans et al. (Eds.): APPROX-RANDOM 2001, LNCS 2129, pp. 159–169, 2001.

from each other arises naturally. Various metrics, such as NNI (nearest-neighbor interchange), SPR (subtree prune and regraft) and TBR (tree bisection and reconnection) for measuring the distance between two phylogenies have been defined [5,4,2]. Many results relating these concepts are presented by Allen and Steel [1]. In particular, they show that the size of a maximum agreement forest of two trees is precisely the TBR-distance between them.

We are concerned here with the problem of finding the size of a maximum agreement forest of two trees, which is known to be $\mathcal{NP}$-hard [3,1]. The formal definition of this problem is given in the next section. We present a 3-approximation algorithm for this problem and show that a previous algorithm by Hein et al. [3], claimed to have performance ratio 3, has performance ratio 4.

The algorithm we shall describe is simple, but the analysis of its approximation ratio is quite long and technical. To give an idea of the analysis we have to introduce many concepts that make the notation somewhat heavy. We hope once the formal definitions are given and the proof is outlined, the reader will be able to complete the details that we shall omit.

In Section 2 we present the basic concepts that will be needed to define the problem and to describe the algorithm and its proof, in Section 3 we outline our approximation algorithm, and in Section 4 we sketch the proof that its approximation ratio is 3 and give a family of instances for which Hein's algorithm attains approximation ratio 4.

2 Basic Definitions

A *phylogenetic tree* consists of an unordered rooted tree, called its *topology*, such that each internal node has two children, and of a set of labels which are mapped one-to-one to the leaves of the tree. If $\mathcal{T}$ is a phylogenetic tree, then $T_{\mathcal{T}}$ denotes its topology, $L_{\mathcal{T}}$ its set of leaves, $S_{\mathcal{T}}$ its set of labels, $f_{\mathcal{T}}$ its one-to-one label-to-leaf mapping, and $r_{\mathcal{T}}$ its root. Since we consider phylogenetic trees as having rooted topologies, these are naturally oriented: we assume the arcs are oriented towards its root.

For each arc e in a phylogenetic tree $\mathcal{T}$, we denote by $l(e)$ its *lower endpoint* (the endpoint of e which is the farthest from the root) and by $u(e)$ its *upper endpoint* (the other endpoint of e). The *lowest common ancestor* of a set of $m \geq 1$ nodes $v_1, \ldots, v_m$ of $\mathcal{T}$, written $\text{lca}_{\mathcal{T}}(v_1, \ldots, v_m)$, is the farthest node from the root that is an ancestor of all v_i, $1 \leq i \leq m$. For each node u in $\mathcal{T}$, we denote by $D(u)$ the set of leaves in $\mathcal{T}$ that are descendants of u.

The concept of phylogenetic tree can be generalized so as to consider topologies with more than one component. Such phylogenies are called *phylogenetic forests*. Each component in the topology of a phylogenetic forest corresponds to a *component* of the forest, with its own topology, root, leaf set, label set and label-to-leaf mapping. The notation adopted for all these objects is the same as that for phylogenetic trees, except that we use $F_{\mathcal{F}}$ instead of $T_{\mathcal{F}}$ to denote the topology of a phylogenetic forest $\mathcal{F}$.

2.1 Restrictions and Agreement Forests

Two phylogenetic forests $\mathcal{G}$ and $\mathcal{H}$ are said to be *isomorphic* if $S_\mathcal{G} = S_\mathcal{H}$, their topologies $F_\mathcal{G}$ and $F_\mathcal{H}$ are isomorphic, and the isomorphism preserves both the roots of the components and the labels of the leaves. If S is a subset of $S_\mathcal{V}$ for some component $\mathcal{V}$ of $\mathcal{G}$, then the *simple restriction* of $\mathcal{G}$ to S, denoted by $\mathcal{G}|S$, is the phylogenetic tree $\mathcal{G}'$ such that $S_{\mathcal{G}'} = S$; $L_{\mathcal{G}'} = f_\mathcal{G}(S) := \{f_\mathcal{G}(a) : a \in S\}$; for each pair of nodes u, v in $\mathcal{G}'$ we have $\mathrm{lca}_{\mathcal{G}'}(u,v) = \mathrm{lca}_\mathcal{G}(u,v)$; and $r_{\mathcal{G}'} = \mathrm{lca}_{\mathcal{G}'}(L_{\mathcal{G}'})$. Let $\mathcal{G}[S]$ denote the minimal subtree of $F_\mathcal{G}$ which connects all leaves with labels in S.

Let $\mathcal{G}$ be a phylogenetic forest and $\mathcal{S} = \{S_i : 1 \leq i \leq m\}$ a family of $m \geq 1$ subsets of $S_\mathcal{G}$ such that for each S_i we have $S_i \subseteq S_\mathcal{V}$ for some component $\mathcal{V}$ of $\mathcal{G}$, and the trees $\mathcal{G}[S_i]$ are pairwise node-disjoint. Then the *restriction* of $\mathcal{G}$ to $\mathcal{S}$, written $\mathcal{G}|\mathcal{S}$, is the phylogenetic forest composed of the node-disjoint components $\mathcal{G}|S_i$. The *size* of $\mathcal{G}|\mathcal{S}$ is the cardinality of $\mathcal{S}$. If $\mathcal{G}' = \mathcal{G}|\mathcal{S}$, then we denote the size of $\mathcal{G}'$ by $|\mathcal{G}'|$. If $S_{\mathcal{G}|\mathcal{S}} = S_\mathcal{G}$, then $\mathcal{G}|\mathcal{S}$ is a *full restriction* of $\mathcal{G}$. An *agreement phylogenetic forest*, abbreviated to an *agreement forest*, of two given phylogenetic forests $\mathcal{G}$ and $\mathcal{H}$ is a phylogenetic forest $\mathcal{F}$ that is isomorphic to a full restriction of $\mathcal{G}$ and to a full restriction of $\mathcal{H}$. An *agreement forest* is said to be *maximum* if it has *minimum* size. The problem of computing a maximum agreement forest of two given phylogenetic trees $\mathcal{T}$ and $\mathcal{U}$ is the *Maximum Agreement Forest (MAF) problem*, abbreviated as the *MAF problem*.

We can prove the following simple claim from the above definitions:

Lemma 1. *If $\mathcal{V}_1$ and $\mathcal{V}_2$ are two simple restrictions of a phylogenetic forest $\mathcal{G}$ such that the trees $\mathcal{G}[S_{\mathcal{V}_1}]$ and $\mathcal{G}[S_{\mathcal{V}_2}]$ are node-disjoint, a is the label of a leaf in $\mathcal{V}_1$, b if the label of a leaf in $\mathcal{V}_2$, and $f_\mathcal{G}(a)$ and $f_\mathcal{G}(b)$ are sibling leaves in $\mathcal{G}$, then either $\mathcal{V}_1$ or $\mathcal{V}_2$ consists of a unique node.*

2.2 Links

Let $\mathcal{G}$ be a phylogenetic forest and $\mathcal{G}'$ be a full restriction of $\mathcal{G}$. We say an arc e of $\mathcal{G}$ is said to be a *link* of $\mathcal{G}$ with respect to $\mathcal{G}'$ if there are $m \geq 1$ components $\mathcal{V}_1, \ldots, \mathcal{V}_m$ of $\mathcal{G}'$ such that $\bigcup_{1 \leq i \leq m} L_{\mathcal{V}_i} = D(l(e))$ in $\mathcal{G}$.

The following lemma gives an important property of links:

Lemma 2. *Let $\mathcal{G}$ be a phylogenetic forest and $\mathcal{G}'$ be a full restriction of $\mathcal{G}$. Let e be an arc of $\mathcal{G}$. If e is not a link with respect to $\mathcal{G}'$, then:*

1. *there is a unique component $\mathcal{V}$ in $\mathcal{G}'$ such that $S_\mathcal{V} \cap D(l(e)) \neq \emptyset$ and $S_\mathcal{V} \setminus D(l(e)) \neq \emptyset$;*
2. *$\mathcal{V}[S_\mathcal{V} \cap D(l(e))]$ and $\mathcal{V}[S_\mathcal{V} \setminus D(l(e))]$ are node-disjoint.*

Observe that the definition of a link depends on the full restriction one is considering. Hereafter, full restrictions will be given by the context and therefore not explicitly written.

2.3 Eliminations

Let $\mathcal{W}$ be a phylogenetic tree, and e an arc of $\mathcal{W}$. In what follows, we define the *elimination* of an arc e, which is the basic operation used in the forthcoming algorithms.

The elimination of an arc e is the operation that yields the phylogenetic forest whose components are $\mathcal{W}|D(l(e))$ and $\mathcal{W}|(S_\mathcal{W} \setminus D(l(e)))$. This operation can be implemented by *removing* arc e from $\mathcal{W}$ (which splits $\mathcal{W}$ into $\mathcal{W}|D(l(e))$ and $\mathcal{W}[S_\mathcal{W} \setminus D(l(e))]$) and by performing a *forced contraction* of arc e' in $\mathcal{W}[S_\mathcal{W} \setminus D(l(e))]$ where $u(e') = u(e)$ (which yields $\mathcal{W}|(S_\mathcal{W} \setminus D(l(e)))$ from $\mathcal{W}[S_\mathcal{W} \setminus D(l(e))]$). Figure 1 illustrates this definition. This operation can be defined for phylogenetic forests as well, by letting all components that do not contain arc e remain the same and applying elimination on the component that contains e.

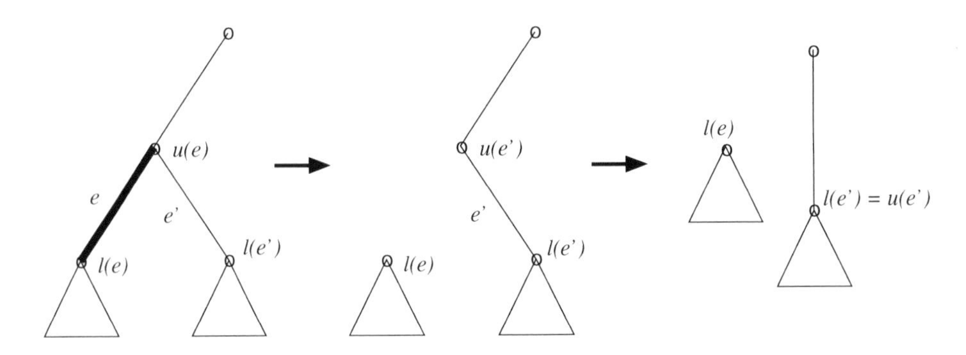

Fig. 1. Elimination of an arc e.

The lemma below exhibits the equivalence between a full restriction of a phylogenetic forest and a sequence of arc eliminations starting from $\mathcal{W}$.

Lemma 3. *Let $\mathcal{W}$ be a phylogenetic forest. Then performing a sequence of m eliminations starting from $\mathcal{W}$ yields a full restriction of $\mathcal{W}$ with size $|\mathcal{W}| + m$. Conversely, let $\mathcal{S}$ be a partition of $S_\mathcal{W}$ such that the full restriction $\mathcal{W}|\mathcal{S}$ is defined. Then $\mathcal{W}|\mathcal{S}$ can be obtained by performing a sequence of $|\mathcal{S}| - |\mathcal{W}|$ eliminations.*

3 Algorithms

The algorithm takes as input two phylogenetic trees $\mathcal{T}$ and $\mathcal{U}$ with $S_\mathcal{T} = S_\mathcal{U}$, and returns an agreement forest of $\mathcal{T}$ and $\mathcal{U}$ with size up to 3 times the size of a maximum agreement forest of $\mathcal{T}$ and $\mathcal{U}$. Our algorithm proceeds by cutting arcs and shrinking pairs of sibling leaves in $\mathcal{T}$ and $\mathcal{U}$ until two isomorphic full restrictions are obtained.

The algorithm is iterative: each iteration starts with two restrictions obtained from $\mathcal{T}$ and $\mathcal{U}$ by the sequence of operations performed up to that iteration by the algorithm. In each iteration, the algorithm searches for a pair of sibling leaves in one of the restrictions and, if it succeeds, it identifies the *case* that is satisfied by this pair in the other (see Figure 2). According to the case, the algorithm then selects and applies on both restrictions a predefined sequence of operations, which we call *transactions* (see Section 3.2), is selected and applied on both restrictions.

An earlier algorithm, designed by Hein, Jiang, Wang and Zhang [3], proceeds mostly within this same framework. In our paper, we refer to this algorithm as Algorithm 1, and ours shall be called Algorithm 2. Section 3.3 outlines both algorithms.

In what follows, we shall consider the iterations numbered 1, 2, ... , and so on. In each iteration i, we label the operations, dividing them into two sequences $(x_i^1, \ldots, x_i^{g_i})$ and $(y_i^1, \ldots, y_i^{h_i})$, where g_i and h_i are the number of operations performed at iteration i on (a full restriction of) $\mathcal{T}$ and (a full restriction of) $\mathcal{U}$ respectively. Let $\mathcal{G}_1 := \mathcal{T}$ and $\mathcal{H}_1 := \mathcal{U}$, and for $i \geq 2$ let $\mathcal{G}_i$ be the full restriction we get from $\mathcal{G}_{i-1}$ after executing the g_{i-1} operations of iteration $i-1$. Let $\mathcal{G}_i^1 := \mathcal{G}_i$, and for $2 \leq j \leq g_i + 1$ let $\mathcal{G}_i^j$ be the full restriction we get by applying operation x_i^{j-1} on $\mathcal{G}_i^{j-1}$. Similarly, we define $\mathcal{H}_i$ and $\mathcal{H}_i^j$ for each $i \geq 1$ and $j \in \{1, \ldots, h_i + 1\}$.

For the analysis of the performance of the algorithm, we consider throughout the iterations two full restrictions, $\mathcal{T}'$ and $\mathcal{U}'$, of $\mathcal{T}$ and $\mathcal{U}$ respectively, which in the beginning of the first iteration are isomorphic to a maximum agreement forest $\mathcal{F}$ of $\mathcal{T}$ and $\mathcal{U}$. For each operation done on $\mathcal{T}$ (respectively $\mathcal{U}$), a "similar" operation is performed on $\mathcal{T}'$ (respectively $\mathcal{U}'$), in order to preserve the invariants of Algorithm 2 to be stated in the sequel. In the same way as we have done for $\mathcal{T}$ and $\mathcal{U}$, for each iteration i we also define for $\mathcal{T}'$ and $\mathcal{U}'$ the sequences of operations $({x'}_i^j : 1 \leq j \leq g_i)$, $({y'}_i^j : 1 \leq j \leq h_i)$, and the corresponding full restrictions $\mathcal{G}_i'$, $\mathcal{H}_i'$, ${\mathcal{G}'}_i^j$ and ${\mathcal{H}'}_i^j$.

Roughly speaking, the initial full restrictions $\mathcal{T}'$ and $\mathcal{U}'$ give a block configuration attained by an optimum agreement forest $\mathcal{F}$ of $\mathcal{T}$ and $\mathcal{U}$ (each block being a component of $\mathcal{F}$). This (fixed) "ideal" block configuration is used to derive some information of the size of the agreement forest returned by the algorithm. Whenever the algorithm performs a cut of an arc e, a "similar" operation is performed on $\mathcal{T}'$ and $\mathcal{U}'$: if e is not a link then this cut splits a block of the current configuration, if e is a link then no new block is derived. With this procedure, we can get hold of the number of the "additional" blocks our algorithm produces.

3.1 Operations

In this section we define the operations that can be applied on the phylogenetic trees $\mathcal{T}$ and $\mathcal{U}$.

Let $\mathcal{W}$ be a phylogenetic tree. The operation of *cutting an arc* e of $\mathcal{W}$ is the elimination of e defined in Section 2.3. The operation of *shrinking a pair of sibling leaves u and v to leaf u* corresponds to the elimination of the arc e such that $l(e) = v$, followed by the removal of the isolated node v. This operation is used whenever a pair of labels which yields a pair of sibling leaves both in $\mathcal{G}_i^j$ and $\mathcal{H}_i^j$ is located.

As Algorithm 2 proceeds performing cuts and shrinkings in $\mathcal{T}$ and $\mathcal{U}$, their full restrictions $\mathcal{T}'$ and $\mathcal{U}'$ must have some of their arcs eliminated as well, if we want to retain as invariants the following properties:

$$\mathcal{G}'^j_i \text{ is a full restriction of } \mathcal{G}_i^j \text{ and } \mathcal{H}'^j_i \text{ is a full restriction of } \mathcal{H}_i^j.$$

For each arc e eliminated, if e is a link, then the invariants are satisfied just by making $\mathcal{G}'^{j+1}_i := \mathcal{G}'^j_i$ and $\mathcal{H}'^{j+1}_i := \mathcal{H}'^j_i$. Otherwise, it can be proved from Lemmas 2 and 3 that for each arc e eliminated that is not a link, there exists an arc in $\mathcal{T}'$ or in $\mathcal{U}'$ respectively whose elimination preserves the invariants.

A similar result is valid for $\mathcal{U}$ and $\mathcal{U}'$. Henceforth, we shall take these eliminations in $\mathcal{T}'$ and $\mathcal{U}'$ for granted at each elimination in $\mathcal{T}$ and $\mathcal{U}$.

When a shrinking is performed and a one-node component is removed in $\mathcal{G}_{i+1}$ and $\mathcal{H}_{i+1}$, we must also remove the corresponding component in $\mathcal{G}'_{i+1}$ and $\mathcal{H}'_{i+1}$. Observe that after shrinkings, $\mathcal{G}_i$ and $\mathcal{H}_i$ are no longer full restrictions of $\mathcal{G}_1$ and $\mathcal{H}_1$ respectively.

3.2 Cases and Transactions

Suppose that at the beginning of iteration i there is at least an arc in $\mathcal{G}_i$. Then $\mathcal{G}_i$ admits at least a pair of sibling leaves. Let a and b be its labels. We consider five cases, each of them defined by a different configuration yielded in $\mathcal{H}_i$ by the leaves $f_{\mathcal{H}_i}(a)$ and $f_{\mathcal{H}_i}(b)$.

If $f_{\mathcal{H}_i}(a)$ and $f_{\mathcal{H}_i}(b)$ belong to the same component in $\mathcal{H}_i$, then the *(a,b)-axis* in $\mathcal{H}_i$ is the (unique) path in $\mathcal{H}_i$ connecting $f_{\mathcal{H}_i}(a)$ and $f_{\mathcal{H}_i}(b)$, and the *stems* of the (a,b)-axis are the arcs e in $\mathcal{H}_i$ such that $u(e)$ belongs to the axis but e does not. The stems are labeled s_1 through s_k according to the order in which they appear along the axis, with stem s_1 being the nearest to $f_{\mathcal{H}_i}(a)$. The cases are shown in Figure 2. Case 1 differs from Case 2 in that there is a fixed limit on the number k of stems: in Case 1 we have $1 \leq k \leq 2$, and in Case 2 we have $k \geq 3$.

The transactions for each case are listed below. We denote by $\textsc{Cut}(\mathcal{G}, v)$ the operation "cut the arc in $\mathcal{G}$ whose lower endpoint is v", and by $\textsc{Srk}(\mathcal{G}, u, v)$ the operation "shrink pair u, v of sibling leaves to leaf u".

Case 1: For each $j \in \{1, \ldots, k\}$ do $\textsc{Cut}(\mathcal{H}_i^j, l(s_j))$.
Case 2: $\textsc{Cut}(\mathcal{G}_i^1, f(a))$; $\textsc{Cut}(\mathcal{G}_i^2, f(b))$; $\textsc{Cut}(\mathcal{H}_i^1, f(a))$; $\textsc{Cut}(\mathcal{H}_i^2, f(b))$.
Case 3: $\textsc{Cut}(\mathcal{G}_i^1, f(a))$; $\textsc{Cut}(\mathcal{G}_i^2, f(b))$; $\textsc{Cut}(\mathcal{H}_i^1, f(a))$; $\textsc{Cut}(\mathcal{H}_i^2, f(b))$.
Case 4: $\textsc{Cut}(\mathcal{G}_i^1, f(b))$.
Case 5: $\textsc{Srk}(\mathcal{G}_i^1, f(a), f(b))$; $\textsc{Srk}(\mathcal{H}_i^1, f(a), f(b))$.

We give below an important lemma, whose proof is straightforward from Lemmas 1 and 2.

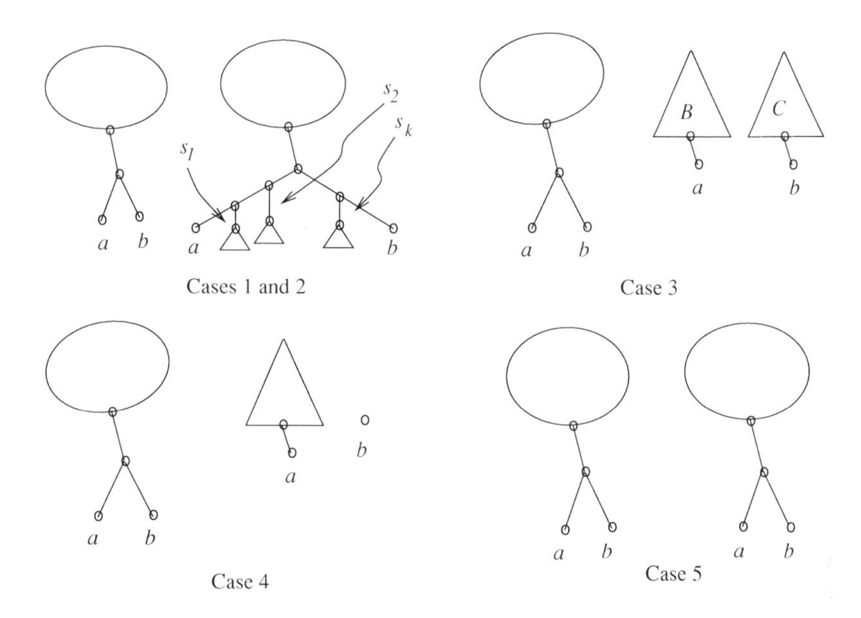

Fig. 2. Cases for Algorithm 2. In each case $\mathcal{G}_i$ is to the left and $\mathcal{H}_i$ is to the right.

Lemma 4. *Let $\mathcal{G}$ and $\mathcal{H}$ be two phylogenetic forests with $S_{\mathcal{G}} = S_{\mathcal{H}}$, and let $\mathcal{G}'$ and $\mathcal{H}'$ be two full restrictions of $\mathcal{G}$ and $\mathcal{H}$ respectively, such that $\mathcal{H}'$ is isomorphic to a full restriction of $\mathcal{G}'$. Suppose that $f_{\mathcal{G}}(a)$ and $f_{\mathcal{G}}(b)$ form a pair of sibling leaves in $\mathcal{G}$.*

1. *If $f_{\mathcal{G}'}(a)$ and $f_{\mathcal{G}'}(b)$ are separated in $\mathcal{G}'$, then at least one of the arcs incident to $f_{\mathcal{H}}(a)$ or $f_{\mathcal{H}}(b)$ is a link.*
2. *If $f_{\mathcal{H}'}(a)$ and $f_{\mathcal{H}'}(b)$ are in the same component of $\mathcal{H}'$, and the (a, b)-axis in $\mathcal{H}$ admits at least one stem, then all of its stems in $\mathcal{H}$ are links.*

3.3 Algorithms 1 and 2

We give below an outline of Algorithm 2. We remark that before the algorithm outputs $\mathcal{H}$ as a solution, those nodes which are removed from $\mathcal{H}$ at shrinkings must be reattached to it in reverse order.

(1) $\mathcal{G}_1 := \mathcal{T}$;
(2) $\mathcal{H}_1 := \mathcal{U}$;
(3) $i := 1$;
(4) While there is a pair of sibling leaves $f_{\mathcal{G}_i}(a)$ and $f_{\mathcal{G}_i}(b)$ in $\mathcal{G}_i$
(5) Find out which case is satisfied by $f_{\mathcal{H}_i}(a)$ and $f_{\mathcal{H}_i}(b)$;
(6) Apply the corresponding transaction;
(7) $i := i + 1$;
(8) Return $\mathcal{H}_i$ with all nodes that were removed at shrinkings reattached.

Algorithm 1 differs from Algorithm 2 by defining otherwise the limit on the number k of stems in Case 1. In Algorithm 1, if $k = 1$ we have Case 1 and if $k \geq 2$ we have Case 2. The transactions applied in each of these cases are the same as before.

It is not difficult to conclude that Algorithm 2 can be implemented to run in polynomial time in the size of the instance. In each iteration i, whenever a pair (a, b) of sibling leaves is considered in $\mathcal{G}_i$, all we need to find out the case satisfied by (a, b) is to find (if existent) the (a, b)-axis in $\mathcal{H}_i$ and check the number k of stems of this axis. According to the value of k we are in one among cases 1, 2 or 5. If no such path exists, then we are in Case 3 or 4. In all these cases, the transactions can be performed in time linear in the size of the corresponding tree. Thus, since the number of iterations is bounded by the number of leaves of the original trees, the claim follows. This rough analysis allows to say that if the original trees have m arcs, then Algorithm 2 can be implemented to run in $O(m^2)$. The same holds for Algorithm 1.

4 Approximation Ratio

Our proof technique uses an accounting system, according to which an initial amount of *credits* is alloted to one of the trees given as input, and this tree (or rather a full restriction of it) is charged by means of *debits* recorded whenever an operation needs payment. The initial amount of credits is related to the size of a maximum agreement forest whereas the debits are related to the number of cuts made by the algorithm, and the goal in this proof technique is to show that the credits suffice to pay the debits recorded. This technique also follows the sketch given in [3], but here we improve the control over the charged elements in the trees. It turns out that some debits recorded can only be paid later; in such situations these debits are attached to special sets of arcs which we call *barriers*. These arc sets not only work as "bags" of postponed credits and as such are fit to be charged for the postponed debits, but they also keep these debits apart from each other, which avoids their clustering and the need for paying great amounts of them at once.

4.1 Barriers

Let B be a node set of a phylogenetic forest $\mathcal{G}$. We denote by $U_{\mathcal{G}}(B)$ the set of arcs whose upper endpoints are the lca's in $\mathcal{G}$ of at least two nodes in B.

Take some $\mathcal{H}_i^j$ with $(i, j) \neq (1, 1)$. Consider the arcs e and e' as shown in Figure 1 for the elimination performed to produce $\mathcal{H}_i^j$. A node set B in $\mathcal{H}_i^j$ with at least two nodes is a *barrier* if:

1. B is a *primary barrier*. That is, $j = 1$; iteration $i - 1$ satisfies Case 2, with $f_{\mathcal{H}'_{i-1}}(a)$ and $f_{\mathcal{H}'_{i-1}}(b)$ in a component of $\mathcal{H}'_{i-1}$ which has at least three leaves; $B = \{l(e) : e \text{ is a stem of the } (a, b)\text{-axis in } \mathcal{H}_{i-1}\}$; or
2. B admits an *antecessor barrier* B'. That is, $j \geq 2$ and there exists a barrier B' in $\mathcal{H}_i^{j-1}$ such that one of the following conditions holds:

(a) $e, e' \in U_{\mathcal{H}_i^{j-1}}(B')$; $B = B' \cap D(l(e))$ or $B = B' \setminus D(l(e))$, the one with the largest cardinality (if there is a tie, choose any of them); or
(b) $e, e' \notin U_{\mathcal{H}_i^{j-1}}(B')$; $B = B'$.

Below we list some properties of barriers.

Lemma 5. *Let $\mathcal{H}_i^j$ be such that $(i, j) \neq (1, 1)$, and let B be a barrier of $\mathcal{H}_i^j$. Then all arcs in $U_{\mathcal{H}_i^j}(B)$ are links.*

Lemma 6. *In some $\mathcal{H}_i^j$ with $(i, j) \neq (1, 1)$, consider two distinct barriers B and C. Let $t_B := \text{lca}_{\mathcal{H}_i^j}(B)$ and $t_C := \text{lca}_{\mathcal{H}_i^j}(C)$. Then $U_{\mathcal{H}_i^j}(B)$ and $U_{\mathcal{H}_i^j}(C)$ are arc-disjoint and if they have a node in common, then either $t_B \in C$ or $t_C \in B$.*

The notion of barrier is fundamental in our proof that Algorithm 2 yields approximation ratio 3 for the maximum agreement forest problem. In the next subsection, we introduce credits and debits and the rules that command them along the iterations of the algorithm, as well as their relation with barriers.

4.2 Credits and Debits

At the beginning of the first iteration, we consider that a certain amount of *credits* is *placed* on the unique component of $\mathcal{U}$. As Algorithm 2 proceeds doing cuts and shrinkings, those operations which must be paid *record* debits equivalent to their due fees, and our proof aims to guarantee that at the end of the algorithm, each debit recorded is paid with a credit out of those placed on $\mathcal{U}$ at the beginning. However, as the algorithm proceeds, the credits are *kept* by the components of $\mathcal{U}$, and they must be *released* before they can be used to pay any recorded debit. Also, as long as they are kept, the credits must be *redistributed* among the components at each elimination performed on $\mathcal{U}$.

The precise rules for debit recording and credit placing, releasing and redistributing are as follows. Let $\mathcal{F}$ be the maximum agreement forest taken as reference in Section 3. At the beginning, we place $3(|\mathcal{F}| - 1)$ credits on $\mathcal{U}$. These credits are released and redistributed when eliminations are performed. We have already seen that each elimination splits a component $\mathcal{W}$ of its argument $\mathcal{H}_i^j$ into two new components, $\mathcal{W}|D(l(e))$ and $\mathcal{W}|(S_\mathcal{W} \setminus D(l(e)))$. Let m be the number of components $\mathcal{W}'$ of $\mathcal{H'}_i^j$ such that $S_{\mathcal{W}'} \subseteq S_\mathcal{W}$, $m^\cap$ be the number of components $\mathcal{W}'$ of $\mathcal{H'}_i^{j+1}$ such that $S_{\mathcal{W}'} \subseteq D(l(e))$, and $m^\setminus$ be the number of components $\mathcal{W}'$ of $\mathcal{H'}_i^{j+1}$ such that $S_{\mathcal{W}'} \subseteq S_\mathcal{W} \setminus D(l(e))$. It can easily be verified that the following rules preserve credits throughout the eliminations.

If we eliminate a link e of a component $\mathcal{W}$ of some $\mathcal{H}_i^j$, then from the $3(m-1)$ credits placed on $\mathcal{W}$, three credits are released, and each of the two resulting components keeps now respectively $3(m^\cap - 1)$ and $3(m^\setminus - 1)$ credits. Otherwise, if we eliminate an arc which is not a link, then no credit is released and each of the resulting components keeps respectively $3(m^\cap - 1)$ and $3(m^\setminus - 1)$ credits.

Among the operations used in Algorithm 2, only cuts record debits. Each cut operation records one debit, which either is paid with credits released in the

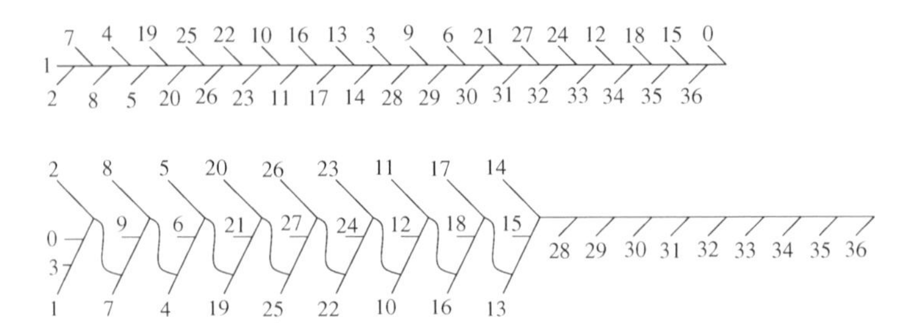

Fig. 3. An example showing that Algorithm 1 attains approximation ratio close to 4.

same iteration at which the cut is performed, or (if the cut arc is not a link) is attached to a suitable barrier. In fact, at each iteration either a credit is released or a primary barrier is created. It turns out that each primary barrier has two debits attached to itself. As these debits are placed on the whole primary barrier rather than on any of its arcs, they are paid as soon as the algorithm begins to eliminate arcs of the barrier.

4.3 An Upper Bound for the Approximation Ratio of Algorithm 2

Since all arcs in a barrier are links (Lemma 5), barriers are suited for receiving postponed debits, that is those debits which could not be handled at the iteration in which they were recorded because no credit was released in that iteration. Also, since distinct barriers are arc-disjoint (Lemma 6), then postponed debits are kept apart, and thus the maximum amount of postponed debits that must be paid at each iteration is bounded. For Algorithm 2 this maximum is 1 since each primary barrier has at least three nodes. For Algorithm 1 we can have primary barriers with only two nodes, so we must guarantee that two postponed debits can be paid.

Our proof consists in verifying that at the end of Algorithm 2, all recorded debits are paid. Once this is done, we have that the number of cuts done by Algorithm 2 in $\mathcal{U}$ is $|\mathcal{H}_i| - 1$, where i denotes the last iteration, and that the number of cuts is equal to the number of recorded debits, which is less than or equal to the number of credits alloted to $\mathcal{U}$ in the beginning, that is $3(|\mathcal{F}| - 1)$; so $|\mathcal{H}_i| \leq 3|\mathcal{F}|$.

A slight modification in this proof technique yields a proof that Algorithm 1 has approximation ratio at most 4. It suffices to place $4(|\mathcal{F}| - 1)$ credits on $\mathcal{U}$ at the beginning of the algorithm and release four credits whenever links are eliminated.

4.4 A Lower Bound for the Approximation Ratio of Algorithm 1

In this section, we exhibit a family of instances which shows that the approximation ratio for Algorithm 1 gets close to 4.

Figure 3 shows an instance of this family. Tree $\mathcal{U}$ (*bottom*) has two parts, one comprising m Cases 1 with $k = 2$ piled in a stack, and the other composed by

m single leaves (in the instance shown above we have $m = 9$). Tree $\mathcal{T}$ (*top*) is constructed so as to make Algorithm 1 handle all Cases 1 in the stack (cutting leaves 1–2–7–8– ... –14 in $\mathcal{U}$ in the example) and then alternate cuts between the remaining leaves out of the stack (0– ... –15) and the single leaves (28– ... –36).

It turns out that for any $m \geq 1$, Algorithm 1 yields an agreement forest with $4m$ components, while there is an agreement forest with at most $m + 2$ components (in Figure 3 consider the agreement forest which contains leaves 1–2–7– ... –14 together with leaves 28–29–30– ... –36 as a single component and the other leaves as isolated nodes).

This family of instances can be modified to show that the approximation ratio of Algorithm 2 is at least 3. It suffices to substitute the Cases 1 with $k = 2$ by Cases 2 with $k = 3$.

5 Final Remarks

The concept of barriers and the proof technique based on an accounting system were useful to prove the approximation ratio of Algorithm 1 and Algorithm 2. The barriers were introduced because of the transactions defined for Case 2. We note, however, that if we can provide a set of transactions (one for each case) with the property that each of them performs at most k cuts and eliminates at least p links, then an approximation ratio k/p can be guaranteed. These ideas will be further explored in a forthcoming paper. We hope that the concept of barrier, in spite of being complicated, may lead to some further improvement.

Acknowledgments

We thank the referee for the suggestions that improved the presentation of this paper, and Cristina G. Fernandes for the helpful discussions.

References

1. B. Allen and M. Steel. Subtree transfer operations and their induced metrics on evolutionary trees. *Submitted to the Annals of Combinatorics*, 2001.
2. B. dasGupta, X. He, T. Jiang, M. Li, J. Tromp, and L. Zhang. On distances between phylogenetic trees. In *Proceedings of the 8th ACM-SIAM Symposium of Discrete Algorithms*, pages 427–436, 1997.
3. J. Hein, T. Jiang, L. Wang, and K. Zhang. On the complexity of comparing evolutionary trees. *Discrete Applied Mathematics*, 71:153–169, 1996.
4. M. Li, J. Tromp, and L. Zhang. On the nearest neighbor interchange distance between evolutionary trees. *Journal on Theoretical Biology*, 182(4):463–467, 1996.
5. D. L. Swofford, G. J. Olsen, P. J. Waddell, and D. H. Hillis. Phylogenetic inference. In D. Hillis, C. Moritz, and B. Mable, editors, *Molecular Systematics*, pages 407–513. Sinauer Associates, 1996.

Near-optimum Universal Graphs for Graphs with Bounded Degrees
(Extended Abstract)

Noga Alon[1,*], Michael Capalbo[2,**], Yoshiharu Kohayakawa[3,***], Vojtěch Rödl[4,†], Andrzej Ruciński[5,‡], and Endre Szemerédi[6,§]

[1] Department of Mathematics, Raymond and Beverly Sackler Faculty of Exact Sciences, Tel Aviv University, Tel Aviv, Israel
[2] Department of Mathematical Sciences, The Johns Hopkins University, 3400 N. Charles Street, Baltimore, MD
[3] Instituto de Matemática e Estatística, Universidade de São Paulo, Brazil
[4] Department of Mathematics and Computer Science, Emory University, Atlanta
[5] Faculty of Mathematics and Computer Science, Adam Mickiewicz University, Poznań, Poland
[6] Department of Computer Science, Rutgers University, NJ

Abstract. Let $\mathcal{H}$ be a family of graphs. We say that G is $\mathcal{H}$-universal if, for each $H \in \mathcal{H}$, the graph G contains a subgraph isomorphic to H. Let $\mathcal{H}(k,n)$ denote the family of graphs on n vertices with maximum degree at most k. For each fixed k and each n sufficiently large, we explicitly construct an $\mathcal{H}(k,n)$-universal graph $\Gamma(k,n)$ with $O(n^{2-2/k}(\log n)^{1+8/k})$ edges. This is optimal up to a small polylogarithmic factor, as $\Omega(n^{2-2/k})$ is a lower bound for the number of edges in any such graph.
En route, we use the probabilistic method in a rather unusual way. After presenting a deterministic construction of the graph $\Gamma(k,n)$, we prove, using a probabilistic argument, that $\Gamma(k,n)$ is $\mathcal{H}(k,n)$-universal. So we use the probabilistic method to prove that an *explicit* construction satisfies certain properties, rather than showing the *existence* of a construction that satisfies these properties.

1 Introduction and Main Result

For a family $\mathcal{H}$ of graphs, a graph G is $\mathcal{H}$*-universal* if, for each $H \in \mathcal{H}$, the graph G contains a subgraph isomorphic to H. Thus, for example, the complete

* Partially supported by a USA-Israeli BSF grant, by the Israel Science Foundation and by the Hermann Minkowski Minerva Center for Geometry at Tel Aviv University.
** Supported by NSF grant CCR98210-58 and ARO grant DAAH04-96-1-0013.
*** Partially supported by MCT/CNPq through ProNEx Proj. 107/97 (Proc. CNPq 664107/1997–4), by CNPq (Proc. 300334/93–1, 468516/2000–0, and 910064/99–7), and by FAPESP (Proj. 96/04505–2).
† Partially supported by NSF grants DMS 0071261 and INT 0072064.
‡ Supported by KBN grant 2 P03A 032 16. Part of this research was done during the fifth author's visit to Emory University.
§ Partially supported by the NSF.

M. Goemans et al. (Eds.): APPROX-RANDOM 2001, LNCS 2129, pp. 170–180, 2001.

graph K_n is $\mathcal{H}_n$-universal, where $\mathcal{H}_n$ is the family of all graphs on at most n vertices. The construction of sparse universal graphs for various families arises in the study of VLSI circuit design, and has received a considerable amount of attention.

For example, as discussed in [5], page 308, universal graphs are of interest to chip manufacturers. It is very expensive to design computer chips, but relatively inexpensive to make many copies of a computer chip with the same design. This encourages manufacturers to make their chip designs configurable, in the sense that the entire chip is prefabricated except for the last layer, and a final layer of metal is then added corresponding to the circuitry of a customer's particular specification. Hence, most of the design costs can be spread out over many customers. We may view the circuitry of a computer chip as a graph, and may also model the problem of designing chips with fewer wires that are configurable for a particular family of applications as designing smaller universal graphs for a particular family of graphs.

Also, as discussed in [12], we may model data structures and circuits as graphs. The problem of designing, say, an efficient single circuit that can be specialized for a variety of other circuits can be viewed as constructing a small universal graph. With these applications in mind, we note that, given a family $\mathcal{H}$ of graphs, it is often desirable to find an $\mathcal{H}$-universal graph with small number of edges.

Motivated by such practical applications, universal graphs for several different families of graphs have been studied by numerous researchers since the 1960s. For example, extensive research exists on universal graphs for forests [4], [7], [8], [9], [10], [13], and for planar and other sparse graphs [1], [3], [4], [6], [11], [16].

Here we construct near-optimum universal graphs for families of bounded-degree graphs. More specifically, for all positive integers k and n, let $\mathcal{H}(k,n)$ denote the family of all graphs on n vertices with maximum degree at most k. By the *size* of a graph we always mean the number of its edges. Several techniques were introduced in [2] to obtain, for fixed k, both randomized and explicit constructions of $\mathcal{H}(k,n)$-universal graphs of size $O(n^{2-\frac{1}{k}}\log^{1/k} n)$, thereby setting a new upper bound for the minimum possible size of an $\mathcal{H}(k,n)$-universal graph. In addition, a (simple) lower-bound of $\Omega(n^{2-\frac{2}{k}})$ was also established. However, closing the gap between the upper and lower bounds was left as an open problem.

Here we almost completely close this gap by presenting an explicit construction of an $\mathcal{H}(k,n)$-universal graph $\Gamma(k,n)$ of size $O(n^{2-\frac{2}{k}}\log^{1+8/k} n)$. We describe the construction of $\Gamma(k,n)$ in the next paragraph.

Construction of $\Gamma(k,n)$: Let us set $q \equiv \log(kn/8\log^4 n)$, and $s = q/k$; so $q = \log n - 4\log\log n + O(1)$, and s is just slightly smaller than $(\log n)/k$. For the sake of simplicity let us omit all floor and ceiling signs, and assume that s and q are integers; this will not affect our arguments. Unless otherwise stated, our logarithms are to the base 2. Let $\Gamma'(k,n) = \Gamma'$ be the graph with the vectors in $\{0,1\}^q$ as its vertex-set; two vertices v and w are adjacent in Γ' if and only if there exist two distinct indices $j', j'' \in \{1, 2, ..., k\}$ such that the $(tk + j)$-th coordinate of v agrees with the $(tk + j)$-th coordinate of w, for all but at most one of the pairs of integers (j, t), where $j = j', j''$ and $t \in \{0, ..., s - 1\}$. To form $\Gamma(k, n)$ from $\Gamma'(k, n)$, replace each vertex v in Γ' with a clique V_v of $64q^4/k = \Theta\big((\log n)^4/k\big)$ vertices, and interconnect each vertex of V_v with each vertex of V_w if and only if the pair vw is an edge of Γ'.

Note that, for each fixed $k \geq 3$, the graph $\Gamma(k, n)$ has size at most

$$\frac{kn}{16 \log^4 n} \binom{k}{2} (2s) 2^{q-2s+2} \left(\frac{64q^4}{k}\right)^2 = O(n^{2-\frac{2}{k}} (\log n)^{1+8/k}),$$

and only about $8n$ vertices. Our main result is the following theorem.

Theorem 1. *The graph $\Gamma(k, n)$ is $\mathcal{H}(k, n)$-universal for all $k \geq 3$, and n sufficiently large.*

The rest of this extended abstract is organized as follows. In §2, we apply a graph embedding technique to prove that $\Gamma(k, n)$ is $\mathcal{H}(k, n)$-universal, provided each member of $\mathcal{H}$ satisfies a certain decomposition property (see Lemma 1, below). This decomposition property is easily satisfied by all graphs in $\mathcal{H}(k, n)$ for k even, and by all graphs in $\mathcal{H}(k, n)$ for k odd with chromatic index k, including all bipartite members of $\mathcal{H}(k, n)$ (see Examples 2.3 and 2.4).

The remaining case, i.e. k odd and H of chromatic index $k + 1$ is, however, quite troublesome. In §3 we sketch a proof of the existence of a suitable decomposition of every graph $H \in \mathcal{H}(k, n)$. Finally, in §4 we show how to turn $\Gamma(k, n)$ into an $\mathcal{H}(k, n)$-universal graph $\Lambda(k, n)$ that has, say, only $(1 + \epsilon)n$ vertices, and still only $O(n^{2-\frac{2}{k}} (\log n)^{1+8/k})$ edges.

The techniques in §2 combine combinatorial and probabilistic ideas, the proofs in §3 are based on tools from matching theory, including Tutte's Theorem and the Gallai-Edmonds Structure Theorem, while the result in §4 is obtained by applying some of the known constructions of expanders and concentrators. In order to make this abstract more complete, we present some of the more technical parts of §3 in an appendix.

2 A Graph Embedding Technique

A graph F is *(m, M)-path-separable* if there exists a collection $\mathcal{P}$ of edge disjoint paths in F, each of length between $2m$ and $4m$ such that for every $E \subseteq E(F)$ which intersects each $P \in \mathcal{P}$, every connected component of $F \setminus E$ has fewer than M vertices. Thus a graph is path-separable if we may pick in it a collection of short edge-disjoint paths with the property that any transversal of the edge sets of these paths breaks up the graph into components of bounded size.

Example 2.1. Every union F of vertex disjoint paths and cycles is $(m, 8m)$-path-separable for all positive integers m. Indeed, partition the edge set of each component of F containing at least $4m$ edges into paths of lengths between $2m$ and $4m$.

Example 2.2. A graph in $\mathcal{H}(k,n)$ obtained from a union of vertex disjoint paths and cycles (called later *units*) by designating one of its components as *the central unit* and connecting some of the other components to the center, each by exactly one edge (called a *spoke*), will be called *a windmill.* It is easy to see that every windmill is $(m, 64km^2)$-path-separable. Again, partition every unit containing at least $4m$ edges into paths of lengths between $2m$ and $4m$. After cutting the paths with a set E, each vertex of the largest piece (of order at most $8m$) in what is left of the central unit can be connected with up to $k-1$ paths from the other units, each of length less than $8m$.

A graph $H \in \mathcal{H}(k,n)$ is *$(2,k,m,M)$-decomposable* if one can find subgraphs $F_1, ..., F_k$ of H, not necessarily all distinct, such that each F_i is (m, M)-path-separable, and each edge of H appears in exactly 2 subgraphs F_i. Let us call $F_1, ..., F_k$ a *$(2,k,m,M)$-decomposition* of H.

Example 2.3. Every graph $H \in \mathcal{H}(k,n)$ of chromatic index k is $(2,k,m,8m)$-decomposable into graphs from $\mathcal{H}(2,n)$. Indeed, let $M_1, ..., M_k$ be matchings that cover the edges of H. Let $F_1 = M_1 \cup M_2, F_2 = M_2 \cup M_3, ..., F_i = M_i \cup M_{i+1}, ..., F_{k-1} = M_{k-1} \cup M_k, F_k = M_k \cup M_1$.

Example 2.4. If k is an even integer, then every graph $H \in \mathcal{H}(k,n)$ is $(2,k,m,$ $8m)$-decomposable into graphs from $\mathcal{H}(2,n)$. This time, by the Petersen Theorem (see, e.g., [15], p. 218), every such graph can be covered by $k/2$ subgraphs $F_1, ..., F_{k/2}$, where $F_i \in \mathcal{H}(2,n)$ for all i. Set $F_{k/2+j} = F_j$, $j = 1, \ldots, k/2$.

As far as we know, for odd k, it is still open as to whether or not every graph $H \in \mathcal{H}(k,n)$ has a $(2,k,m,O(m))$-decomposition. However, for all integers k and m, we will prove in the next section that every such graph has a $(2,k,m,64km^2)$-decomposition. This, and Lemma 1, will imply Theorem 1. Indeed, $64sq = 64ks^2$, since we set s to be q/k.

In the remainder of this section we prove Lemma 1.

Lemma 1. *If $H \in \mathcal{H}(k,n)$ is $(2,k,s,64sq)$-decomposable, then $\Gamma(k,n) \supset H$.*

Proof of Lemma 1. Let $F_1, ..., F_k$ be a $(2,k,s,64sq)$-decomposition of H. Define F_i for all $i = k+1, ..., q$, by setting $F_{tk+j} = F_j$ for each $j \in \{1, ..., k\}$, and each $t \in \{1, ..., s-1\}$. Trivially, for each edge $e \in E(H)$, there are two distinct indices $j', j'' \in \{1, ..., k\}$ such that $e \in F_{tk+j}$ for each $j = j', j''$ and $t \in \{0, 1, ..., s-1\}$.

Let $\mathcal{P}_i$ be a family of paths which exhibits the $(s, 64sq)$-path-separability of F_i, $i = 1, \ldots, q$. The following fact is crucial.

Claim 2 *There exist subsets $E_i \subseteq F_i$, $i = 1, \ldots, q$, such that*
(i) for all $1 \le i < j \le q$, we have $E_i \cap E_j = \emptyset$, and
(ii) for all $i = 1, \ldots, q$, we have $E_i \cap P \ne \emptyset$, for each $P \in \mathcal{P}_i$.

Proof of Claim 2. Consider an auxiliary bipartite graph B with the paths from (the multiset) $\bigcup_i^q \mathcal{P}_i$ on one side (red vertices) and the edges of H on the other (blue vertices), where the edges of B connect the edges of H with the paths they belong to. In this graph, the degree of every red vertex is at least $2s$ (the length of the path), while the degree of every blue vertex is at most $2s$ (since every edge of H belongs to exactly $2s$ graphs F_i and, for given i, to at most one path from $\mathcal{P}_i$). Hence, by Hall's matching theorem, one can assign to each path a different edge. The edges assigned to the paths of $\mathcal{P}_i$ form the desired set E_i, $i = 1, \ldots, q$. □

Continuing with the proof of Lemma 1, let $E_i \subseteq F_i$, $i = 1, ..., q$, satisfy (i) and (ii) of Claim 2, and let $L_i = F_i \setminus E_i$ for each $i = 1, ..., q$. Then, clearly,

(a) for each edge $e \in E(H)$, there exist two distinct indices $j', j'' \in \{1, 2, ..., k\}$ such that $e \in L_{kt+j}$ for all but at most one of the pairs of integers (j, t), where $j = j', j''$ and $t \in \{0, ..., s-1\}$, and

(b) each connected component of each L_i has at most $64sq$ vertices.

Recall that $\Gamma(k, n) = \Gamma$ is constructed by blowing up the vertices of another graph $\Gamma'(k, n) = \Gamma'$. Now, we will show the existence of an embedding $f : V(H) \to V(\Gamma') = \{0, 1\}^q$ such that

(I) if $xy \in H$, then $f(x) = f(y)$, or $f(x)f(y) \in \Gamma'$, and

(II) $|f^{-1}(v)| \leq 64q^4/k$ for each $v \in \Gamma'$.

This will prove that H is a subgraph of Γ.

For each $i = 1, ..., q$, let $\mathcal{C}_i$ denote the set of connected components of L_i, and let a function $f_i : \mathcal{C}_i \to \{0, 1\}$ be given. We now specify f: for each $x \in V(H)$, let $f(x)$ be such that the i-th coordinate of $f(x)$ is $f_i(C_i(x))$, where $C_i(x)$ is the connected component of L_i that contains x. Observe that if $xy \in L_i$, then clearly, x and y are in the same connected component of L_i, and the i-th component of $f(x)$ equals the i-th component of $f(y)$. Hence, by (a) and the construction of Γ', if $xy \in H$ then $f(x)f(y) \in \Gamma'$, unless $f(x) = f(y)$. Consequently, f satisfies condition (I).

It remains to show that there exists such an f with $|f^{-1}(v)| \leq 64q^4/k$ for all $v \in V(\Gamma')$. We apply the probabilistic method. Let each f_i be chosen randomly according to the uniform distribution on $\{0, 1\}^{\mathcal{C}_i}$. Then f is also random, but not necessarily uniform on $V(\Gamma')^{V(H)}$. To avoid this problem, we split $V(H)$ suitably, being guided by the following elementary observation.

Claim 3 *Let each $f_i : X_i \to V$ be drawn uniformly at random, $i = 1, \ldots, q$. Let Y be a set of vectors in $X_1 \times \ldots \times X_q$, such that no two vectors in Y have a common coordinate. Then, letting y_i ($i = 1, \ldots, q$) denote the i-th coordinate of each $y \in X_1 \times \ldots \times X_q$, the function $f : Y \to V^q$, defined by $f(y) = (f_1(y_1), \ldots, f_q(y_q))$ is also drawn according to the uniform distribution on $(V^q)^Y$.* □

Let H' be the graph obtained from H by connecting every two vertices x and y which, for some $i = 1, \ldots, q$, are in the same connected component of L_i. If Y is an independent set in H', then, by Claim 3, $f|_Y$ is distributed uniformly on $V(\Gamma')^Y$. As the degree of H' is smaller than $q(64qs) = 64sq^2$, we can partition the vertices of H into $r = 64sq^2 = 64q^3/k$ sets $Y_1, Y_2, \ldots, Y_r$, each independent in H', and so, for each $j = 1, \ldots, r$, $f|_{Y_j}$ is distributed uniformly on $V(\Gamma')^{Y_j}$.

In fact, by applying the Hajnal-Szemerédi Theorem [14] to H', we can ensure that $Y_1, ..., Y_r$ have all equal cardinality (to within 1). So each Y_j has cardinality

$n/64sq^2$. Since $V(\Gamma') = V$ has cardinality $2^q = kn/8\log^4 n$, which is at least $kn/10q^4$ for large enough n, it follows that $|Y_j|/|V| \le 5sk/32 = 5q/32$.

To confirm condition (II), it suffices to show that, for each fixed $j \in \{1, \dots, r\}$, with probability at least $1 - o(1/r)$, $|f^{-1}(v) \cap Y_j| \le q$ for all $v \in V$. Thus, the following simple, probabilistic fact is just what we need.

Claim 4 *Let Y and V be two sets with $|Y| \le 5q|V|/32$ and $|V| = 2^q$. If a function $f : Y \to V$ is chosen uniformly at random, then*

$$\mathrm{Prob}(\exists v \in V : |f^{-1}(v)| > q) = o(1/q^3).$$

Proof of Claim 4. The probability in question can be bounded from above by

$$|V|\binom{|Y|}{q}|V|^{-q} < |V|\left(\frac{15}{32}\right)^q = \left(\frac{15}{16}\right)^q = o\left(\frac{1}{q^3}\right).$$

□

To finish the proof of Lemma 1 we apply Claim 4 r times, with $Y = Y_j$, $j = 1, \dots, r$, and $V = V(\Gamma')$. □

3 Windmill Decomposition of Graphs

In this section we prove the following proposition which together with Lemma 1 completes the proof of Theorem 1.

Proposition 1. *For each $k \ge 3$, every graph $H \in \mathcal{H}(k, n)$ is $(2, k, s, 64sq)$-decomposable*

In view of Examples 2.2–2.4, Proposition 1 is a simple corollary of the next result and the fact that $64sq = 64s^2k$. Recall the definition of a windmill given in Example 2.2. We say that $H \in \mathcal{H}(k, n)$ is $(2, k)$*-decomposable into windmills* if there exist subgraphs $F_1, ..., F_k$ of H, not necessaily all distinct, such that

(i) each edge of H appears in exactly two of the F_i's, and

(ii) each F_i is a vertex-disjoint collection of windmills.

In this case, the collection $F_1, ..., F_k$ is called a $(2, k)$*-decomposition of H into windmills.*

Proposition 2. *For each odd $k \ge 3$, every graph H of maximum degree at most k is $(2, k)$-decomposable into windmills.*

Proof. We first present a construction of subgraphs $W, F_2, ..., F_{(k-1)/2}$ of H. Next we prove that each F_i is a vertex-disjoint collection of windmills (see Lemma 2, below), and that W is $(2, 3)$-decomposable into windmills (see Lemma 2 and Lemma 3).

The construction of $W, F_2, ..., F_{(k-1)/2}$. Let us assume, without loss of generality, that H is k-regular, as H is a subgraph of a k-regular graph (that may have a larger vertex-set). We further assume that H is connected. Recall that a *Tutte* *set* in H is a set S of vertices such that if $H - S$ has m connected components with an odd number of vertices, then the size of the maximum matching in H is $\frac{1}{2}(|V(H)| - m + |S|)$. Let S be a maximal Tutte set of H, and let $\mathcal{C} = \{C_1, ..., C_m\}$ denote the set of odd connected components of $H \setminus S$. For any $C \in \mathcal{C}$, and any subgraph $F \subseteq H$, let $\delta_F(C)$ denote the number of edges in F with exactly one endpoint in C.

Using the Gallai-Edmonds Structure Theorem (see, e.g., [15], pp. 94-95) and Hall's Theorem, one can prove that there exists a collection M^* of vertex-disjoint stars of H that satisfies the following properties.
(i) Each $s \in S$ is a center of a star $\chi_s \in M^*$, where the χ_s's, $s \in S$, are such that

(†) each such χ_s has at least one edge,
(a) no χ_s contains any vertices of $S \setminus \{s\}$,
(b) no χ_s contains any vertices in any even connected component of $H \setminus S$,
(c) for each $C \in \mathcal{C}$, there is exactly one edge e_C that is incident to a vertex in C, and also belongs to some χ_s, and
(d) for each $s \in S$, there is at most one $C \in \mathcal{C}$ such that $\delta_H(C) \geq k$ and $V(\chi_s) \cap V(C)$ is nonempty.

(ii) The subgraph of M^* induced by the vertices of H not belonging to any χ_s as in (i) is a perfect matching.

Note that

(*) $M^*[V']$ has maximum degree at most 1 for any set V' of vertices disjoint from S.

Each vertex in H has degree at least 1 in M^*; let $F'_1, ..., F'_{(k-1)/2}$ be subgraphs with maximum degree at most 2 such that $\bigcup_j F'_j = H \setminus M^*$. Such subgraphs exist by the Petersen Theorem (cf. Example 2.4). For each $j \geq 2$ in its turn, we now construct F_j from F'_j as follows. For each $C \in \mathcal{C}$ such that $\delta_{F'_j}(C) = 0$, add e_C to F'_j unless it already belongs to $F_{j'}$ for some $2 \leq j' < j$, and call the resulting graph F_j. Note that

(**) $\delta_{F_j}(C) \leq 1$ if $\delta_{F'_j}(C) = 0$, for each $C \in \mathcal{C}$.

Let $W = H \setminus F_2 \cup ... \cup F_{(k-1)/2}$. (Note that $W \setminus M^* = F'_1$.)

The next two lemmas describe the structure of the subgraphs W, F_2,..., $F_{(k-1)/2}$. Their proofs are given in the appendix.

Lemma 2. *The graphs $W, F_2, ..., F_{(k-1)/2}$ satisfy the following two conditions.*
(i) We can partition $V(W) = V(H)$ into sets $V_0, ..., V_t$ such that

(A) for each $j \in \{1, ..., t\}$, there is at most one edge e_j in W with exactly one endpoint in V_j (the edges e_j will be called the parting edges *of W), and*

(B) each $W[V_i]$ has maximum degree 3, and a matching M_i that saturates all vertices of degree 3 in $W[V_i]$.

(ii) Each F_i is a vertex-disjoint collection of windmills.

Lemma 3. *Let H' be a graph in $\mathcal{H}(3,n)$ that contains a matching M that saturates each vertex of degree 3 in H'. Then there exist three subgraphs F_1, F_2, F_3, such that*

(i) each edge of H' appears in exactly two of the F_i's;

(ii) F_1, F_2 have maximum degree 2, and F_3 is a collection of vertex disjoint windmills.

We now use Lemmas 2 and 3 to finish the proof of Proposition 2. Take two copies of each of the graphs $F_2, ..., F_{(k-1)/2}$ to obtain $F_2, ..., F_{k-2}$, which are each vertex-disjoint collections of windmills, by Lemma 2. Therefore, to prove Proposition 2, all we need to show is that W is $(2,3)$-decomposable into windmills, say, F_1, F_{k-1}, and F_k, and thus obtain a $(2,k)$-decomposition of H into vertex-disjoint windmills.

To this end we use Lemma 3. Let $V_0, V_1, ...$ be as in Lemma 2 (i). For each $W[V_j]$, let $Y_{j,1}, Y_{j_2}, Y_{j_3}$ be a $(2,3)$-decomposition of $W[V_j]$ into windmills, such that each $Y_{j,1}$ and $Y_{j,2}$ have maximum degree 2; such graphs exist by Lemma 3. Let E' denote the set of parting edges in W. Note that $F_1 = E' \cup (\bigcup_j Y_{j,1})$ and $F_{k-1} = E' \cup (\bigcup_j Y_{j,2})$ is a vertex-disjoint collection of windmills, and $F_k = \bigcup Y_{j,3}$ is also a vertex-disjoint collection of windmills, and that each edge of W appears in at least 2 of the graphs F_1, F_{k-1}, F_k. This completes the proof of Proposition 2. □

4 Universal Graphs with Fewer Vertices

In this section, we sketch a construction of an $\mathcal{H}(k,n)$-universal graph $\Lambda(k,n) = \Lambda$, which still has $O(n^{2-\frac{2}{k}}(\log n)^{1+8/k})$ edges, but only has $(1+\epsilon)n$ vertices, for any fixed $\epsilon > 0$.

Let us write $V(\Gamma(k,n)) = V$, and let $\Omega = (V, Q, E)$ be a bipartite graph of bounded degree such that $|Q| = (1+\epsilon)n$, and $|N(X)| \geq |X|$ for each subset $X \subset V$ such that $|X| \leq n$. It is well-known that such an Ω, usually called a concentrator, exists, and can be constructed explicitly using the known constructions of bounded-degree expanders. We now construct $\Lambda(k,n)$, which has Q as its vertex-set. Let ν and ν' be vertices in Q. The edge $\nu\nu' \in \Lambda$ if and only if there exist vertices $v, v' \in V$ such that $vv' \in \Gamma(k,n)$, and $v\nu, v'\nu' \in \Omega$. We have $|E(\Lambda)| \leq |E(\Gamma)|\Delta(\Omega)^2 = O(|E(\Gamma)|)$.

The following theorem can be easily deduced from Theorem 1.1.

Theorem 5. *$\Lambda(k,n)$ is $\mathcal{H}(k,n)$-universal for all $k \geq 3$, and n sufficiently large.*

Proof. Let $H \in \mathcal{H}(k,n)$. Then, by Theorem 1.1, $H \subset \Gamma(k,n)$. By the expanding property of Ω and by Hall's Theorem, Ω has a matching f between $V(H)$ and a subset of Q. Thus, if $xy \in H$ then $f(x)f(y) \in \Lambda$. □

A More Details for Section 3

Proof of Lemma 2: We first prove (ii), namely, show that each F_i is indeed a vertex-disjoint collection of windmills. As F_i' is a collection of vertex-disjoint cycles and paths, each connected component of F_i not containing an edge of $F_i \setminus F'_i$ is either a path or a cycle. Thus, it remains to show that L is a windmill, for each L that is a connected component of F_i containing an edge $e = xy$ in $F_i \setminus F'_i$. The edge e is of the form e_C, for some $C \in \mathcal{C}$; let us assume without loss of generality that $y \notin C$, but $x \in C$. Note that (a') both x and y cannot be in the same connected component of F'_i, since otherwise $\delta_{F'_i}(C) > 0$, and $e \notin F_i$. Similarly, (b') the connected component L'_x of F'_i containing x must be contained in C. But (a'), (b') and (**), together with the fact that each connected component of F'_i is either a path or cycle, imply that L must be a windmill. Thus Lemma 2 (ii) follows.

We now show that W satisfies (i) of Lemma 2.

Claim 3.1: For each $C \in \mathcal{C}$, the quantity $\delta_H(C)$ is an odd integer. So if $\delta_H(C) < k$, then $\delta_H(C) \leq k - 2$.

Claim 3.1 follows from the fact that H is k-regular, with k an odd integer, and that C has an odd number of vertices.

Claim 3.2: For each $C \in \mathcal{C}$, and each i, the quantity $\delta_{F'_i}(C)$ is an even integer. So if $\delta_{F'_i}(C) > 0$, then $\delta_{F'_i}(C) \geq 2$.

Claim 3.2 follows from the fact that each vertex in C has degree exactly 1 in M^*, so each vertex in C has degree exactly $k - 1$ in $H \setminus M^*$, and so exactly 2 in each F'_i.

Claim 3.3: All but one of s's neighbors in $M^* \cap W$ are in some $C \in \mathcal{C}$ such that $\delta_W(C) \leq 1$.

Proof of Claim 3.3: From the definition of M^* and Claim 3.1, all but one of s's neighbors in M^* are in some $C \in \mathcal{C}$ such that $\delta_H(C) \leq k - 2$. But by definition of the F_i's, and Claim 3.2, for each such C, either $\delta_{H\setminus(F_2\cup\ldots\cup F_i)}(C) \leq \delta_{H\setminus(F_2\cup\ldots\cup F_{i-1})}(C) - 2$ for each i, or $e_C \in F_i$, and therefore, $e_C \notin W$, and so Claim 3.3 follows.

Claim 3.4: Let V_0 be the set of vertices v such that either $v \in S$, or $v \in C \in \mathcal{C}$ such that $\delta_W(C) > 1$, or v is in any even-sized connected component of $H \setminus S$. Then (1) each vertex in $W[V_0]$ has degree at most 3, and (2) $W[V_0]$ has a matching covering all vertices of degree 3 in $W[V_0]$.

Proof of Claim 3.4: By Claim 3.3, $M^*[V_0]$ is a matching. Since $W[V_0] \setminus M^* \subseteq F'_1$, which has maximum degree 2, both (1) and (2) follow.

Claim 3.5: For each $C_{i_j} \in \mathcal{C}' = \{C_{i_1}, ..., C_{i_l}\} \subseteq \mathcal{C}$ such that $\delta_W(C_{i_j}) \leq 1$ for each $j \in \{1, .., l\}$, let V_j denote the set of vertices of C_{i_j}. The graph $W[V_j]$ has a matching that saturates all vertices of degree 3 in $W[V_j]$.

Proof of Claim 3.5: $M^*[V_j]$ is a matching by (*), and $W[V_0] \setminus M^* \subseteq F'_1$, which has maximum degree 2.

Lemma 2 follows from Claim 3.6.

Claim 3.6: W satisfies (i) of Lemma 2.

Proof of Claim 3.6: Use Claims 3.4 and 3.5. □

Proof of Lemma 3: The idea is to find a subset M_1 of M, and a matching M_2 that is a subset of $H' \setminus M$, such that $F_2 = (M \setminus M_1) \cup ((H' \setminus M) \setminus M_2)$ has maximum degree 2, and $F_3 = (H' \backslash M) \cup M_1$ is a vertex-disjoint collection of windmills. Then let

$$F_1 = M \cup M_2;\ F_2 = (M \backslash M_1) \cup ((H' \backslash M) \backslash M_2);\ F_3 = (H' \backslash M) \cup M_1.$$

Each edge of H' appears in exactly two of the F_i's, and hence this will imply Lemma 3.

We now describe how to find M_1 and M_2. Let us add edges to $H' \setminus M$ to obtain a graph F_3, consisting of vertex-disjoint windmills, such that

(i) the units of the windmills in F_3 are the connected components of $H' \setminus M$, and

(ii) each odd cycle C of $H' \setminus M$ such that each $v \in C$ has degree 3, is a unit of a windmill in F_3 with at least two units.

To find such an F_3, contract each connected component C of $H' \backslash M$ to a single vertex v_C; call the resulting graph G, and take any subgraph of G with the fewest possible edges such that each such v_C has positive degree if v_C has positive degree in G; the resulting graph corresponds to such an F_3. Let the set of edges that are the spokes of the windmills in F_3 be M_1; note that $M_1 \subseteq M$, and that $F_3 = (H' \setminus M) \cup M_1$. Note also that each odd cycle of $H' \setminus M$ contains a vertex v such that v has degree exactly 2 in $H' \setminus M_1$, and degree exactly zero in $M \setminus M_1$.

Let us now specify $M_2 \subseteq H' \backslash M$. Let C be a connected component of $H' \backslash M$. If C is a path or an even cycle, let $M_2 \cap C$ be any matching such that $C \setminus M_2$ is also a matching. If C is an odd cycle, let v be a vertex in C that has degree exactly 2 in $H' \setminus M_1$, and let M_2 be a matching of C such that the only vertex of C that has degree 2 in $C \setminus M_2$ is v.

One can check that the maximum degree of each vertex in F_1 and in F_2 is 2; Indeed, F_1 is the union of two matchings. Each vertex on a path or even cycle C of $H' \setminus M$ has degree at most 1 in $M \setminus M_1$, and degree at most 1 in $(H' \setminus M) \setminus M_2$, while each vertex on an odd cycle C of $H' \setminus M$ having degree 2 in $(H' \setminus M) \backslash M_2$ has degree 0 in $M \backslash M_1$. Because each vertex of H' has degree at most 2 in $H' \setminus M$, each vertex of H' has degree at most 2 in F_2. Finally, as we have already established that F_3 is a collection of vertex-disjoint windmills, Lemma 3 follows. □

References

1. N. Alon and V. Asodi, Sparse universal graphs, *Journal of Computational and Applied Mathematics*, to appear.
2. N. Alon, M. Capalbo, Y. Kohayakawa, V. Rödl, A. Ruciński, and E. Szemerédi, Universality and tolerance, *Proceedings of the 41st IEEE Annual Symposium on FOCS*, pp. 14–21, 2000.
3. L. Babai, F. R. K. Chung, P. Erdős, R. L. Graham, J. Spencer, On graphs which contain all sparse graphs, *Ann. Discrete Math.*, 12 (1982), pp. 21–26.
4. S. N. Bhatt, F. Chung, F. T. Leighton and A. Rosenberg, Universal graphs for bounded-degree trees and planar graphs, *SIAM J. Disc. Math.* 2 (1989), 145–155.

5. S. N. Bhatt and C. E. Leiserson, How to assemble tree machines, *Advances in Computing Research*, F. Preparata, ed., 1984.
6. M. Capalbo, A small universal graph for bounded-degree planar graphs, *SODA* (1999), 150–154.
7. F. R. K. Chung and R. L. Graham, On graphs which contain all small trees, *J. Combin. Theory Ser. B*, 24 (1978) pp. 14–23.
8. F. R. K. Chung and R. L. Graham, On universal graphs, *Ann. New York Acad. Sci.*, 319 (1979) pp. 136–140.
9. F. R. K. Chung and R. L. Graham, On universal graphs for spanning trees, *Proc. London Math. Soc.*, 27 (1983) pp. 203–211.
10. F. R. K. Chung, R. L. Graham, and N. Pippenger, On graphs which contain all small trees II, *Proc. 1976 Hungarian Colloquium on Combinatorics*, 1978, pp. 213–223.
11. M. Capalbo and S. R. Kosaraju, Small universal graphs, *STOC* (1999), 741–749.
12. F. R. K. Chung, A. L. Rosenberg, and L. Snyder, Perfect storage representations for families of data structures, *SIAM J. Alg. Disc. Methods.*, 4 (1983), pp. 548–565.
13. J. Friedman and N. Pippenger, Expanding graphs contain all small trees, *Combinatorica*, 7 (1987), pp. 71–76.
14. A. Hajnal and E. Szemerédi, Proof of a conjecture of Erdős, in *Combinatorial Theory and its Applications*, Vol. II (P. Erdős, A. Rényi, and V. T. Sós, eds.), Colloq. Math Soc. J. Bolyai 4, North Holland, Amsterdam 1970, 601–623.
15. L. Lovász and M. D. Plummer, *Matching Theory*, North Holland, Amsterdam (1986).
16. V. Rödl, A note on universal graphs, *Ars Combin.*, 11 (1981), 225–229.

On a Generalized Ruin Problem

Kazuyuki Amano[1], John Tromp[2], Paul M.B. Vitányi[3], and Osamu Watanabe[4]

[1] GSIS, Tohoku University, Japan
ama@ecei.tohoku.ac.jp
[2] CWI, The Netherlands
tromp@cwi.nl
[3] CWI and University of Amsterdam, The Netherlands
paulv@cwi.nl
[4] Dept. of Math. & Comp. Sci., Tokyo Inst. of Tech., Japan
watanabe@is.titech.ac.jp

Abstract. We consider a natural generalization of the classical ruin problem to more than two parties. Our "ruin" problem, which we will call the (k, I)-game, starts with k players each having I units as its initial capital. At each round of the game, all remaining k' players pay $1/k'$th unit as *game fee*, play the game, and one of the players wins and receives the combined game fees of 1 unit. A player who cannot pay the next game fee goes *bankrupt*, and the game terminates when all players but one are bankrupt. We analyze the length of the game, that is, the number of rounds executed until the game terminates, and give upper and lower bounds for the expected game length.

1 Introduction

We define a notion of "multiparty" ruin problem, generalizing the classical ruin problem to more than two parties. In this initial study we analyze the expected length of the associated play until all but one player (the final winner that takes all) are ruined. Specifically, for a given k_0 and I_0, (k_0, I_0)-*game* starts with k_0 players each having I_0 units as its capital. The game is divided into discrete rounds, and each round is executed as follows.

(1) One of the remaining players wins and receives 1 unit. Here the winning probability of each remaining player is $1/k$, where k is the number of remaining players.
(2) All players (including the winner) lose $1/k$ unit. In this way, the total amount of capital is kept the same.
(3) A player whose capital becomes $< 1/k$ is declared *bankrupt*. Players who are not bankrupt are called *alive*. Once a player goes bankrupt, he leaves the game and his remaining capital is evenly divided between the remaining players. If there are more than one players whose capital is smaller than $1/k$ in the same round, then the bankruptcies are handled from the lowest indexed player on.

The game terminates when only one player is alive. This last remaining player is the winner. The *length* of the game is the number of rounds executed until

M. Goemans et al. (Eds.): APPROX-RANDOM 2001, LNCS 2129, pp. 181–191, 2001.

the termination of the game. The sequence of rounds, starting with the round following the round that created the sth bankruptcy up to and including the round creating the $(s+1)$th bankruptcy, is called the *sth epoch*. The game starts with the 0th epoch.

One might also consider a variation of this game where each player first pays 1 unit and the winning player receives all of them; in that case we do not need to worry about rational numbers. The ruin problem, however, changes nontrivially under this altered game rule. We will discuss the difference in the last section.

Our generalized "ruin" game is defined in [WY97] as *Monopolist Game* in order to study a neural network updating rule due to von der Malsburg [vdM73] that plays a key role in explaining the development of the orientation selectivity in the brain. It is also considered as a model for randomized agents working under a certain constraint (i.e., keeping the total amount of capital the same). In fact, a variation of this model has been shown useful for some random selection procedure [DGW98]. In [DWY98] a probabilistic analysis resulting in a plausibility argument (but no proof) for the average game length was given.

Note that the game with two players is a classical "ruin" problem [Fel57], and our general game can be viewed as a multi-dimensional random walk with absorbing barriers. As far as the authors are aware, this multiparty ruin problem has not been considered before. Special cases, for example [It73], and a related ruin problems consisting of games with three players A, B, C where A plays with C and B plays with C but A doesn't play with B have been studied before in probability theory [IM98]. The problem in this paper, while natural in the computer science setting, does not seem to have been studied before in the random walk and ruin literature, as perusal of the most recent literature failed to turn up evidence, see for example [Fel57,Asm00].

We simulated this game, and it is somewhat interesting to see that the game always terminates quite quickly. By a bit more careful experiments, one finds evidence that the average game length may be proportional to $(k_0 I_0)^2$ (Figure 1). In this paper, we derive lower and upper bounds for the average game length that are close to $(k_0 I_0)^2$.

2 Notations

In the following, the symbols k and I indicate variables. Actual values of these variables are indicated by subscripting them: for example, k_0 and I_0 denote the initial number of players and the initial amount of capital of each player, respectively, at the start of the 0th epoch. Thus, we consider (k_0, I_0)-games throughout this paper.

All possible histories of the (k_0, I_0)-game are represented by a, possibly infinite, directed tree $\mathcal{T}_0$. Every directed edge of $\mathcal{T}_0$ represents the outcome of a throw with a fair k-side coin, and hence the winner, at the corresponding round of the game. Therefore, a path from the root to a vertex v of $\mathcal{T}_0$ represents a sequence of winners of the game. Such a path is called a *game path*. We can label v with the sequence of outcomes leading up to it.

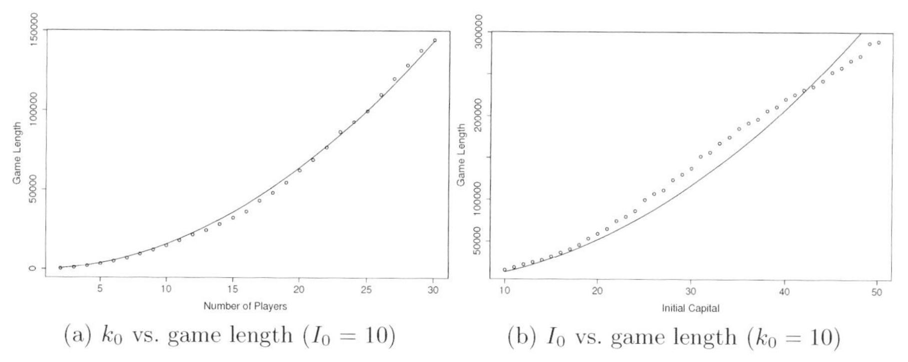

(a) k_0 vs. game length ($I_0 = 10$) (b) I_0 vs. game length ($k_0 = 10$)

These two figures show the results of our experiments on the relation between the average game length and (a) the number of players (k_0), and (b) each player's initial capital (I_0). For each parameter setting, we run the game 1000 times and take the average. Lines indicate (a) length $= 160k_0^2$, and (b) length $= 130I_0^2$.

Fig. 1. The Average Game Length

With every vertex v of $\mathcal{T}_0$, we naturally associate a *wealth vector* $(w_1, ..., w_k)$ that represents the capital of each player at (the game configuration corresponding to) v. A player becomes *bankrupt* at v when his capital decreases below $1/k$ for the first time, where k is the number of alive players just before v. It may be the case that a bankrupt player i still possesses some w units; but when computing a wealth vector, bankrupt player's capital w_i is set to 0 while his remaining wealth w is equally distributed among the remaining players. We use $w(v)$ to denote the wealth vector associated with v and $k(v)$ to denote the number of players alive at v. Note that v has $k(v)$ outgoing edges. (In some rare situation where there are more than one players whose capital is just below $1/k$, the notion of bankrupt players may change depending on the order of determining bankruptcy because the capital of bankrupt players is distributed among remaining players. According to our game rule, the bankruptcies are handled from players with smaller indices first.)

With every vertex v of $\mathcal{T}_0$, we associate the probability of reaching v as the probability that the sequence of winners specifies the game path τ from the root to v. Let $P(v)$ denote this probability. The probability that some event occurs in the game is formally specified as the sum of $P(v)$ for all v comprising the event.

In our analysis we sometimes consider a subtree of the game tree $\mathcal{T}_0$ rooted at some vertex v of $\mathcal{T}_0$, which we denote by $\mathcal{T}_v$. This subtree $\mathcal{T}_v$ represents a partial game stating from (the game configuration corresponding to) v. For a vertex u of $\mathcal{T}_v$, we use $P_v(u)$ to denote the probability of reaching to u in this partial game, the probability that the partial game path from v to u occurs in the game.

Finally, let us state a formula computing the capital of each player after t rounds. Consider a t round execution of the (partial) game (starting from some vertex v). We assume that no player went bankrupt during the execution, and let

k denote the number of participating players. For an alive player i, let s_i denote the number of times that the player i wins during the t round execution, and let w_i and w'_i be respectively the player i's capital before and after the execution. Then we have the following relation.

$$w'_i = w_i + s_i - t/k. \quad (1)$$

3 Upper Bound Analysis

We consider a (k_0, I_0)-game for sufficiently large I_0. Let $W_0 = k_0 I_0$, i.e., the total amount of capital, which is fixed throughout the game. From the evidence supplied by our experiments, we can conjecture that the expected length of the game is proportional to W_0^2. This we cannot quite prove, but we show that the expected length of the game is $O(W_0^2 \ln \ln k_0)$. We start by estimating the number of rounds that is sufficient to have someone go bankrupt from a given game configuration. More specifically, we prove the following lemma.

Lemma 1. *Let v be a vertex of the game tree $\mathcal{T}_0$, and let $k = k(v)$, the number of players alive at v. For $t \geq 1$, define $P_{\rm one}(v, t)$ be the probability that no player gets bankrupt during a t round execution of the partial game starting from v. Then for every positive integer r, we have*

$$P_{\rm one}\left(v, \frac{rc_0 W_0}{(k-1)\ln k}\right) < 2^{-r}.$$

Where c_0 is defined by $c_0 = 3c'_0$ with some integer c'_0 such that $(0.993)^{c'_0} < 1/2$. For example, we may choose $c_0 = 200$.

Proof. Let v and k be a vertex of the game tree $\mathcal{T}_0$ and $k(v)$ respectively. For a given t, consider the partial game tree $\mathcal{T}_{v,t}$ under v of depth t, which represents the t round partial game starting from v. (The tree may have some path that terminates earlier than the tth round.) Let X be the set of leaves x of $\mathcal{T}_{v,t}$ with $k(x) = k$, that is, no player gets bankrupt from v to x. Then $P_{\rm one}(v, t)$ is defined formally as $P_v(X)$ $(= \sum_{x \in X} P_v(x))$. We will bound this $P_{\rm one}(v, t)$ for $t = rc_0 W_0/((k-1)\ln k)$.

However, instead of considering the partial game tree $\mathcal{T}_{v,t_0}$, we first analyze a similar but much simpler tree obtained by considering a modified game. In the modified game, which we call a *pseudo game*, no one gets bankrupt; that is, even a player with a negative capital still participates in the game. Let $\widetilde{\mathcal{T}}$ denote the game tree of depth t for this pseudo game starting from (the game configuration corresponding to) v. Notice that this tree is k-regular, i.e., every vertex of the tree has k outgoing edges and that it has k^t leaves. For a vertex x in the tree, we define $\widetilde{P}_v(x)$ to be the probability of reaching to x in the pseudo game from v. ($\widetilde{P}_v(x)$ is simply $k^{-t'}$, where t' is the number of rounds from v to x.) Let $\widetilde{X}$ be the set of leaves x of $\widetilde{\mathcal{T}}$ whose wealth vector $w(x)$ has a negative element, that is, there exists some player with a negative capital at x. We define

$\widetilde{Q}_{\text{one}}(v,t) = \widetilde{P}_v(\widetilde{X})$ $(= \|\widetilde{X}\|/k^t)$, that is, the probability that a leaf of $\widetilde{\mathcal{T}}$ has a player with a negative capital. For this $\widetilde{Q}_{\text{one}}(v,t)$, it is easy to prove the following bound.

Claim.

$$\widetilde{Q}_{\text{one}}\left(v, \frac{3W_0^2}{(k-1)\ln k}\right) > 0.007.$$

Proof. We consider the partial game tree $\widetilde{\mathcal{T}}$ of depth $t_0 = 3W_0^2/((k-1)\ln k)$. Let I be the set of players whose capital at v is at most twice the average, i.e., $I = \{i \mid w_i \le 2W_0/k\}$ where w_i denotes the capital of the player i at v. It is easy to see that $|I| \ge k/2$ since if not, there would be too many rich players.

We first show that, for a player i in I, the probability that the player i's capital becomes negative at a leaf of $\widetilde{\mathcal{T}}$ is $\Omega(1/k)$.

Since no one gets bankrupt in a pseudo game, we can use the formula (1) for computing i's capital at every vertex in $\widetilde{\mathcal{T}}$. Thus, we have

$$\begin{aligned} s_i < t_0/k - 2W_0/k \Leftrightarrow 0 &> 2W_0/k + s_i - t_0/k \\ \Rightarrow 0 &> w_i + s_i - t_0/k \\ &\Rightarrow \text{the player } i \text{ has a negative capital at a leaf of } \widetilde{\mathcal{T}}, \end{aligned}$$

where s_i is the number of times that the player i wins during a t_0 round execution of the game.

Since the probability that each player wins in the game is always $1/k$, s_i follows the binomial distribution with mean $\mu = t_0/k$ and variance $\sigma^2 = t_0(k-1)/k^2$ $(= 3W_0^2/(k^2 \ln k))$. Thus, the probability that $s_i < t_0/k - 2W_0/k$ is estimated as follows.

$$\begin{aligned} \Pr\left\{ s_i < \frac{t_0}{k} - \frac{2W_0}{k} \right\} &\ge \Pr\left\{ s_i < \frac{t_0}{k} - \sqrt{\frac{3W_0^2}{k^2 \ln k} \cdot (3/2)\ln k} \right\} \\ &= \Pr\left\{ s_i < \mu - \sigma\sqrt{(3/2)\ln k} \right\} \\ &= \frac{1}{\sqrt{2\pi}} \int_{-\infty}^{-\sqrt{(3/2)\ln k}} e^{-z^2/2} dz \\ &\ge \frac{1}{\sqrt{2\pi}} e^{-(3/4)\ln k} \cdot \frac{1}{\sqrt{(3/2)\ln k}} \left(1 - \frac{1}{(3/2)\ln k}\right) \\ &\ge \frac{0.015}{k}. \end{aligned}$$

Here we used the approximation by the normal distribution, which holds if $t_0 = 3W_0^2/((k-1)\ln k)$ is sufficiently large.

Note that, the probability that a player has a positive capital at the leaf level decreases under the condition that some other players have a positive capital. Hence, for probability that given players have a positive capital at a leaf of $\widetilde{\mathcal{T}}$, we have, for example, that $\Pr\{\text{both } i_1 \text{ and } i_2 \text{ are positive}\} \le \Pr\{i_1 \text{ is positive}\}$

$\times$ Pr$\{i_2$ is positive$\}$. By using this, the probability that a leaf of $\widetilde{\mathcal{T}}$ has a player with a negative capital is bounded as follows.

$$\begin{aligned}\widetilde{Q}_{\text{one}}(v,t_0) &\geq \Pr\left\{\exists i \in I\,[\text{the player } i \text{ has a } \textit{negative} \text{ capital at a leaf of } \widetilde{\mathcal{T}}]\right\}\\ &= 1 - \Pr\left\{\forall i \in I\,[\text{the player } i \text{ has a } \textit{positive} \text{ capital at a leaf of } \widetilde{\mathcal{T}}]\right\}\\ &\geq 1 - \prod_{i\in I} \Pr\left\{\text{the player } i \text{ has a } \textit{positive} \text{ capital at a leaf of } \widetilde{\mathcal{T}}\right\}\\ &\geq 1 - \left(1 - \left(\frac{0.015}{k}\right)\right)^{k/2} > 0.007.\end{aligned}$$

This completes the proof of the claim. □

We continue the proof of Lemma 1. Let v and k be a vertex of the game tree $\mathcal{T}_0$ and $k(v)$ respectively. First for $t_0 = 3W_0^2/((k-1)\ln k)$, we consider the partial game tree $\mathcal{T}_{v,t_0}$ under v of depth t_0, and let X' be the set of vertices x in $\mathcal{T}_{v,t_0}$ such that someone gets bankrupt at x for the first time in the partial game from v. We argue that $P_v(X') > 0.007$.

Consider again the pseudo game tree $\widetilde{\mathcal{T}}$ from v of depth t_0. As before, let $\widetilde{X}$ be the set of leaves x of $\widetilde{\mathcal{T}}$ that has some player with a negative capital. For every $x \in \widetilde{X}$, we can find a vertex x' on a path from v to x such that some player's capital gets smaller than $1/k$ for the first time from v on the path. Let $\widetilde{X}'$ be the set of such vertices. Notice that $\widetilde{X}'$ covers all paths from v to some leaf $x \in \widetilde{X}$; that is, for every path from v to some leaf $x \in \widetilde{X}$, there exists some x' in $\widetilde{X}'$ on the path. Thus, $\widetilde{P}_v(\widetilde{X}') \geq \widetilde{P}_v(\widetilde{X})$. Note also that for every $x \in \widetilde{X}'$, no one gets bankrupt in the original game from v to the vertex before x. This means that there is no difference between the original and pseudo game on the game path from v to x. Hence, we have $X' = \widetilde{X}'$, and furthermore, $P_v(X') = \widetilde{P}_v(\widetilde{X}')$. Therefore, we have $P_v(X') = \widetilde{P}_v(\widetilde{X}') \geq \widetilde{P}_v(\widetilde{X}) = \widetilde{Q}_{\text{one}}(v,t_0) > 0.007$, by Claim 3.

Consider the partial game tree $\mathcal{T}_{v,t_0}$ again. Let X_1 be the set of leaves x of $\mathcal{T}_{v,t_0}$ such that no one gets bankrupt from v to x. Clearly, a leaf x of $\mathcal{T}_{v,t_0}$ is in X_1 if and only if it has no ancestor vertex in X'. Thus, $P_v(X_1) = 1 - P_v(X') < 0.993$. Now for a vertex x in X_1, we consider the partial game tree $\mathcal{T}_{x,t_0}$ of depth t_0 starting from x. Then by the same argument, we have $P_x(X_{1,x}) < 0.993$, where $X_{1,x}$ is the set of leaves y of $\mathcal{T}_{x,t_0}$ such that no one gets bankrupt from x to y. Thus, we have $P_v(X_2) < (0.993)^2$, where X_2 is defined as $X_2 = \cup_{x\in X_1} X_{1,x}$. Note that X_2 is the set of leaves x in the partial game tree $\mathcal{T}_{v,2t_0}$ such that no one gets bankrupt from v to x. Similarly, for $j \geq 1$, we have $P_v(X_j) < (0.993)^j$, where X_j is the set of vertices x in the partial game tree $\mathcal{T}_{v,jt_0}$ such that no one gets bankrupt from v to x. Here recall that $rc_0't_0 = rc_0W_0^2/((k-1)\ln k)$; thus, $P_v(X_{rc_0'})$ is $P_{\text{one}}(v, rc_0W_0^2/((k-1)\ln k))$, the probability that we want to bound. On the other hand, we have $P_v(X_{rc_0'}) < (0.993)^{rc_0'}$ and $(0.993)^{c_0'} < 1/2$ from our choice of c_0', we have the bound of the lemma. □

From this lemma, we can easily derive our upper bound for the expected total number of rounds of a (k_0, I_0)-game. For every k, $2 \leq k \leq k_0$, let T_k be the random variable denoting the number of rounds executed in the $(k_0 - k)$th epoch of the game, i.e., the period from the round when the number of players becomes k to the round when the number of players becomes $k - 1$. Then the expected total rounds of (k_0, I_0)-game is computed as $\mathrm{Exp}(T_{k_0} + T_{k_0-1} + \cdots + T_2)$. From the above lemma, we have the following upper bound.

Theorem 2. *For c_0 as above and k, $2 \le k \le k_0$, we have $\mathrm{Exp}(T_k) \le 2c_0W_0^2/((k-1)\ln k)$. Therefore, we have*

$$\begin{aligned}
&\textit{The average length of } (k_0, I_0)\textit{-game} \\
&\quad = \mathrm{Exp}(T_{k_0} + T_{k_0-1} + \cdots + T_2) \le 2c_0W_0^2 \left(\sum_{k=2}^{k_0} \frac{1}{(k-1)\ln k} \right) \\
&\quad \le 2c_0W_0^2 \ln\ln k_0.
\end{aligned}$$

Proof. Consider k, $2 \le k \le k_0$. In Lemma 1, we showed that $P_{\mathrm{one}}(v, rc_0W_0^2/((k-1)\ln k)) < 2^{-r}$ for every integer $r \ge 1$, which means that the probability that $T_k > rc_0W_0^2/((k-1)\ln k)$ is less than 2^{-r} for every integer $r \ge 1$. Thus,

$$\mathrm{Exp}(T_k) \le \frac{c_0W_0^2}{(k-1)\ln k} \cdot \left(1 \cdot 2^{-1} + 2 \cdot 2^{-2} + 3 \cdot 2^{-3} + \cdots\right) = \frac{2c_0W_0^2}{(k-1)\ln k}.$$

□

In order to see how the total rounds of (k_0, I_0)-game is distributed around its average, let us consider a set of random variables $\widehat{T}_k$, $2 \le k \le k_0$, that are mutually independent and each $\widehat{T}_k$ takes the value $rc_0W_0^2/((k-1)\ln k)$ with the probability 2^{-r} for each integer $r \ge 1$. Let $\widehat{T} = \sum_{k=2}^{k_0} \widehat{T}_k$. Intuitively, we may consider that $\widehat{T}$ bounds the total rounds of the game. In fact, the average of $\widehat{T}$ is bounded in the same way as Theorem 2; that is, $\mathrm{Exp}(\widehat{T}) \le 2c_0W_0^2 \ln\ln k_0$. For this $\widehat{T}$, we can easily compute its variance in the following way.

$$\begin{aligned}
\mathrm{Var}(\widehat{T}) &= \mathrm{Var}\left(\sum_{k=2}^{k_0} \widehat{T}_k \right) = \sum_{k=2}^{k_0} \left(\mathrm{Exp}(\widehat{T}_k^2) - \mathrm{Exp}(\widehat{T}_k)^2 \right) \\
&= \sum_{k=2}^{k_0} \left(6 \cdot \left(\frac{c_0W_0^2}{(k-1)\ln k} \right)^2 - 4 \cdot \left(\frac{c_0W_0^2}{(k-1)\ln k} \right)^2 \right) \\
&\le 2(c_0W_0^2)^2 \cdot \sum_{k=1}^{\infty} \frac{1}{k^2} \le d_0(c_0W_0^2)^2.
\end{aligned}$$

Thus, the probability that $\widehat{T}$ is much larger than its estimated upper bound, say, larger than $3c_0W_0^2 \ln\ln k_0$ (for sufficiently large W_0), is very small. We believe the situation is the same for the actual total rounds of (k_0, I_0)-game. Unfortunately, though, what we have been able to prove formally is the following probability bound.

Theorem 3. *For every integer $r \geq 1$, the probability that the total rounds of (k_0, I_0)-game exceeds $rc_0W_0^2 \ln k_0 \ln\ln k_0$ is less than 2^{-r}.*

Proof. If the total rounds of the game exceeds $rc_0W_0^2 \ln k_0 \ln\ln k_0$, then there exists some k, $2 \leq k \leq k_0$, for which we have $T_k > (r \ln k_0)c_0W_0^2/((k-1)\ln k)$. On the other hand, by Lemma 1, the probability that $T_k > (r \ln k_0)c_0W_0^2/((k-1)\ln k)$ holds is less than $2^{-r}/k_0$. Thus, the probability that such an event occurs for some k, $2 \leq k \leq k_0$, is less than 2^{-r}. □

4 Lower Bound Analysis

Here again we consider (k_0, I_0)-game for sufficiently large I_0, and let $W_0 = k_0I_0$, i.e., the total amount of capital. We will show that the expected game length is $\Omega(W_0^2/\log W_0)$. More specifically, we will show that, for every integer $r \geq 1$, the probability that the game terminates within $c_1W_0^2/(2r \log k_0 + 2\log W_0)$ rounds is smaller than 2^{-r}.

We first consider the game with two players.

Lemma 4. *Consider $(2, I)$-game, the game with two players each having I units as its initial capital. Then for every $r \geq 1$, we have*

$$\Pr\left\{ \textit{the length of the game} \leq \frac{2I^2}{3\ln 2} \cdot \frac{1}{r + 2\log I + 2} \right\} \leq 2^{-r}.$$

Proof. For $t \geq 1$, suppose that the game terminates at the tth round. This implies that $|s_1 - t/2| \geq I$; i.e., we have either $s_1 - t/2 \leq -I$ or $s_1 - t/2 \geq I$, where s_1 is the number of rounds where the player 1 wins. On the other hand, note that $t/2$ is the expectation of s_1. Hence, the Chernoff bound gives us the following bound.

$$\Pr\{\, |s_1 - t/2| \geq I \,\} \leq 2e^{-\frac{2I^2}{3t}}.$$

Now we analyze the probability of the event E_1 that the game terminates within $t_1 = 2I^2/\{(3\ln 2)(r + 2\log I + 2)\}$ steps. Note that the event *implies* that $|s_1 - t/2| \geq I$ holds for *some* t, $1 \leq t \leq t_1$. Thus, $\Pr\{E_1\}$ is bounded as follows, which proves the lemma.

$$\begin{aligned}\Pr\{E_1\} &\leq \sum_{t=1}^{t_1} \Pr\{\, |s_1 - t/2| \geq I \,\} \leq \sum_{t=1}^{t_1} 2e^{-\frac{2I^2}{3t}} \\ &\leq t_1 \cdot 2e^{-\frac{2I^2}{3t_1}} = t_1 \cdot \frac{2^{-r}}{2I^2} \leq 2^{-r}.\end{aligned}$$

□

By using this lemma, we can show the following probabilistic lower bound for the length of (k_0, I_0)-game.

Theorem 5. *For $r \geq 1$, the probability that a (k_0, I_0)-game terminates within $c_1W_0^2/(2r\log k_0 + 2\log W_0)$ rounds is less than 2^{-r}, where $c_1 = 1/(6\ln 2)$.*

Proof. For $r \geq 1$, let $t'_1 = c_1 W_0^2/(2r \log k_0 + 2 \log W_0)$. Consider the game tree $\mathcal{T}_0$ and the set V of its leaves v at depth $\leq t'_1$; that is, each of such leaves correspond to some execution of the game that terminates within t'_1 rounds. We would like to show that $P(V) \leq 2^{-r}$. For this, we fix some i_0 and i_1, and focus on the set U of leaves $v \in V$ such that the players i_0 and i_1 are the last two players. Since there are $\binom{k_0}{2} < k_0^2$ choices of pairs, we have $P(V) < k_0^2 P(U)$. Thus, to get the desired bound, it is enough to show that $P(U) \leq 1/(k_0^2 2^r)$.

Consider some leaf v in U, and let τ be a *winner sequence*, the sequence of winners in the game path from the root to v. By removing winners other than i_0 and i_1 from τ, we obtain a winner sequence τ', which is a winner sequence for the game with two players. In fact, it is easy to show that τ' defines a game path for $(2, (k_0 I_0)/2)$-game that terminates at v. Note that the same τ' is obtained from some other $v' \in U$. Let $U' \subseteq U$ be the set of such leaves. It is also easy to see that the probability of the winner sequence τ' occurs in the two player game (which is $2^{-|\tau'|}$) is the same as the probability of U' in $\mathcal{T}_0$. Hence we have

$$\Pr(U) \leq \Pr\{\ (2, (k_0 I_0)/2)\text{-game terminates within } t'_1 \text{ rounds }\}.$$

Here note that

$$t'_1 = \frac{c_1 W_0^2}{2r \log k_0 + 2 \log W_0} = \frac{(k_0 I_0)^2}{(6 \ln 2)(2r \log k_0 + 2 \log(k_0 I_0/2) + 2)}$$
$$= \frac{2(k_0 I_0/2)^2}{3 \ln 2} \cdot \frac{1}{2r \log k_0 + 2 \log(k_0 I_0/2) + 2}.$$

Thus, by using Lemma 4, the probability that a $(2, (k_0 I_0)/2)$-game terminates within t'_1 rounds is at most $2^{-2r \log k_0} = 2^{-r}/k_0^2$, which gives us the desired bound for $P(U)$. □

5 Concluding Remarks

The "ruin" game studied here is defined based on the neural network training rule proposed in [vdM73]. But one can consider similar constrained games. For example, we may consider the following variation of the game: at each round, (1) every player (of the remaining k players) must pay 1 unit as a game fee at the beginning of the round, and (2) the winner of the round takes all these k units. This rule seems more natural (as a game) since we do not need to worry about noninteger units.

It seems like the game does not change so much by this change. In fact, we have similar experimental results on the length of the game under this new rule (Figure 2). When analyzing the game length more carefully, however, we would soon notice the difference. In the original game, each player's capital changes relatively smoothly when the number of players is large, whereas under the new game rule, it changes drastically when the number of players is large. This difference leads several differences in the behavior of two types of game. For example, as shown in Figure 3, the length of the 0th epoch differs between

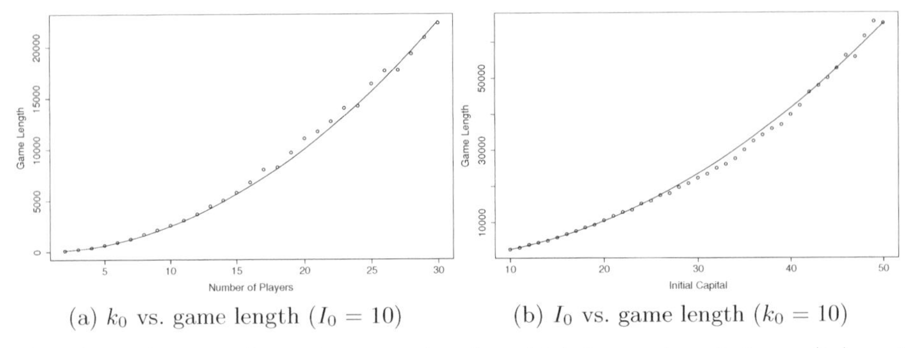

(a) k_0 vs. game length ($I_0 = 10$) (b) I_0 vs. game length ($k_0 = 10$)

The relation between the average game length and (a) the number of players (k_0), and (b) each player's initial capital (I_0). For each parameter setting, we run the game 1000 times and take the average. Lines indicate (a) length $= 25k_0^2$, and (b) length $= 26I_0^2$.

Fig. 2. The Average Game Length of the New Monopolist Game

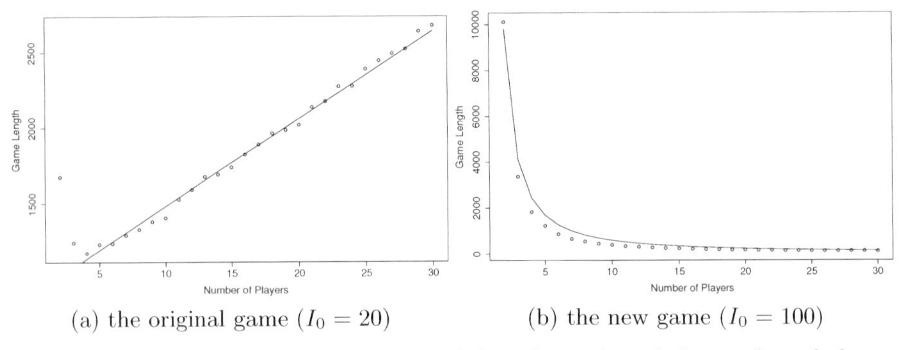

(a) the original game ($I_0 = 20$) (b) the new game ($I_0 = 100$)

The relation between the average length of the 0th epoch and the number of players k_0 in (a) the original game, and (b) in the new game. For each parameter setting, we run the game 1000 times and take the average. Lines indicate (a) length $= 58k_0 + 900$, and (b) length $= 13600/(k_0 \ln k_0)$.

Fig. 3. The Average Length of the 0th Stage

two games. What is more important for our analysis is the fact that the length of the last epoch dominates the total game length under the new game rule. Thus, the analysis for the two player case, which is for the simple ruin problem, is sufficient for the new game rule. On the other hand, what makes our analysis a bit more difficult is that we had to consider all epochs in the analysis of the original game.

Acknowledgments

K. Amano and O. Watanabe were partially supported by a Grant-in-Aid for Scientific Research on Priority Areas "Discovery Science" from the Ministry of

Education, Science, Sports and Culture of Japan. The work of J. Tromp and P.M.B. Vitányi was supported in part by the EU fifth framework project QAIP, IST–1999–11234, the NoE QUIPROCONE IST–1999–29064, the ESF QiT Programme and the EU Fourth Framework BRA Neuro COLT II Working Group EP 27150.

References

Asm00. S. Asmussen, *Ruin Probabilities*, World Scientific, 2000.

DGW98. Carlos Domingo, Ricard Gavaldà, and Osamu Watanabe, Practical algorithms for on-line sampling, in *Proc. of the First International Conference on Discovery Science*, Lecture Notes in Artificial Intelligence 1532, 150–162 (1998).

DWY98. C. Domingo, O. Watanabe, T. Yamazaki, A role of constraint in self-organization, in *Proc. Workshop on Randomization and Approximation Techniques in Computer Science (RANDOM98)*, Lecture Notes in Computer Science 1518, Springer-Verlag, 307–318 (1998).

Fel57. W. Feller, *An Introduction to Probability Theory and Its Applications*, Wiley, New york, 1957.

It73. Y. Itoh, On a ruin problem with interaction, *Ann. Inst. Statist. Math.* 25, 635–641 (1973).

IM98. Y. Itoh and H. Maehara, A variation to the ruin problem, *Math. Japonica* 47 (1), 97–102 (1998).

vdM73. C. von der Malsburg, Self-organization of orientation sensitive cells in the striate cortex, *Kybernetik* 14, 85–100 (1973).

WY97. O. Watanabe and T. Yamazaki, Orientation selectivity: An approach from theoretical computer science, TR97-0008, Comp. Sci. Dept., Tokyo Institute of Technology, Tokyo, November 1997, http://www.cs.titech.ac.jp/TR/tr97.html.

On the b-Partite Random Asymmetric Traveling Salesman Problem and Its Assignment Relaxation

Andreas Baltz*, Tomasz Schoen**, and Anand Srivastav

Mathematisches Seminar, Christian-Albrechts-Universität zu Kiel,
Ludewig-Meyn-Str. 4, D-24098 Kiel, Germany
{aba,tos,asr}@numerik.uni-kiel.de

Abstract. We study the relationship between the value of optimal solutions to the random asymmetric b-partite traveling salesman problem and its assignment relaxation. In particular we prove that given a $bn \times bn$ weight matrix $W = (w_{ij})$ such that each finite entry has probability p_n of being zero, the optimal values $bATSP(W)$ and $AP(W)$ are equal (almost surely), whenever np_n tends to infinity with n. On the other hand, if np_n tends to some constant c then $\mathbb{P}[bATSP(W) \neq AP(W)] > \epsilon > 0$, and for $np_n \to 0$, $\mathbb{P}[bATSP(W) \neq AP(W)] \to 1$ (a.s.). This generalizes results of Frieze, Karp and Reed (1995) for the ordinary asymmetric TSP.

Keywords: Asymmetric b-partite TSP, Assignment Problem

1 Introduction

The b-partite TSP is a variant of the traveling salesman problem with natural applications in robotics [2],[6]. For a given integer $b \geq 2$ we consider a graph $G = (V, E)$ whose vertices are partitioned into b distinct sets $V_1, \ldots, V_b$ of equal size. Each edge $e \in E$ is assigned a non-negative weight w_e which is finite if and only if e connects a vertex $v \in V_i$ to a vertex $v' \in V_{i+1}$, $i \in \{1, \ldots, b\}$ $(V_{b+1} := V_1)$. The aim is to find a Hamiltonian cycle C such that $\sum_{e \in E(C)} w_e$ is minimum. A straight-forward reduction proves that the b-partite TSP is no easier than the TSP. In particular, the *asymmetric* b-partite TSP, where G is interchanged with a b-partite *di*graph, does not allow for any constant factor approximation in polynomial time, unless $P = NP$. However, this hardness result applies in the worst case only. There is a very simple polynomially solvable relaxation called the *assignment problem* which can be exploited to obtain near optimal tours on average. Balas and Toth [5] carried out 400 computational experiments where they solved both the TSP and its assignment relaxation for problems with $50 \leq n \leq 250$ vertices and edge weights drawn independently from

* Supported by the graduate school "Effiziente Algorithmen und Mehrskalenmethoden", Deutsche Forschungsgemeinschaft.

** Also member of Adam Mickiewicz University, Poznań, Poland.

M. Goemans et al. (Eds.): APPROX-RANDOM 2001, LNCS 2129, pp. 192–201, 2001.

a uniform distribution of the integers over the intervals $[1, 100]$ and $[1, 1000]$. As the outcome of these experiments the average weight of the assignment was found to be 99.2% of the average optimal TSP tour length. Miller and Pekny [7] managed to obtain optimal solutions to random TSP instances on up to 500,000 vertices by scanning the optimal solutions of the assignment problem for a cyclic permutation. Frieze, Karp and Reed [3] confirmed these observations in a probabilistic analysis. In this paper we turn to the analysis of b-partite random instances. Denote by AP(W) and bATSP(W) the optimal values of solutions to the assignment problem and the b-partite TSP, respectively, specified by a $bn \times bn$ weight matrix W and let $I := \bigcup_{i=1}^{n} V_i \times V_{i+1} \cup \bigcup_{i=1}^{n} V_{i+1} \times V_i$ be the set of indices for which the entries of W are finite. We will prove the following results.

Theorem 1. *Let (Z_n) be a sequence of random variables over the nonnegative reals and let $\alpha \geq 0$. Let $W = (w_{ij})$ be a $bn \times bn$ matrix such that $w_{ij} = \infty$ for $(i, j) \notin I$ and w_{ij} is drawn independently from the same distribution as Z_n for $(i, j) \in I$. If $p_n := \mathbb{P}[Z_n \leq \frac{\alpha AP(W)}{bn}]$ satisfies $n \cdot p_n \to \infty$ as $n \to \infty$ then $bATSP(W) \leq (1 + \alpha)AP(W)$ almost surely.*

Corollary 1. *Let (Z_n) be a sequence of random variables over the nonnegative reals and let $p_n := \mathbb{P}[Z_n = 0]$. Let $W = (w_{ij})$ be a $bn \times bn$ matrix such that $w_{ij} = \infty$ for $(i, j) \notin I$ and w_{ij} is drawn independently from the same distribution as Z_n for $(i, j) \in I$. If $n \cdot p_n \to \infty$ as $n \to \infty$ then $bATSP(W) = AP(W)$ (a.s.).*

Theorem 2. *Let $W = (w_{ij})$ be a $bn \times bn$-matrix where, for $(i, j) \in I$, w_{ij} is independently and uniformly drawn from $\{0, 1, \ldots, \lfloor cn \rfloor\}$ and $w_{ij} = \infty$ for $(i, j) \notin I$.*

(a) $\mathbb{P}[AP(W) \neq bATSP(W)] \not\to 0$ as $n \to \infty$, if $c \in \mathbb{R}_{>0}$.
(b) $\mathbb{P}[AP(W) \neq bATSP(W)] \longrightarrow 1$ as $n \to \infty$, if $c \xrightarrow{n \to \infty} \infty$.

Theorem 1 shows the approximation behavior of the assignment relaxation. Corollary 1 and Theorem 2 are generalizations of two theorems of Frieze, Karp and Reed [3] for the ordinary asymmetric TSP. Our proof is based on two observations:

- the number of light edges is large, since $np_n \to \infty$.
- the weights are independently and identically distributed, hence an optimal assignment may be viewed as consisting of b random bijections (in the case of the ordinary TSP *one* random permutation suffices) which induce only $O(\log n)$ cycles (a.s.).

We will show how to construct a (near) optimal assignment with a small number of cycles and how to combine these into a tour via light edges. While Frieze, Karp and Reed may rely on known facts about random permutations, we have to establish similar results for unions of b random bijections.

2 Preliminaries

As a tool from probability theory the following two large deviation inequalities of Angluin and Valiant [1] will be useful. Note that, for small expectations, Theorem 3 is stronger than the standard Chernoff-bound.

Theorem 3 (Angluin-Valiant). *Let $X_1, \ldots, X_n$ be independent random variables with $0 \le X_i \le 1$ for all i and set $X := \sum_{i=1}^{n} X_i$. Then for every $0 < \beta \le 1$:*

$$(i)\ \mathbb{P}[X > (1+\beta)\mathbb{E}[X]] \le e^{-\frac{\beta^2 \mathbb{E}[X]}{3}}, \quad (ii)\ \mathbb{P}[X < (1-\beta)\mathbb{E}[X]] \le e^{-\frac{\beta^2 \mathbb{E}[X]}{2}}.$$

Moreover we will need the following theorem of Walkup [8].

Theorem 4 (Walkup). *Let $P(n,d)$ be the probability of a matching in a graph selected uniformly from the class $G(n,d)$ of directed bipartite graphs with n nodes in each class and outward degree d at each node. Then $P(n,1) \le 3n^{1/2}(2/e)^n$, $1 - P(n,2) \le 1/5n$, $1 - P(n,d) \le \frac{1}{122}(d/n)^{(d+1)(d-2)}$. In particular $P(n,1) \to 0$ but $P(n,2) \to 1$ as $n \to \infty$.*

For a directed graph $G = (V,E)$, $V = \{1, \ldots, t\}$, $E \subseteq V \times V$, with weight function $w : E \to \mathbb{R}_{\ge 0}$ we define an associated bipartite graph $G' = (V', E')$ as $V' = \{x_1, \ldots, x_t, y_1, \ldots, y_t\}$, $\{x_i, y_j\} \in E' \Leftrightarrow (i,j) \in E$ with edge weights $w'(\{x_i, y_j\}) := w(i,j)$. Since any Hamiltonian cycle in G corresponds to a perfect matching in G', whereas each perfect matching in G' corresponds to a cycle cover of G, we see that the *assignment problem (AP)* of finding a minimum weight perfect matching in G' is a relaxation of the (b-partite) asymmetric TSP. An instance on n vertices of the above problems for the ordinary case can be specified by an $n \times n$ matrix $W = (w_{ij})$, where w_{ij} represents the weight of the edges $(i,j) \in E$ and $\{x_i, y_j\} \in E'$, respectively. In the b-partite case we view the digraph G as joining sets $V_i := \{(i-1) \cdot n + 1, \ldots, i \cdot n\}$, $i \in \{1, \ldots, b\}$, where V_i is connected to V_{i-1} and V_{i+1} ($V_0 := V_b$, $V_{b+1} := V_1$). We forbid other than these edges by defining the entries w_{ij} of our $bn \times bn$ weight matrix W as being ∞ for $(i,j) \in V^2 \setminus (V_i \times V_{i+1} \cup V_{i+1} \times V_i)$. Recall that AP($W$) and bATSP(W) denote the optimal values of solutions to instances specified by W and $I := \bigcup_{i=1}^{n} V_i \times V_{i+1} \cup \bigcup_{i=1}^{n} V_{i+1} \times V_i$ is the set of indices for which the entries of W are finite.

3 Proofs

We will give a detailed proof of Theorem 1 and a brief proof sketch of Theorem 2. Complete proofs will appear in the full version of this paper.

Proof of Theorem 1. Let G denote the complete digraph on V with edge-weights given by W. We will use the abbreviation $\beta := \frac{\alpha \mathrm{AP}(W)}{bn}$ and distinguish between "light" and "heavy" edges depending on whether or not their weights exceed β. Consider for each $(i,j) \in I$ an indicator variable

$$z_{ij} := \begin{cases} 1, & w_{ij} \le \beta \\ 0 & \text{otherwise.} \end{cases}$$

The z_{ij} are independent, and each z_{ij} is equal to 1 with probability $p := p_n$. The light edges indicated by the z_{ij} are crucial for our construction. Note that once we observed the weight of an edge it may no longer be considered random. To make our construction work we thus have to be careful not to observe too many light edges too early. Therefore we decompose the graph induced by the z_{ij} into separate random subgraphs we can consider independently of each other. For the sake of this, let $h \in \mathbb{R}_{>0}$ be such that $(1-h)^5 = 1-p$. For all $k \in \{1,\ldots,5\}$, $(i,j) \in I$ let z_{ij}^k be independent indicator random variables satisfying $\mathbb{P}[z_{ij}^k = 1] = h$ and define $\hat{z}_{ij} := \max\left\{z_{ij}^k \mid k \in \{1,\ldots,5\}\right\}$. Then the $\hat{z}_{ij}$ are independent and $\mathbb{P}[\hat{z}_{ij} = 1] = 1 - \mathbb{P}[\hat{z}_{ij} = 0] = 1 - \prod_{k=1}^{5} \mathbb{P}[z_{ij}^k = 0] = 1 - (1-h)^5 = p$. So we can identify z_{ij} with $\hat{z}_{ij}$ and view the z_{ij} as being generated by the above construction. For $k \in \{1,\ldots,5\}$ let G_k be a digraph on $\{1,\ldots,bn\}$ with $(i,j) \in E(G_k) \Leftrightarrow (i,j) \in I$ and $z_{ij}^k = 1$. We call the edges of G_3, G_4, and G_5 out-, in-, and patch-edges respectively. Let $D := (X \cup Y, F)$, $X = \{x_1,\ldots,x_{bn}\}$, $Y = \{y_1,\ldots,y_{bn}\}$ be a bipartite digraph where

$$F := \{(x_i,y_j) \mid \{x_i,y_j\} \in E(G_1')\} \cup \{(y_i,x_j) \mid \{y_i,x_j\} \in E(G_2')\}.$$

The random graphs D, G_3, G_4, G_5 are completely independent. For $v \in X \cup Y$ we denote by $N^+(v) := \{w \in X \cup Y \mid (v,w) \in F\}$ the set of "out-neighbors" of v, and we call $\deg^+(v) := |N^+(v)|$ the "out-degree" of v. Then $\delta := nh$ is the expected out-degree of each vertex in G_i, $i \in \{1,\ldots,5\}$.

Our first aim is to cover those vertices by a minimum weight perfect matching that are likely to be incident with an edge of nonzero weight in an optimal assignment for G'. Clearly, the vertices of small out-degree in D are candidates for this "troublesome" set $T' \subseteq X \cup Y$. Therefore we put $T'_{-1} := \{v \in X \cup Y \mid \deg^+(v) \le \delta/2\}$, determine a minimum-weight matching M_0 in G' that covers T'_{-1}, and set $T'_0 := \bigcup M_0$. We also consider vertices as troublesome if many of their neighboring edges have been observed. Hence we proceed to construct T' inductively as follows: for $i \ge 1$ we choose $v \in X \cup Y \setminus T'_{i-1}$ such that $|N^+(v) \cap T'_{i-1}| \ge \delta/4$ and we let M_i be a minimum-weight matching (in G') covering $T'_{i-1} \cup \{v\}$. Since T'_{i-1} is the vertex set of a matching there is a vertex $w \in X \cup Y \setminus (T'_{i-1} \cup \{v\})$ covered by M_i, and we can define $T'_i := T'_{i-1} \cup \{v,w\}$. Our construction terminates after r steps with a set $T' := T'_r$ and a perfect matching $M := M_r$ in $G_{|_{T'}}$, the subgraph of G induced by T'.

Lemma 1. $|T'| \le 3ne^{-\delta/8}$ *(a.s.)*.

Proof. Let us assume that we have already shown the following two claims:

- **Claim 1:** $|T'_{-1}| \le ne^{-\delta/8}$
- **Claim 2:** For $S \subseteq X \cup Y$ with $|S| \le 3ne^{-\delta/8}$ we have

$$|\,(E(G_1') \cup E(G_2')) \cap \mathcal{P}_2(S)| < 2|S| \text{ (a.s.)}.$$

Then $|T'_0| \le 2|T'_{-1}| \le 2ne^{-\delta/8}$ by Claim 1. By construction of T'_i we have for each $i \in \mathbb{N}_{\le r}$: $|T'_i| = |T'_0| + 2i$ and $|\,(E(G_1') \cup E(G_2')) \cap \mathcal{P}_2(T'_i)| \ge i\delta/4$. Assume for a contradiction that $|T'| > 3ne^{-\delta/8}$. It follows that $|T' \setminus T'_0| > ne^{-\delta/8}$, so $r \geq r_0 := \lfloor ne^{-\delta/8}/2 \rfloor$. But now, $|T'_{r_0}| \leq 2ne^{-\delta/8} + ne^{-\delta/8} = 3ne^{-\delta/8}$ and $|(E(G'_1) \cup E(G'_2)) \cap \mathcal{P}(T'_{r_0})| \geq r_0\delta/4 = \Omega(\delta ne^{-\delta/8}) > 6ne^{-\delta/8}$ for n large (since $\delta = nh$ grows with n) $\geq 2|T'_{r_0}|$, which contradicts Claim 2.

For a proof of Claim 1 let $v \in X \cup Y$. We have

$$\mathbb{P}[\deg^+(v) \leq \delta/2] = \mathbb{P}[\delta - \deg^+(v) \geq \delta/2] = \mathbb{P}[e^{\delta - \deg^+(v)} \geq e^{\delta/2}].$$

By Markov's inequality, $\mathbb{P}[e^{\delta - \deg^+(v)} \geq e^{\delta/2}] \leq e^{\delta/2} \cdot \mathbb{E}[e^{-\sum_{i=n+1}^{2} nz^1_{1i}}]$ (assuming w.l.o.g. that v corresponds to vertex 1 in G_1). Hence

$$\begin{aligned}\mathbb{P}[\deg^+(v) \leq \delta/2] &\leq e^{\delta/2} \prod_{i=n+1}^{2} n\mathbb{E}[e^{-z^1_{1i}}] = e^{\delta/2}\left((1-h) + h \cdot e^{-1}\right)^n \\ &= e^{\delta/2}\left(1 + \frac{\delta(e^{-1}-1)}{n}\right)^n \leq e^{\delta/2}e^{\delta(e^{-1}-1)} = e^{\delta(1/e-1/2)},\end{aligned}$$

since $\delta = hn$ and $\left(1+\frac{a}{n}\right)^n \leq e^a$ for all $a \in \mathbb{R}$. Hence $\mathbb{E}[|T_{-1}|] \leq 2bne^{\delta(1/e-1/2)}$ and $P[|T_{-1}| > ne^{-\delta/8}] \leq \frac{\mathbb{E}[|T_{-1}|]}{ne^{-\delta/8}} \leq 2be^{\delta(1/e-1/2+1/8)} = O(e^{-\frac{\delta}{141}}) \xrightarrow{n\to\infty} 0$.
Turning to Claim 2, let $S \subseteq X \cup Y$ with $|S| \leq 3ne^{-\delta/8}$ and let Z abbreviate $|(E(G'_1) \cup E(G'_2)) \cap \mathcal{P}_2(S)|$. Then $\mathbb{E}[Z] < \binom{|S|}{2} \cdot 2h < |S|^2h$ and

$$P[Z \geq 2|S|] < \frac{|S|^2h}{2|S|} = |S|h/2 \leq \frac{3}{2}nhe^{-\delta/8} = \frac{3}{2}\delta e^{-\delta/8} = \frac{3}{2}e^{\ln\delta - \delta/8} \xrightarrow{n\to\infty} 0.$$

□

Let $T'_X := T' \cap X$, $T'_Y := T' \cap Y$ be the subsets of troublesome vertices in the partition classes X and Y, and consider the subgraph of D induced by $(X \setminus T'_X) \cup (Y \setminus T'_Y)$. We claim that this subgraph contains a perfect matching $\hat{M}$ (a.s.). Since all edges in D are light, we obtain an assignment of weight $\leq$ $\mathrm{AP}(W) + |\hat{M}| \cdot \beta \leq (1+\alpha) \cdot \mathrm{AP}(W)$ by combining M and $\hat{M}$.

Lemma 2. $D_{|(X\setminus T'_X)\cup(Y\setminus T'_Y)}$ *contains a perfect matching* $\hat{M}$ *(a.s.).*

Proof. By construction of T', for every $i \in \{1,\ldots,b\}$ each $x \in \{x_j \mid j \in V_i\} \setminus T'_X$ has at least $\delta/4$ out-neighbors in $\{y_j \mid j \in V_{i+1}\} \setminus T'_Y$. Moreover, these neighbors can be considered as being chosen independently at random. Clearly, similar statements hold for the $y \in Y \setminus T'_Y$. By Theorem 4, $D_{|\{x_j \mid j \in C_i\} \cup \{y_j \mid j \in C_{i+1}\}}$ contains a perfect matching $\hat{M}_i$. Hence $\hat{M} := \bigcup_{i=1}^{b} \hat{M}_i$ is a perfect matching in $D_{|(X\setminus T'_X)\cup(Y\setminus T'_Y)}$. □

Our next aim is to show that the subgraph of G corresponding to $M \cup \hat{M}$ contains only few cycles and is thus not too far away from a salesman tour. In G the matching $M \cup \hat{M}$ corresponds to a set σ of pairs $\in V^2$ which we can view as a mapping on V. We claim that σ can be considered as the union of b

random bijections $\sigma_i : V_i \to V_{i+1}$. To see this, we define the equivalence class $[W]$ of a matrix W to consist of all matrices obtained by (repeatedly) permuting the n columns of W indicated by any V_i. Since the probability that W has two identical columns is negligibly small we can assume that $[W]$ is a set of $(n!)^b$ distinct and equally likely matrices $W^\pi = (w_{ij}^\pi)$, where $\pi = \bigcup_{i=1}^b \pi_i$ for permutations $\pi_i : V_i \to V_i$, $i \in \{1, \ldots, b\}$, and $w_{ij}^\pi := w_{i\pi(j)}$. If σ is an assignment for W obtained by the above algorithm then $\pi^{-1} \circ \sigma$ is an assignment of the same weight for W^π. Clearly, $\sigma_i := \pi_{i+1}^{-1} \circ \sigma_{|V_i}$ ranges over all bijections $V_i \to V_{i+1}$ if π_i ranges over all permutations. Hence it only remains to show that we can assume that $\pi \circ \sigma$ is constructed by the above algorithm for W^π.For this we slightly alter the procedure in the following way: for each $i \in \{1, \ldots, b\}$ we permute the columns of W indicated by V_i into lexicographic order, perform the described steps on the ordered matrix to construct T'_X, T'_Y and σ, and permute back. Now we are ready to prove the following.

Claim 3:

(a) σ has at most $2 \ln n$ cycles (a.s.),
(b) for all $k \in \{1, \ldots, n\}$ there are less than $(np)^k$ cycles of length bk (a.s.),
(c) there are at most $\frac{n}{\sqrt[3]{np}}$ vertices on cycles of length at most $b\sqrt{n/p}$ (a.s.).

Proof. (a) σ can be seen as the union of b random bijections $\sigma_i : V_i \to V_{i+1}$ generated by a succession of bn decisions: first we choose $\sigma_1(1)$; next we select $\sigma_2(\sigma_1(1))$ up to $\sigma_b(\sigma_{b-1}(\ldots(\sigma_1(1))\ldots))$; if $\sigma_b(\ldots(\sigma_1(1))\ldots) \neq 1$ we proceed by choosing $\sigma_1(\sigma_b(\ldots(\sigma_1(1))\ldots))$, otherwise we select $\sigma_1(2)$ and continue with $\sigma_2(\sigma_1(2))$, and so forth. In each bth step we have the chance of completing a cycle. Let

$$J_i := \begin{cases} 1, & \text{if a cycle is completed in step } bi, \\ 0 & \text{otherwise.} \end{cases}$$

The J_i, $i \in \{1, \ldots, n\}$, are independent random variables with $\mathbb{P}[J_i = 1] = 1/(n - i + 1)$, $\mathbb{P}[J_i = 0] = 1 - 1/(n - i + 1)$, and $\sum_{i=1}^n J_i$ counts the overall number of cycles in σ. Hence we have

$$\begin{aligned} \mathbb{E}[\#\text{ cycles in } \sigma] &= \mathbb{E}\left[\sum_{i=1}^n J_i\right] = \sum_{i=1}^n \mathbb{P}[J_i = 1] = \sum_{i=1}^n \frac{1}{n - i + 1} \\ &= \sum_{i=1}^n \frac{1}{i} = (1 + o(1)) \ln n. \end{aligned}$$

Using the first Angluin-Valiant inequality we conclude

$$\mathbb{P}[\#\text{ cycles in } \sigma > 2 \ln n] = O(e^{-\ln n/3}) \xrightarrow{n\to\infty} 0.$$

(b) Let $k \in \{1, \ldots, n\}$. For each cycle γ of length bk let

$$J_\gamma := \begin{cases} 1, & \text{if } \gamma \subseteq \sigma, \\ 0 & \text{otherwise.} \end{cases}$$

We have

$$\begin{aligned}\mathbb{E}[\#\text{ cycles of length } bk \text{ in } \sigma] &= \mathbb{E}\left[\sum_{\gamma} J_\gamma\right] \\ &= (\#\text{ cycles of length } bk \text{ in } I)\cdot \mathbb{P}[J_\gamma = 1] \\ &= \binom{n}{k}^b (k-1)!k!^{b-1}\cdot \frac{((n-k)!)^b}{(n!)^b} = \frac{1}{k}.\end{aligned}$$

Markov's inequality yields $\mathbb{P}[\#\text{cycles of length } bk > (pn)^k] \leq \frac{1}{k(np)^k} \xrightarrow{n\to\infty} 0$.
(c) Similar to (b) we calculate

$$\mathbb{E}[\#\text{ vertices on cycles of length } \leq b\sqrt{n/p}] = \sum_{k=1}^{\sqrt{n/p}} \frac{1}{k}\cdot bk = b\sqrt{n/p},$$

so, by Markov's inequality, $\mathbb{P}[(\text{c}) \text{ is false}] \leq \frac{b}{\sqrt[6]{np}} \xrightarrow{n\to\infty} 0$. □

Let $T := \{i \in V \mid x_i \in T' \text{ or } y_i \in T'\}$, $T_j := T \cap V_j$ be the sets of troublesome vertices in G. So far we did not consider the weight of any edge with both its endpoints in $V \setminus T$. Hence each such edge still has the probability h of being an out-, in-, or patch-edge. We will use these edges to eliminate the cycles of σ. To succeed in this we have to make sure that no cycle has too many vertices in T.

Lemma 3. *No cycle of σ has more than $\frac{1}{20}$ of its vertices in T (a.s.).*

Proof. By Lemma 1, $|T| \leq 3ne^{-\delta/8}$ (a.s.), so $t := |T_i| \leq \frac{3}{b}ne^{-\delta/8}$ (a.s.) for all $i \in \{1,\ldots,b\}$. Let Z denote the number of cycles in σ with at least $\frac{1}{20b}$ of their vertices in T_i. Similar to the proof of Claim 3 (b) we estimate $\mathbb{E}[Z] = o(1)$. □

Let us start with eliminating "small" cycles of σ, i.e. cycles on less than $b\sqrt{n/p}$ vertices. Let C be a small cycle of length l in σ. We describe an algorithm for removing C via the out- and in-edges. Once the in- or out-edges incident with a vertex have been observed, they may no longer be considered random. Therefore we mark such a vertex as "dirty". At the beginning all vertices apart from those in T are clean. Let r_C, s_C be two clean vertices from C forming an edge. W.l.o.g. we can assume that this edge is directed from r_C to s_C. By a "near-cycle-cover" we mean a digraph θ consisting of a directed path P_θ between two clean vertices and a cycle cover extending over those vertices that are not on P_θ. We grow a rooted tree with near-cycle-covers as nodes. As the root we take the near-cycle-cover obtained by deleting (r_C, s_C) from $(V, M \cup \hat{M})$. The tree is grown as follows: let θ be any node of the tree with path $s_\theta \to r_\theta$. We consider out-edges of the form (r_θ, s) such that s and its predecessor r in θ are clean. (r_θ, s) is "successful" if either s lies on a cycle of length $\geq b\sqrt{n/p}$ or if s is on P_θ and the paths $s_\theta \to r$ and $s \to r_\theta$ both have length $\geq b\sqrt{n/p}$. For each successful edge we create a successor of θ by removing (r, s) and inserting (r_θ, s). s is marked dirty immediately afterwards. When all successful edges have been considered r_θ

is marked dirty, too. We stop growing the tree when we have produced $\sqrt{n \ln n}$ leaves. This construction implies that throughout the algorithm the starting vertex of our paths stays the same, namely s_C.

Claim 4. Provided that we marked at most $o(n)$ vertices as dirty, we succeed to produce a tree with $\sqrt{n \ln n}$ leaves, thereby marking at most $O(\sqrt{n \ln n})$ additional vertices as dirty (a.s.).

Proof. We have proved as Claim 3 (c) that the number of vertices on small cycles is at most $\frac{n}{\sqrt[3]{np}} (= o(n))$. Since for each node θ in our tree the path in the associated cycle cover ends in some clean vertex, we can view the number of successors of θ as a random variable with binomial distribution $B(n - o(n), h)$. The marking steps in our algorithm ensure that paths belonging to distinct nodes θ_1, θ_2 end in distinct vertices. Thus the corresponding random variables are independent. Let Z_t denote the number of nodes in the tth level of the tree, and let $\hat{\delta}$ abbreviate $(n - o(n))h$. An easy induction shows that

$$\mathbb{P}\left[Z_t \notin [\hat{\delta}^{t-1}/2, 2\hat{\delta}^{t-1}]\right] < 2 \sum_{j=1}^{t-1} e^{-\frac{\hat{\delta}^j}{8}} \text{ for } t \geq 2.$$

Let $k \in \mathbb{N}$ be such that $\frac{\hat{\delta}^k}{2} \geq \sqrt{n \ln n} \geq \frac{\hat{\delta}^{k-1}}{2}$. The probability that we fail to produce $\sqrt{n \ln n}$ leaves is bounded by

$$\mathbb{P}\left[Z_{k+1} \notin [\hat{\delta}^k/2, 2\hat{\delta}]\right] < 2 \sum_{j=1}^{k+1} e^{-\frac{\hat{\delta}^j}{8}} < 2 \sum_{j=1}^{\infty} e^{-\frac{\hat{\delta}^j}{8}} = \frac{2e^{-\hat{\delta}/8}}{1 - e^{-\hat{\delta}^8}} = o(1).$$

□

Let $\theta_1, \ldots, \theta_m$ denote the $\sqrt{n \ln n}$ leaves of the tree thus constructed. Each θ_j is a near-cycle-cover containing a path which starts at the clean vertex s_C and ends at some vertex r_{θ_j}. We take each leaf θ_j as the root of a new tree T_{θ_j}. These trees T_{θ_j} are grown simultaneously in the following fashion using in-edges: initially s_C is the single starting vertex of all paths corresponding to the roots θ_j. Let $\Sigma_1 := \{s_C\}$. If (r, s) is an edge in G such that r and s both are clean and (r, s_C) is an in-edge, we construct a successor for every θ_j by inserting (r, s_C) and deleting (r, s). r is immediately marked as dirty; s_C is marked dirty when all in-edges ending at s_C have been considered. Note that this construction ensures that for each tree T_{θ_j} the sets Σ_2^j of starting vertices of the new paths in level 2 are identical. Now suppose we have grown the trees up to a level $t - 1$ and Σ_{t-1} is the single set of starting vertices for the paths belonging to the current level of each tree. Determining the direct successors for each node in level $t - 1$ in the same way as above maintains the condition $\Sigma_t^j = \Sigma_t = \Sigma_t^{j'}$ for all j, j'. We stop growing the trees when we have reached a depth k with $\frac{\hat{\delta}^k}{2} \geq \sqrt{n \ln n} \geq \frac{\hat{\delta}^{k-1}}{2}$ as above. By the same calculations as in the proof of Claim 4 we obtain that every tree has $\sqrt{n \ln n}$ leaves almost surely. Moreover, since the edges we used for growing the trees are identical for each tree, a total number of at most $O(\sqrt{n \ln n})$ additional vertices is marked dirty. So the overall number of dirty vertices is $O(\sqrt{n \ln n}) = o(n)$. Note that the above construction did not care about the sizes of the cycles and paths. Therefore we now delete those nodes from each T_{θ_j} that contain a small path or a small cycle. We call T_{θ_j} "good" if it still has $\sqrt{n \ln n}$ leaves and "bad" otherwise. Since the number of vertices that can cause the creation of small cycles or paths is $o(n)$ we can repeat the proof of Claim 4 to see that $\mathbb{P}\left[T_{\theta_j} \text{ is bad}\right] \leq \frac{2e^{-\hat{\delta}/8}}{1-e^{-\hat{\delta}/8}}$. Thus the expected number of bad trees is at most $\sqrt{n \ln n} \cdot \frac{2e^{-\hat{\delta}/8}}{1-e^{-\hat{\delta}/8}}$ and by Markov's inequality

$$\mathbb{P}\left[\text{there exist more than } \frac{\sqrt{n \ln n}}{2} \text{ bad trees}\right] \leq \frac{4e^{-\hat{\delta}/8}}{1-e^{-\hat{\delta}/8}} = o(1).$$

So with probability $1 - o(1)$ more than $\sqrt{n \ln n}/2$ trees are good. Each good tree T_{θ_j} has $\sqrt{n \ln n}$ leaves each of which consists of cycles plus a long path from distinct starting vertices b to r_{θ_j}. Since for distinct θ_j, θ'_j the ending vertices $r_{\theta_j}, r_{\theta'_j}$ are distinct, too, we have created an overall number of $(n \ln n)/2$ distinct long paths. Note that neither the out-edges of the ending vertices nor the in-edges of the starting vertices have been observed so far. Hence the probability that we cannot close any path via an in- or out-edge is smaller than

$$(1-h)^{2\frac{n \ln n}{2}} = \left(1 + \frac{-hn}{n}\right)^{n \ln n} \leq e^{-hn \ln n}.$$

But this means that with probability at least $1 - e^{-hn \ln n}$ we can eliminate the small cycle C. Since the number of cycles is less than $2 \ln n$, the probability that we fail to remove all small cycles is at most $2 \ln n \cdot e^{-hn \ln n} = o(1)$. Being left with cycles $C_1, C_2, \ldots, C_l$ each having at least $b\sqrt{n/p}$ vertices, so that $l \leq 2\sqrt{np}$, we want to use the patch-edges to combine these cycles into a single tour. Cycles C, C' can be patched together if we find edges $(v, w) \in C$, $(v', w') \in C'$ such that (v, w') and (v', w) are patch-edges. All edges, except those incident with T have probability h to be patch edges. So the probability that C, C' cannot be patched together is at most

$$\begin{aligned}
&(1-h^2)^{\left(\frac{|C|}{b} - |T_i|\right) \cdot \left(\frac{|C'|}{b} - |T_i|\right)} \leq (1-h^2)^{\frac{|C||C'|}{b} - |T_i|(|C| + |C'|)} \\
&\leq (1-h^2)^{\frac{bn}{p} - 3ne^{-\delta/8} \cdot 2b\sqrt{n/p}} \leq (1-h^2)^{\frac{bn}{p} - 6bn^2 e^{-\delta/8}} \\
&= (1 - \delta^2/n^2)^{n^2\left(\frac{b}{np} - 6be^{-\delta/8}\right)} \leq e^{-\frac{\delta^2 b}{np} + 6\delta^2 be^{-\delta/8}} \\
&= e^{-\frac{nh^2 b}{p} + 6be^{2 \ln \delta - \delta/8}} = e^{-npb\left(\frac{h}{p}\right)^2 + o(1)}.
\end{aligned}$$

Since $\frac{h}{p} = \frac{h}{1-(1-h)^5} > \frac{1}{5}$ for all h we can upper bound the above expression by $e^{-\frac{npb}{25} + o(1)}$. Now the probability that all cycles can be patched together is at least $1 - le^{-\frac{npb}{25} + o(1)} \geq 1 - 2\sqrt{np}e^{-\frac{npb}{25} + o(1)} = 1 - o(1)$. We used $o(n)$ additional light edges to eliminate all cycles. Hence the weight of the tour is

$$\mathrm{AP}(W)(1+\alpha)+o(n)\cdot\frac{\alpha\mathrm{AP}(W)}{bn}\to\mathrm{AP}(W)(1+\alpha)\text{ for }n\to\infty,$$

concluding the proof of Theorem 1. □

Proof Sketch of Theorem 2. Call an edge e of G "forced" if it is contained in *every* optimal assignment of G, and call e "positive" if it is contained in *at least* one optimal assignment and $w(e)\geq 1$. A "forced" resp. "positive" cycle is a cycle consisting of forced resp. positive edges. Let $\mathcal{F}_e$ and $\mathcal{P}_e$ be the sets of weight functions such that e is a forced or a positive edge, respectively. For every $w\in\mathcal{P}_e$ we see that $w' := w|_{E\setminus\{e\}}\cup\{(e,w(e)-1)\}\in\mathcal{F}_e$, so $|\mathcal{F}_e|\geq|\mathcal{P}_e|$. Let σ be an optimal assignment. We have $\mathbb{P}[\mathrm{AP}(W)\neq b\mathrm{ATSP}(W)]=\mathbb{P}[$every optimal Assignment contains a cycle of length $<bn]\geq\mathbb{P}[\sigma$ contains a forced cycle of length $<bn]\geq\mathbb{P}[\sigma$ contains a positive cycle of length $t<bn]$. Since σ can be seen as being induced by b *random* bijections, we can estimate the latter probability to be at least $\frac{t}{t+2b}(1+o(1))e^{-t/c}$ for every $t\in\{b,2b,\ldots,(n-1)b\}$, giving the claim of (a) for $t=b$ and of (b) for $t=\theta(\sqrt{c})$. □

4 Open Problems

Recently, Frieze and Sorkin [4] proved that for random weight matrices $W\in[0,1]^{n\times n}$, $\mathrm{ATSP}(W)-\mathrm{AP}(W)\leq c_0\frac{(\ln n)^2}{n}$ and $\mathbb{E}[\mathrm{ATSP}(W)-\mathrm{AP}(W)]\geq\frac{c_0}{n}$, where c_0 and c_1 are absolute constants. It would be interesting to have similar results for the b-partite TSP.

References

1. Angluin, D. Valiant, L.G.: Fast probabilistic algorithms for Hamiltonian circuits and matchings. J. Comp. Sys. Sci. **18**: (1979) 155–194
2. Chalasani, P., Motwani, R., Rao, A.: Approximation Algorithms for Robot Grasp and Delivery. In: Proceedings 2nd International Workshop on Algorithmic Foundations of Robotics (WAFR). Toulouse, France (1996)
3. Frieze, A., Karp, M., Reed, B.: When is the Assignment Bound Tight for the Asymmetric Traveling-Salesman Problem? SIAM J. on Computing **24** (1995) 484–493
4. Frieze, A., Sorkin, G.B.: The probabilistic relationship between the assignment and asymmetric traveling salesman problems. In: Proceedings ACM-SIAM Symposium on Discrete Algorithms (SODA). Washington D.C. (2001) 652–660
5. Lawler, E.L., Lenstra, J.K., Rinnooy Kan, A.H.G., Shmoys, D.G. (editors): The Traveling Salesman Problem – A Guided Tour of Combinatorial Optimization. John Wiley & Sons (1985)
6. Michel, C., Schroeter, H., Srivastav, A.: Approximation algorithms for pick-and-place robots. Preprint (2001)
7. Miller, D.L., Pekny, J.F.: Exact solution of large asymmetric traveling salesman problems. Science **251** (1991) 754–762
8. Walkup, D.W.: Matchings in random regular bipartite graphs. Discrete Mathematics **31** (1980) 59–64

Exact Sampling in Machine Scheduling Problems

Sung-woo Cho and Ashish Goel

Department of Computer Science
University of Southern California
{sungwooc,agoel}@cs.usc.edu

Abstract. Analysis of machine scheduling problem can get very complicated even for simple scheduling policies and simple arrival processes. The problem becomes even harder if the scheduler and the arrival process are complicated, or worse still, given to us as a black box. In such cases it is useful to obtain a *typical* state of the system which can then be used to deduce information about the performance of the system or to tune the parameters for either the scheduling rule or the arrival process. We consider two general scheduling problems and present an algorithm for extracting an *exact* sample from the stationary distribution of the system when the system forms an ergodic Markov chain. We assume no knowledge of the internals of the arrival process or the scheduler. Our algorithm assumes that the scheduler has a natural monotonic property, and that the job service times are geometric/exponential. We use the Coupling From The Past paradigm due to Propp and Wilson to obtain our result. In order to apply their general framework to our problems, we perform a careful coupling of the different states of the Markov chain.

1 Introduction

Consider a system consisting of multiple machines where jobs arrive according to some probabilistic process, get scheduled on a machine using some scheduling rule, and leave when they have been serviced. Analysis of such a system can get very complicated even for simple scheduling policies and simple arrival processes [4]. The problem becomes even harder if the scheduler and the arrival process are complicated, or worse still, given to us as a black box. In such cases it is useful to obtain a *typical* state of the system. More formally, if the entire system forms an ergodic Markov chain [11], then we would like to have a sample from the stationary distribution of the system. A set of such samples could then be used to deduce information about the performance of the system or to tune the parameters for either the scheduling rule or the arrival process.

In this paper we study scheduling problems where the system forms an ergodic Markov chain. We will not assume any knowledge of the internal structure of the scheduler or the arrival process; we will treat them as black-boxes that we are allowed to query. However we do need to assume that the scheduling rule is monotonic, where a monotonic scheduler must have the following property. Suppose the scheduler assigns a job i to machine j when the system is in state s. Consider another state s' where machine j has the same load as in state s and

M. Goemans et al. (Eds.): APPROX-RANDOM 2001, LNCS 2129, pp. 202–210, 2001.

all other machines have at least as much load as in state s. Then the scheduler must assign job i to the same machine j even if the system is in state s' instead of s. Intuitively, we can not force the scheduler to change its decision to assign a job to machine j by increasing the load on the other machines. We will further assume that each job gets executed on a single machine, that the service times for all jobs are i.i.d. geometric or exponential variables, and that each machine has a finite buffer size. Under these fairly general assumptions we exhibit an efficient algorithm to find a sample from the stationary state of the system. The algorithm simulates the Markov process for an expected time $O(T_{mix} \ln C)$ where T_{mix} is the mixing time for the Markov chain and C is the total buffer size of the machines. It is important to note that the algorithm does not need to know the mixing time.

Our algorithm uses an elegant tool called Coupling From The Past (CFTP) due to Propp and Wilson [12]. CFTP is essentially a variant of the Markov Chain Monte Carlo method [5,9] and a natural extension of the work on approximate and exact sampling for specific Markov chains [2,10,9]. In order to use CFTP efficiently, we exhibit a coupling that allows different states of the Markov chain to evolve in a *monotonic* fashion [12]. A complete survey of work in this area is beyond the scope of this paper. While Markov Chain Monte Carlo methods have been used widely for problems arising from combinatorics (e.g. [3,1]), physics (e.g. [6,8]), and statistics (e.g. [2]), their use for traditional optimization problems such as scheduling has been limited. We hope that our work will lead to more applications of this flavour.

Section 2 presents a formal definition of the problem. Section 3 presents our algorithm and does a running time analysis. Section 4 presents simulation results for our algorithm, and illustrates its use for optimizing a simple scheduler.

2 Problem Definition

We are given m machines, $M_1, \cdots, M_m$. Let c_i denote the capacity of machine M_i. We define C as the sum of the capacities of all machines, that is $C = \sum_{i=1}^{m} c_i$. There are t types of jobs. Different job types may have a different arrival process, but we assume that the service times for the jobs are i.i.d. geometric random variables with mean $1/\mu$.

We are also given a scheduler that assigns arriving jobs to machines. However some machines cannot process certain types of jobs.[1] If a job of type i cannot be processed by M_j, then the scheduler cannot assign that job to M_j. This assignability can be represented as a matrix $V_{(t \times m)}$ as follows.

$V_{i,j} = 1$ if a job of type i can be assigned to machine M_j.
$= 0$ otherwise.

If there is no space available in any machine for the arriving job, scheduler must discard the job. The scheduler may also do admission control. We will not assume any knowledge of the arrival process or the scheduler and will merely use them as black boxes.

[1] This is analogous to the 1-∞ case in machine scheduling [7].

The state s_t of the system at time t is defined as the number of jobs in every machine; let $l_i(s_t)$ denote the number of jobs in machine M_i. The state s_t is an m-dimensional vector, i.e. $\{l_1(s_t), l_2(s_t), \cdots, l_m(s_t)\}$. The state evolves only when a job arrives or leaves, and we define a random mapping $\phi(s_t)$ such that $s_{t+1} = \phi(s_t)$. In this paper, we will assume that the random mapping ϕ defines an Ergodic Markov chain. It is important to note that ϕ depends on the arrival process, the scheduler, and the mean service time $1/\mu$.

Our goal is to efficiently extract a sample from the stationary distribution of this Ergodic Markov chain. We will consider the following two scenarios.

Scenario 1: Each machine has a FIFO buffer. That is, only the job in the front of queue is processed.
Scenario 2: Jobs in each machine are processed simultaneously.

Our algorithm will work for a large class of schedulers that we call *monotonic schedulers.*

Definition 2.1 *We define $\leq$ over state space S such that for x and y in S, $x \leq y$ iff $l_i(x) \leq l_i(y)$ for all machines M_i.*

Definition 2.2 *Let $\Gamma_1(i)$ and $\Gamma_2(i)$ represent the state of the system after arrival of a job of type i in state s_1 and s_2 respectively. A scheduler is said to be monotonic if $s_1 \leq s_2 \Rightarrow \Gamma_1(i) \leq \Gamma_2(i)$ for all job types i.*

We claim that this is a natural class of schedulers to study for the following reason. Suppose that there are two states $\hat{L}$ and $\hat{U}$ such that $\hat{L} \leq \hat{U}$. Consider the situation that a job k is assigned to a machine M_j in the state $\hat{L}$. If $l_j(\hat{L}) < l_j(\hat{U})$, the monotonicity property will be maintained because $l_j(\Gamma_1) \leq l_j(\Gamma_2)$. Suppose that $l_j(\hat{L}) = l_j(\hat{U})$. It seems natural to require that if the load on machine j in $\hat{L}$ is kept the same and all other loads are either kept the same or increased, then the scheduler must still assign the job to machine j. This would result in monotonicity of the scheduler.

3 Applying CFTP to Our Scheduling Problems

3.1 Coupling the Evolution of Different States

We would like to couple the evolution of all the states in the Markov Chain to allow an efficient application of the CFTP technique. A state evolves only when a job arrives or leaves. The other factors – number of machines, capacities of machines, scheduling strategy, and so on – are identical for any state. Hence we just have to distribute job arrivals and departures carefully. Let the random mapping $\phi(s_t)$ be described as a function of the state s_t and a hidden random seed U_t. The random seed affects both the arrival and departure of jobs. If we use the same random seed for evolution from different states, we will obtain the same job arrivals at time t for all sample paths. Therefore all we need to take care of are the job-departure events. Recall (from section 2) that the service times for the jobs are geometrically distributed. We will couple the departure events by associating departure events with machines rather than with jobs, as outlined below.

Scenario 1: Suppose that a departure event happens from the machine M_i at time t. Let $\delta^{(i)}(t)$ be a geometric random variable with mean $1/\mu$. The next departure event for machine i happens at time $t + \delta^{(i)}(t)$. At this time, if the queue in M_i is not empty then the job in the front of the queue leaves the machine, otherwise nothing happens. This ensures that the service times of the jobs are geometrically distributed with mean $1/\mu$ because of the memoryless property of geometric random variables. The evolution of different states is coupled by providing them with identical values of $\delta^{(i)}(t)$.

Scenario 2: In scenario 1, we made m Poisson processes. In scenario 2, we make C Poisson processes. Suppose that a departure event happens from the j-th "slot" of machine M_i at time t. Let $\delta^{(i,j)}(t)$ be a geometric random variable with mean $1/\mu$. The next departure event from this slot happens at time $t + \min_j\{\delta^{(i,j)}(t)\}$. Suppose that there are k jobs in machine M_i at time t. If $j \leq k$ then one job in the buffer leaves the machine, otherwise nothing happens. This process ensures that the service times of the jobs are geometrically distributed with mean $1/\mu$. Like scenario 2, the evolution of different states is coupled by providing them with identical values of $\delta^{(i,j)}(t)$.

The above coupling has been carefully devised to allow us to prove monotonicity of our Markov chain, as explained in the next section.

3.2 Monotonicity of the Coupled Markov Chain

CFTP is particularly efficient for monotonic Markov chains as defined by Propp and Wilson [12].

Definition 3.1 *Suppose that the state space S of our problem admits a natural partial ordering $\leq$, with a minimum element $\hat{0}$ and a maximum element $\hat{1}$. A Markov chain is defined over S in terms of the transition probability matrix P. For elements x and y in S, let the function $\phi(i)$ be a random mapping of S to S itself with $Pr[\phi(x) = y] = P_{x,y}$. We say that the Markov chain is monotonic iff $x \leq y \Rightarrow \phi(x) \leq \phi(y)$.*

The following theorem is now immediate; we sketch the proof of the theorem for completeness.

Theorem 1. *Given a monotonic scheduler and the coupling described above, our Markov chain is monotonic in both scenarios.*

Proof. We will state the proof for the first scenario; the proof for the other scenario is similar. Let $\hat{L}$ and $\hat{U}$ be two states such that $\hat{L} \leq \hat{U}$. Let Γ_1 and Γ_2 be the next states starting from $\hat{L}$ and $\hat{U}$, respectively, after one event. We need to consider

1. Departure Events: Let the departure event correspond to the j-th machine. We know that $l_j(\hat{L}) \leq l_j(\hat{U})$. If $l_j(\hat{L}) = 0$ then $l_j(\Gamma_1) = 0 \leq l_j(\Gamma_2)$. On the other hand, if $l_j(\hat{L}) > 0$ then so is $l_j(\hat{U})$ i.e. a job will depart from machine j in both steps. Now $l_j(\Gamma_2) = l_j(\hat{U}) - 1 \geq l_j(\hat{L}) - 1 = l_j(\Gamma_1)$. Thus, in both cases, $l_j(\Gamma_2) \geq l_j(\Gamma_1)$. Since the loads on the other machines are unchanged, this implies that $\Gamma_1 \leq \Gamma_2$.
2. Arrival Events: Since the scheduler is monotonic, we are guaranteed that $\Gamma_1 \leq \Gamma_2$.

We can now use the CFTP paradigm for monotonic Markov chains as defined by Propp and Wilson [12] to obtain a sample from the stationary distribution. In this scheme, the evolutions of the states $\hat{0}$ and $\hat{1}$ in the Markov chain are simulated from time $-\tau$ to 0. If the two processes are in the same state at time 0, then this common state is output as the sample from the stationary distribution. If not, then τ is doubled and the process is repeated. It is important to reuse the same random numbers at time $-t$ during all the iterations of the algorithm.[2] The main intuition is that when the sample paths starting from the bottom state $\hat{0}$ and the top state $\hat{1}$ converge, they sandwich the entire state space in between. The reader is referred to an excellent description of the process by Propp and Wilson [12] for more detail.

We are going to use the state where all machines are empty as $\hat{0}$ and the state where all machines are filled to capacity as $\hat{1}$. Since departures are associated with machines rather than jobs, the random number reuse is easy to ensure. The following theorem follows from a general theorem due to Propp and Wilson [12] regarding the correctness of the CFTP paradigm.

Theorem 2. *The algorithm outlined above samples* exactly *from the stationary distribution of the ergodic Markov chain defined by the scheduler and the arrival process.*

3.3 Running Time Analysis

Theorem 3. *Define*

$$\Phi_{t_1}^{t_2}(x) = \phi_{t_2-1}(\phi_{t_2-2}(\cdots(\phi_{t_1}(x, U_{t_1}), U_{t_1+1}), \cdots, U_{t_2-2}), U_{t_2-1}).$$

Let T^* be the smallest t such that $\Phi^0_{-t}(\hat{0}) = \Phi^0_{-t}(\hat{1})$. Define $\bar{d}(k) = \max_{\pi_1,\pi_2} ||\pi_1^k - \pi_2^k||$ for particular k, where π^k is the distribution governing the Markov chain at time k when started at time 0 in a random state governed by the distribution π. The mixing time threshold T_{mix} is defined to be the smallest k for which $\bar{d}(k) \leq 1/e$. Then,

$$E[T^*] \leq 2T_{mix}(1 + \ln C).$$

Proof. Propp and Wilson proved that $E[T^*] \leq 2T_{mix}(1 + \ln l)$ where l is the length of the longest chain in the partially ordered state space [12]. Suppose

[2] More formally, we have to use the same instance F_{-t} of the random mapping ϕ at time $-t$ during all the iterations of the algorithm.

that there are l distinct states such that $s_0 \leq s_1 \leq \cdots \leq s_l$. For l to be the length of the longest chain, s_0 should be $\hat{0}$ and s_l should be $\hat{1}$. The number of jobs of each machine in s_{i+1} is greater than or equal to the number of jobs in corresponding machine of s_i. Since the two states are distinct, s_{i+1} must have at least one more job than s_i. Recall that the difference between the number of jobs in $\hat{0}$ and $\hat{1}$ is C. Thus there can be at most C distinct states between $\hat{0}$ and $\hat{1}$, and l cannot exceed C.

We will now explore the implications of the above theorem for the running time of our scheme for scenario 1 (the implications for scenario 2 can be similarly obtained). Let λ be the total arrival rate for all the job types. Since there are m machines, there are an average of μm departure events and λ arrival events during each step. We can store all the arrival/departure events in a priority queue, which can be implemented using heaps. This allows us to find the next event to be processed in time $O(\log n)$. Using the fact that T^* has an exponential tail, we can claim that the expected running time is $\tilde{O}(T_{mix}(1 + \ln C)(\lambda + \mu m))$ – we omit the details from this abstract.

Note that we do not need to define Markov chains to make a random sample using our scheme. In fact it is very hard to define Markov chains sometimes because
1) State space may be too large $(\Pi_{i=1}^{m} c_i)$.
2) The scheduler may be too complicated, or given as a black box.

It is interesting (and a testament to the efficacy of the CFTP paradigm) that we can get an *exact* sample from the stationary distribution of the Markov chain without any knowledge of or assumptions about the structure of the Markov chain beyond ergodicity and monotonicity of the scheduler, and that we can prove good bounds on the expected running time of our algorithm.

3.4 Extensions

The above algorithm can be adapted in a straight forward fashion to

The continuous case: This is the case where the arrival process is continuous and the service times are exponential rather than geometric.

Randomized Schedulers: The notion of monotonicity that we require is that of vector stochastic dominance. More formally, for each state s and job type i there must be a random variable $X(i, s)$ such that
1. $\mathbf{Pr}[X(i, s) = t]$ is the same as the probability that the state changes from s to t as a result of the scheduling of job i.
2. $s_1 \leq s_2 \Rightarrow X(i, s_1) \leq X(i, s_2)$.

Such a scheduler will allow us to couple the evolution of states so that the resulting Markov chain is monotonic.

4 Simulation Results

In this section we will present simulation results only for the first scenario. As mentioned in the introduction, we can tune a black box scheduler using the extracted samples. We will also present an example to tune a black box random scheduler.

We will not show the detail of our code in this paper. However let us explain the simulation setting. We have used a simple(Min-load) scheduler for the first scenario. When a job arrives, the scheduler looks at the loads of all machines to which the job can be assigned, and assigns the job to the minimum loaded machine among them. This is a simple monotonic scheduling strategy. We need some input parameters as follows.

1. average arrival rate
2. average processing time
3. number of machines
4. capacities of machines
5. job types distribution
6. job type assignability matrix

We will not specify the job types in detail in this abstract. We simulated the case where there are 10 machines and 10 job types. We set the total arrival rate equal to 10 times the service rate. Figure 1 shows the running time of our program in seconds as the capacities of machines are increased. The basic case is that each machine has a capacity between 1 and 15. We multiplied the capacities by n where n is between 1 and 500. The running time seems almost linear in the multiplication constant n.

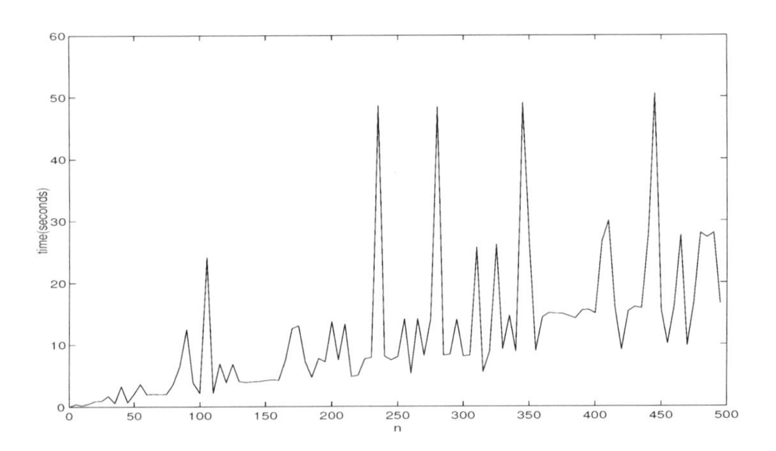

Fig. 1. Running time of our program for scenario 1

Since we can extract a random sample efficiently using our scheme, we can use it to tune a random scheduler. We illustrate this process for a simple random scheduler that assigns a job according to assigning probability matrix $\tilde{V}$ such that $\tilde{V}_{i,j}$ is the probability that a job type i is assigned to a machine j. Of course $\tilde{V}_{i,j} = 0$ if job type i can not be scheduled on machine j. We illustrate a very

```
Tune_Rand_Schd(Ṽ: assigning probability matrix, t: tolerance rate)

1. while True
2.     Extract k samples.
3.     f = total number of full machines in the k samples
4.     if f/k > t
5.        Improve Ṽ.
6.     else
7.        break
```

Fig. 2. A simple optimizer for random scheduler

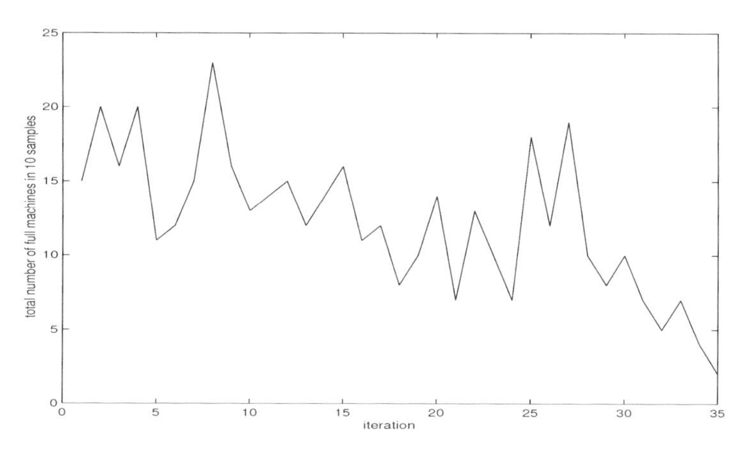

Fig. 3. Evolution of the random scheduler

simple optimizer in figure 2. Assuming that there are not too many arriving jobs and each machine has enough capacity, this optimizer will tune the assigning probability matrix for random scheduler. We simulated the case that we have 10 machines, 10 job types, [1-15] capacities and 10% tolerance rate. Figure 3 shows the evolution of the random scheduler. As the tuning process progresses, the number of machines with completely filled up buffers goes down, indicating an improvement in the performance of the scheduler. Notice that the tuning process does not involve any knowledge of the arrival rates.

Acknowledgements

Ashish Goel would like to thank Chandra Chekuri for introducing him to the problem and for several useful discussions.

References

1. D.J. Aldous. A random walk construction of uniform spanning trees and uniform labelled trees. *SIAM Journal on Discrete Mathematics*, 3(4):450–465, 1990.

2. S. Asmussen, P. Glynn, and H. Thorisson. Stationary detection in the initial transient problem. *ACM trans. on modeling and comp. simulation*, 2(2):130–157, 1992.
3. A. Broder. Generating random spanning trees. *30th Annual Symposium on Foundations of Computer Science*, pages 442–447, 1989.
4. C. Chekuri, K. Ramanan, P. Whiting, and L. Zhang. Blocking probability estimates in a partitioned sector TDMA system. *Fourth International Workshop on Discrete Algorithms and Methods for Mobile Computing and Communications (Dial M for Mobility)*, 2000.
5. P. Diaconis and L. Saloff-Coste. What do we know about the metropolis algorithm? *Twenty-Seventh Annual ACM Symposium on the Theory of Computing*, pages 112–129, 1995.
6. J. Fill. An interruptible algorithm for perfect sampling via Markov chains. *Annals of Applied Probability*, 8(1):131–162, 1998.
7. D. Hochbaum. *Approximation algorithms for NP-hard problems.* PWS Publishing Company, 1997.
8. M. Huber. Exact sampling and approximate counting techniques. *30th ACM Symposium on the Theory of Computing*, pages 31–40, 1998.
9. M. Jerrum and A. Sinclair. The Markov chain Monte Carlo method: an approach to approximate counting and integration. *In "Approximation Algorithms for NP-hard Problems," D.S.Hochbaum ed.*, 1996.
10. L. Lovasz and P. Winkler. Exact mixing in an unknown Markov chain. *Electronic Journal of Combinatorics, 2, paper #R15*, 1995.
11. R. Motwani and P. Raghavan. *Randomized Algorithms.* Cambridge University Press, 1995.
12. J.G. Propp and D.B. Wilson. Exact sampling with coupled Markov chains and applications to statistical mechanics. *Random Structure & Algorithms*, 9:223–252, 1996.

On Computing Ad-hoc Selective Families

Andrea E.F. Clementi[1], Pilu Crescenzi[2], Angelo Monti[3], Paolo Penna[1], and Riccardo Silvestri[3]

[1] Dipartimento di Matematica, Università di Roma "Tor Vergata"
{clementi,penna}@mat.uniroma2.it

[2] Dipartimento di Sistemi e Informatica, Università di Firenze
piluc@dsi.unifi.it

[3] Dipartimento di Scienze dell'Informazione, Università di Roma "La Sapienza",
{monti,silver}@dsi.uniroma1.it

Abstract. We study the problem of computing ad-hoc selective families: Given a collection $\mathcal{F}$ of subsets of $[n] = \{1, 2, \ldots, n\}$, a *selective family for* $\mathcal{F}$ is a collection $\mathcal{S}$ of subsets of $[n]$ such that for any $F \in \mathcal{F}$ there exists $S \in \mathcal{S}$ such that $|F \cap S| = 1$. We first provide a polynomial-time algorithm that, for any instance $\mathcal{F}$, returns a selective family of size $O((1 + \log(\Delta_{max}/\Delta_{min})) \cdot \log |\mathcal{F}|)$ where Δ_{max} and Δ_{min} denote the maximal and the minimal size of a subset in $\mathcal{F}$, respectively. This result is applied to the problem of broadcasting in radio networks with known topology. We indeed develop a broadcasting protocol which completes any broadcast operation within $O(D \log \Delta \log \frac{n}{D})$ time-slots, where n, D and Δ denote the number of nodes, the maximal eccentricity, and the maximal in-degree of the network, respectively. Finally, we consider the combinatorial optimization problem of computing broadcasting protocols with minimal completion time and we prove some hardness results regarding the approximability of this problem.

1 Introduction

Selective Families. The notion of *selective family* has been introduced in [6]. Given a positive integer n, a family $\mathcal{S}$ of subsets of $[n] = \{1, 2, \ldots, n\}$ is said to be (n, h)-selective if and only if, for any subset $F \subseteq [n]$ with $|F| \leq h$, there is a set $S \in \mathcal{S}$ such that $|F \cap S| = 1$. This notion is an essential tool exploited in [6,7,8,9] to develop distributed broadcasting algorithms in radio networks with unknown topology. In particular, [6] provide a polynomial-time algorithm that, for any integer ℓ, computes a $(2^{\ell}, 2^{\lceil \ell/6 \rceil})$-selective family of size $O(2^{5\ell/6})$: Notice that the size of the selective family is a key parameter since, as we will see later, it determines the completion-time of the corresponding broadcasting protocol. Better constructions of selective families have been introduced in [7]. The best known (non-constructive) upper bound has been proved in [9] where it is shown that there exists an (n, h)-selective family of size $O(h \log n)$. This upper bound is almost tight since, in the same paper, it is shown that, any (n, h)-selective family has size $\Omega(h \log \frac{n}{h})$.

M. Goemans et al. (Eds.): APPROX-RANDOM 2001, LNCS 2129, pp. 211–222, 2001.

Radio Networks and Broadcasting. A *radio network* is a set of radio stations that are able to communicate by transmitting and receiving radio signals. A transmission range is assigned to each station s and any other station t within this range can directly (i.e. by one *hop*) receive messages from s. Communication between two stations that are not within their respective ranges can be achieved by *multi-hop* transmissions. In this paper, we will consider the case in which radio communication is structured into synchronous *time-slots*, a paradigm commonly adopted in the practical design of protocols [3,10,16]. A radio network can be modeled as a directed graph $G(V, E)$ where an edge (u, v) exists if and only if u can send a message to v in one hop. The nodes of a radio network are processing units, each of them able to perform local computations. It is also assumed that every node is able to perform *all* its local computations required for deciding the next send/receive operation during the current time-slot. In every time-slot, each node can be *active* or *non-active*. When it is active, it can decide to be either *transmitter* or *receiver* : in the former case the node transmits a message along all of its outgoing edges while, in the latter case, it tries to recover messages from all its incoming edges. In particular, the node can recover a message from one of its incoming edges if and only if this edge is the only one bringing in a message. When a node is non-active, it does not perform any kind of operation.

One of the fundamental tasks in network communication is the *broadcast* operation. It consists in transmitting a message from one source node to all the other nodes of the network. A broadcasting protocol is said to have *completed broadcasting* when all nodes, reachable from the source, have received the source message (notice that when this happens, the nodes not necessarily stop to run the protocol since they might not know that the operation is completed). We also say that a broadcasting protocol *terminates* in time t if, after the time-slot t, all the nodes are in the non-active state (i.e. when all nodes stop to run the protocol). According to the network model described above, a broadcasting protocol operates in time-slots synchronized by a global clock: At every time-slot, each active node decides to either transmit or receive, or turn into the non-active state.

Selective Families and Broadcasting in Unknown Topology. Given a radio network with n nodes, maximal in-degree Δ, and unknown topology (in the sense that nodes know nothing about the network but their own label), the existence of a (n, Δ)-selective family of size m implies the existence of a distributed broadcasting protocol in the network, whose completion time is $O(nm)$. The protocol operates in n phases of m time-slots each. During each phase, at time-slot j the nodes (whose labels are) in the j-th set of the selective family which are *informed* (that is, which have already received the source message) transmit the message to their out-neighbors. Even though no node knows the topology of the network, the definition of a selective family implies that, at the end of each phase, at least one new node has received the source message (hence, n phases are sufficient): This is due to the fact that *for any* subset F of $[n]$, with $|F| \leq \Delta$, there exists at least one subset of the selective family whose intersection with F contains exactly one element. In particular, for any i with $1 \leq i < n$, let R_i be the set of nodes that have received the source message after the first i phases and assume that there are still nodes to be informed. Then, there exists a non-informed node x_i which is an out-neighbor of at least one node in R_i: Let F_i be the set of (labels of the) nodes $r \in R_i$ such that x_i is an out-neighbor of r and let S_j be a subset in the selective family such that $|F_i \cap S_j| = 1$ (notice that S_j *must exist independently of the topology of the network*). Hence, x_i will certainly receive the source message at the latest during the j-th time-slot of the $(i+1)$-th phase since during this time-slot *exactly one* of the informed in-neighbors of x_i transmits the source message to x_i[1].

Ad-hoc Selective Families and Broadcasting in Known Topology. The connection between selective families and broadcasting in radio networks with unknown topology justifies the fact that *each* subset of the domain $[n]$ has to be selected by at least one subset in the family: Indeed, since the topology of the network is not known, in order to be sure that F_i is selected by at least one subset in the family, the family itself has to select every possible subset of the nodes. In this paper, we assume that *each node knows the network topology and the source node* which is a well-studied assumption concerning the communication process in radio networks [4,5,11,1,15]. It is then easy to see that the family to be computed has to select only those sets of nodes which are at distance l from the source node and which are the in-neighbors of a node at distance $l+1$, for any distance l. This observation leads us to the following definition of *ad-hoc selective family*: Given a collection $\mathcal{F} = \{F_1, F_2, \ldots, F_m\}$ of subsets of $[n]$, a family $\mathcal{S} = \{S_1, S_2, \ldots, S_k\}$ of subsets of $[n]$ is said to be *selective for* $\mathcal{F}$ if, for any F_i, there exists S_j such that $|F_i \cap S_j| = 1$. In Fig.1 it is shown how a radio network determines the sets F_i to be selected: For example, for $l = 2$, there are three sets to be selected, that is, F_2 which is the in-neighborhood of node 4, F_3 which is the in-neighborhood of node 5, and F_4 which is the in-neighborhood of node 6.

Our (and Previous) Results. The main result of this paper is a polynomial-time algorithm that, for any collection $\mathcal{F}$, returns a selective family for $\mathcal{F}$ of size $O((1+\log(\Delta_{max}/\Delta_{min}))\cdot\log|\mathcal{F}|)$ where Δ_{max} and Δ_{min} denote the maximal and the minimal size of a subset in $\mathcal{F}$, respectively. The proof of our result is based on the probabilistic method [2]: We first perform a probabilistic analysis of randomly computing a selective family according to a specific probability distribution and we, subsequently, apply the method of the conditional probabilities in order to de-randomize the probabilistic construction.

The above efficient construction of ad-hoc selective families is then used to derive a broadcasting protocol having completion time $O(D \log \Delta \log(n/D))$, where D denotes the maximal eccentricity of the network (that is, the longest distance from a node to any other node). The protocol is efficiently constructible

[1] Actually, this is not always true since the set of informed nodes might have been changed before the j-th time-slot of the $(i+1)$-th phase: However, this would imply that at least one other new node has been informed during the $(i+1)$-th phase.

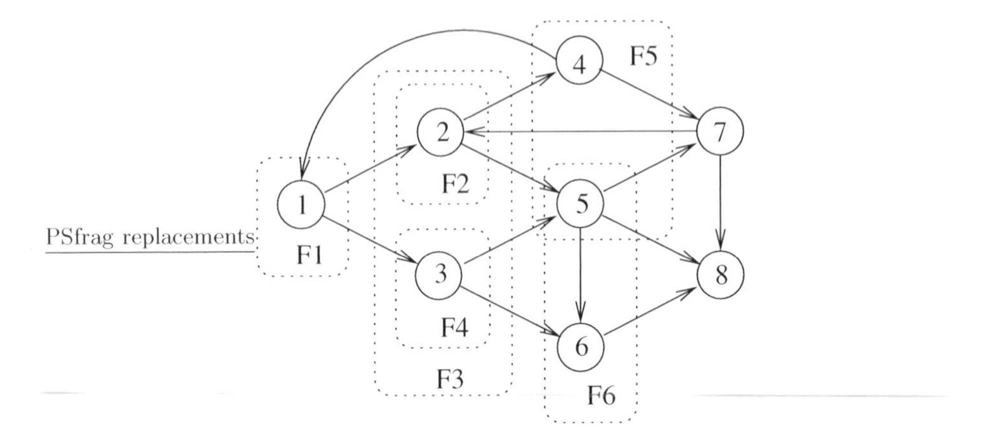

Fig. 1. The collection of sets to be selected corres

(i.e. it can be constructed in deterministic polynomial time in the size of the network) and easy-to-implement. Its completion time is better than the $O(D \log^2 n)$ bound obtained in [5] whenever $\Delta = O(1)$ and $D = \Theta(n)$, or $\Delta = o(n^\alpha)$, or $n/D = o(n^\alpha)$ (for every positive constant $\alpha > 0$). Furthermore, our bound implies that the $\Omega(\log^2 n)$ lower bound, shown in [1] in the case in which D is a constant value greater than or equal to 2, only holds when $\Delta = \Omega(n^\alpha)$, for some positive constant $\alpha > 0$. In [11], an $O(D + \log^5 n)$ upper bound is proved. The efficient construction of the protocol with such a completion time relies on a de-randomization of the well-known distributed randomized protocol in [3]: However, it is not clear whether this de-randomization can be done efficiently.

The *communication complexity* of our protocol, that is, the maximum number of messages exchanged during its execution, turns out to be $O(n \log \Delta \log(n/D))$, since each node sends at most $O(\log \Delta \log(n/D))$ messages. Moreover, when the source is unknown our protocol technique works in $O(D \log \Delta \log n)$ time.

The probabilistic argument used in order to efficiently construct small ad-hoc selective families will then be applied in order to develop a polynomial-time approximation algorithm for the MAX POS ONE-IN-k-SAT problem: Given a set of clauses with each clause containing exactly k literals, all positive, find a truth-assignment to the Boolean variables that 1-in-k satisfies the maximal number of clauses, where a clause is *1-in-k satisfied* if exactly one literal in the clause is assigned the value true. According to [17,13], this problem is NP-hard. Furthermore, from the approximation algorithm for the more general maximum constraint satisfaction problem (MAX CSP), it is known that the problem is $\frac{2^k}{k}$-approximable [14]. The performance ratio of our algorithm is bounded by $\frac{4(k-1)}{k} \leq 4$ (notice that the performance ratio is bounded by a constant, that is, 4 which does not depend on the value of k).

Finally, we investigate the combinatorial optimization problem of computing a broadcasting protocol with minimal completion time, called MIN BROADCAST.

This problem is NP-hard [4]: However, the reduction, which starts from the exact-3 cover problem, yields radio networks with $D = 2$. We instead introduce a new reduction starting from the problem of computing ad-hoc selective families of minimal size, called MIN SELECTIVE FAMILY. This reduction allows us to show that, for any fixed $D \geq 2$, if MIN D-BROADCAST (i.e., the problem restricted to radio networks of maximal eccentricity D) is r-approximable, then MIN SELECTIVE FAMILY is $\frac{rD-1}{D-1}$-approximable. Since the latter is not r-approximable, for any $r < 2$, we obtain that MIN BROADCAST cannot be approximated within a factor less than $2 - 1/D$ (unless P = NP).

2 Efficient Construction of Ad-hoc Selective Families

This section provides an efficient method to construct selective families of small size. More precisely, we prove the following

Theorem 1 *There exists an algorithm that, given a collection $\mathcal{F}$ of subsets of $[n]$, each of size in the range $[\Delta_{min}, \Delta_{max}]$, computes a selective family $\mathcal{S}$ for $\mathcal{F}$ of size $O((1 + \log(\Delta_{max}/\Delta_{min})) \cdot \log |\mathcal{F}|)$. The time complexity of the algorithm is $O(n^2|\mathcal{F}| \log |\mathcal{F}| \cdot (1 + \log(\Delta_{max}/\Delta_{min})))$.*

Proof. The proof consists of two main steps. We first show the *existence* of the selective family $\mathcal{S}$ by using a *probabilistic* construction. Then, an efficient algorithm that de-randomizes this construction is presented.

Probabilistic Construction. Without loss of generality, we can assume that $\Delta_{min} \geq 2$. For each $i \in \{\lceil \log \Delta_{min} \rceil, \ldots, \lceil \log \Delta_{max} \rceil\}$, consider a family $\mathcal{S}_i$ of l sets (the value of l is specified later) in which each set is constructed by randomly picking every element of $[n]$ independently, with probability $\frac{1}{2^i}$.
Fix a set $F \in \mathcal{F}$ and consider a set $S \in \mathcal{S}_i$, where i is the integer such that $\frac{1}{2} \leq \frac{|F|}{2^i} < 1$; then it holds that

$$\Pr[|F \cap S| = 1] = \frac{|F|}{2^i}\left(1 - \frac{1}{2^i}\right)^{|F|-1} > \frac{|F|}{2^i}\left(1 - \frac{1}{2^i}\right)^{2^i} \geq \frac{|F|}{4 \cdot 2^i} \geq \frac{1}{8} \qquad (1)$$

where the second inequality is due to the fact that $\left(1 - \frac{1}{t}\right)^t \geq \frac{1}{4}$ for $t \geq 2$. We then define the family $\mathcal{S}$ as the union of the families $\mathcal{S}_i$, for each $i \in \{\lceil \log \Delta_{min} \rceil, \ldots, \lceil \log \Delta_{max} \rceil\}$. Clearly, $\mathcal{S}$ has size $O((1 + \log(\Delta_{max}/\Delta_{min})) \cdot l)$. The probability that $\mathcal{S}$ does not select F is upper bounded by the probability that $\mathcal{S}_i$ does not select F. The sets in $\mathcal{S}_i$ have been constructed independently, so, from Eq. 1, this probability is at most $\left(1 - \frac{1}{8}\right)^l \leq e^{-\frac{l}{8}}$. Finally, we have that

$$\Pr[\mathcal{S} \text{ is not selective for } \mathcal{F}] \leq \sum_{F \in \mathcal{F}} \Pr[\mathcal{S} \text{ doesn't select } F] \leq \sum_{F \in \mathcal{F}} e^{-\frac{l}{8}} = |\mathcal{F}| e^{-\frac{l}{8}}.$$

The last value is less than 1 for $l > 8 \log |\mathcal{F}|$. Hence, such an $\mathcal{S}$ exists.
De-randomization. The de-randomization is obtained by applying the "greedy" criterium yielded by the method of the *conditional probabilities* [12].

Let us represent any subset $S \subseteq [n]$ as a binary sequence $\langle s_1, \ldots, s_n \rangle$ where, for any $i \in [n]$, $s_i = 1$ if and only if $i \in S$. Let $i \in [n]$, let $F \in \mathcal{F}$ of size Δ, and let $\langle s_1, \ldots, s_{i-1} \rangle$ be any sequence of $i-1$ bits (i.e., any subset of the first $i-1$ elements of $[n]$). Then, define the (conditional) probabilities

$$Y_i(F) = \Pr\left[|F \cap \langle s_1, \ldots, s_{i-1}, 1, x_{i+1}, \ldots, x_n \rangle| = 1\right]$$
$$N_i(F) = \Pr\left[|F \cap \langle s_1, \ldots, s_{i-1}, 0, x_{i+1}, \ldots, x_n \rangle| = 1\right]$$

where, for any $k = i+1, \ldots, n$, x_k is a bit chosen independently at random with

$$\Pr[x_k = 1] = 1/\Delta.$$

The algorithm relies on the following

Lemma 2 *It is possible to compute both $Y_i(F)$ and $N_i(F)$ in $O(n)$ time.*

Proof. Let us define $S_i = \langle s_1, \ldots, s_{i-1}, 0, \ldots, 0 \rangle$, and $I_i = \{i, i+1, \ldots, n\}$. Define also $\delta_i(F) = |F \cap I_i|$. If $\delta_i(F) = 0$, then it is easy to verify that

$$Y_i(F) = N_i(F) = \begin{cases} 1 & \text{if } |F \cap S_i| = 1, \\ 0 & \text{otherwise.} \end{cases}$$

If, instead, $\delta_i(F) > 0$ then two cases may arise

- **Case** $i \in F$. Then, it holds that

$$Y_i(F) = \begin{cases} 0 & \text{if } |F \cap S_i| \geq 1, \\ \left(1 - \frac{1}{\Delta}\right)^{\delta_i(F)-1} & \text{otherwise,} \end{cases}$$

$$N_i(F) = \begin{cases} 0 & \text{if } |F \cap S_i| \geq 2 \\ & \text{or } \ \delta_i(F) = 1 \ \wedge \ |F \cap S_i| = 0, \\ \left(1 - \frac{1}{\Delta}\right)^{\delta_i(F)-1} & \text{if } |F \cap S_i| = 1, \\ \frac{\delta_i(F)-1}{\Delta}\left(1 - \frac{1}{\Delta}\right)^{\delta_i(F)-2} & \text{otherwise.} \end{cases}$$

- **Case** $i \notin F$. Then, it holds that

$$Y_i(F) = N_i(F) = \begin{cases} 0 & \text{if } |F \cap S_i| \geq 2, \\ \left(1 - \frac{1}{\Delta}\right)^{\delta_i(F)} & \text{if } |F \cap S_i| = 1, \\ \frac{\delta_i(F)}{\Delta}\left(1 - \frac{1}{\Delta}\right)^{\delta_i(F)-1} & \text{otherwise.} \end{cases}$$

The proof is completed by observing that all the computations required by the above formulas can be easily done in $O(n)$ time. □

Figure 2 shows the algorithm $\texttt{greedy}_{\texttt{MSF}(\Delta)}$ that finds the desired selective family when all subsets in $\mathcal{F}$ have the same size. As for the general case, the algorithm must be combined with the technique in the probabilistic construction that splits $\mathcal{F}$ into a logarithmic number of families, each containing subsets having "almost" the same size. A formal description of this generalization will be given in the full version of the paper. However, we observe here that the time complexity of the general algorithm is $O(1 + \log(\Delta_{max}/\Delta_{min}))$ times the time complexity of $\texttt{greedy}_{\texttt{MSF}(\Delta)}$.

Input $\mathcal{F} = \{F_1, \ldots, F_m\}$
$\mathcal{F}' := \mathcal{F}$
$j := 0$
While $\mathcal{F}' \neq \emptyset$ **Do** /* Construct the jth selector S_j */
  For each $i = 1, \ldots, n$ **Do**
    For each $F \in \mathcal{F}'$ **Do** compute $Y_i(F)$ and $N_i(F)$ (using Lemma 2)
    $Y_i := \sum_{F \in \mathcal{F}'} Y_i(F)$
    $N_i := \sum_{F \in \mathcal{F}'} N_i(F)$
    If $Y_i \geq N_i$ **Then** $s_i := 1$ **Else** $s_i := 0$.
  End (For)
  $j := j + 1$
  $S_j := \langle s_1, \ldots, s_n \rangle$
  $\mathcal{F}' := \mathcal{F}' - \{F \in \mathcal{F}' : |F \cap S_j| = 1\}$
End (While)
Return $\mathcal{S} = \{S_1, \ldots, S_j\}$.

Fig. 2. Algorithm $\texttt{greedy}_{\texttt{MSF}(\Delta)}$.

Lemma 3 *Let $\mathcal{F}$ be a family of subsets of $[n]$, each of size Δ. Then, Algorithm $\texttt{greedy}_{\texttt{MSF}(\Delta)}$ (with input $\mathcal{F}$) computes a selective family $\mathcal{S}$ for $\mathcal{F}$ of size $O((1 + \log(\Delta_{max}/\Delta_{min})) \log |\mathcal{F}|)$ in time $O(n^2 |\mathcal{F}| \log |\mathcal{F}|)$.*

Proof. We first prove that, at each iteration of the **While** loop, the computed subset S_j selects at least 1/8 of the remaining subsets of $\mathcal{F}$, i.e., $\mathcal{F}'$.
Let B be a subset of $[n]$ randomly chosen according to the following probability function: For each $i \in [n]$, $i \in B$ with probability $1/\Delta$. Let $E(B)$ denote the expected number of subsets F in $\mathcal{F}'$ such that $|F \cap B| = 1$.
For any $i \in [n]$ and for any bit sequence $b_1, \ldots, b_i$, let $E(B|b_1, \ldots, b_i)$ be the expected number of subsets F in $\mathcal{F}'$ such that $|F \cap B| = 1$, where $B = \langle b_1, \ldots, b_n \rangle$ is the random completion of the sequence $b_1, \ldots, b_i$ such that, for any $i + 1 \leq l \leq n$, $b_l = 1$ with probability $1/\Delta$.

Let $s_1, \ldots, s_n$ be the choices made by the **For** loop.

Claim 1 *For any $i = 1, \ldots, n$, $E(B|s_1, \ldots, s_i) \geq E(B)$.*

Proof. The proof is by induction on i. For $i = 1$, by definition, we have that

$$E(B) = \frac{1}{\Delta} E(B|1) + \left(1 - \frac{1}{\Delta}\right) E(B|0)$$

So, $E(B) \leq \max\{E(B|1), E(B|0)\} = \max\{Y_1, N_1\}$, and s_1 is chosen so that $E(B|s_1) = \max\{Y_1, N_1\}$. We now assume that the claim is true for $i - 1$. Then, s_i is chosen so that

$$\begin{aligned} E(B|s_1, \ldots, s_i) &= \max\{Y_i, N_i\} \\ &= \max\{E(B|s_1, \ldots, s_{i-1}, 1), E(B|s_1, \ldots, s_{i-1}, 0)\}. \end{aligned}$$

It also holds that

$$E(B|s_1,\ldots,s_{i-1}) = \frac{1}{\Delta}E(B|s_1,\ldots,s_{i-1},1) + \left(1-\frac{1}{\Delta}\right)E(B|s_1,\ldots,s_{i-1},0)$$
$$\leq \max\{E(B|s_1,\ldots,s_{i-1},1), E(B|s_1,\ldots,s_{i-1},0)\}.$$

By combining the above inequalities with the inductive hypothesis, we get

$$E(B|s_1,\ldots,s_i) \geq E(B|s_1,\ldots,s_{i-1}) \geq E(B).$$

□

Let us observe that, $E(B|s_1,\ldots,s_n)$ is equal to the number of subsets in $\mathcal{F}'$ that are selected by $S = \langle s_1,\ldots,s_n \rangle$. From Claim 1, this number is at least $E(B)$. Moreover, from Eq. 1, it holds that $E(B) \geq |\mathcal{F}'|/8$. Finally, from Lemma 2, it follows that the time complexity of $\mathtt{greedy}_{\mathtt{MSF}(\Delta)}$ is $O(n^2|\mathcal{F}|\log|\mathcal{F}|)$. □

3 Two Applications of Theorem 1

The Broadcast Protocol. For any possible source node $s \in V$, let $L_i(s)$ be the set of nodes whose distance from s is i. For each node in $L_{i+1}(s)$ let us consider the set of its in-neighbors belonging to $L_i(s)$; let $\mathcal{F}_i(s)$ be the family of all such sets. Then, let $\mathcal{S}_i$ be an arbitrarily ordered selective family for $\mathcal{F}_i(s)$.

Description of Protocol BROAD**.** The protocol consists of D phases. The goal of phase i is to inform nodes at distance i from the source.
– In the first phase the source sends its message.
– The i-th phase, with $i \geq 2$, consists of $|\mathcal{S}_{i-1}|$ time-slots. At time-slot j of the i-th phase a node v sends the source message if and only if the following two conditions are satisfied:

- v belongs to the j-th set of $\mathcal{S}_{i-1}$;
- v has been informed for the first time during phase $i-1$.

All the remaining nodes work as receivers.

Theorem 4 *Protocol* BROAD *completes (and terminates) a broadcast operation on an n-node graph of maximum eccentricity D and maximum in-degree Δ within $O(D\log\Delta\log\frac{n}{D})$ time-slots. Moreover, the cost of the protocol is $O(n\log\Delta\log\frac{n}{D})$.*

Proof. To show the correctness (and the performances) of the protocol we prove the following

Claim 2 *all the nodes at distance i from the source s are informed, for the first time, during phase i.*

Sketch of the proof. The proof is by induction on the distance i. For $i = 1$ the claim is obvious. We thus assume that all nodes at distance i have received the source message, for the first time, during phase $i-1$. Let consider a node v at distance $i+1$ and let F_v be the set of all its in-neighbors at distance i from the source. Since F_v belongs to $\mathcal{F}_i(s)$ and $\mathcal{S}_i$ is selective for $\mathcal{F}_i(s)$, there will be a time-slot in phase $i+1$ in which only one of the nodes in F_v transmits the source message so that v will correctly receive it. Notice that, by the inductive hypothesis, any in-neighbor of v that is not in F_v has not been informed in phase i, so it does not transmit during phase $i+1$. □

Since the graph has maximum in-degree Δ, the size of any subset in $\mathcal{F}_i(s)$ is at most Δ. Hence, from Theorem 1, we have that $|\mathcal{S}_i| \le c \log \Delta \log |\mathcal{F}_i(s)|$, for some constant $c > 0$. The total number of time-slots required by the protocol is thus

$$1 + \sum_{i=1}^{D-1} c \log \Delta \log |\mathcal{F}_i(s)| = 1 + c \log \Delta \log \prod_{i=1}^{D-1} |\mathcal{F}_i(s)| \le 1 + c \log \Delta \log \prod_{i=1}^{D-1} \frac{n}{D}$$

where the last inequality is due to the facts that $\sum_{i=1}^{D-1} |\mathcal{F}_i(s)| \le n$ and that $\prod_{i=1}^{D-1} |\mathcal{F}_i(s)|$ is maximized when all the $|\mathcal{F}_i(s)|$ are equal. It thus follows that BROAD has $O(D \log \Delta \log \frac{n}{D})$ completion time.
As for the cost of the protocol, it suffices to observe that once a node has acted as transmitter during a phase, after that phase it can turn into the inactive state forever. □

Remark. If we require a protocol that works for any source, we need to select a bigger set of families, i.e., the families $\mathcal{F}_i = \cup_{s \in V} \mathcal{F}_i(s)$, $i = 1 \ldots n-1$. By applying the same arguments of the above proof, we can easily obtain a broadcast protocol having $O(D \log \Delta \log n)$ completion time.

Approximation of the MAX POS ONE-IN-k-SAT *Problem.* An (even) simplified version of the algorithm $\texttt{greedy}_{\texttt{MSF}(\Delta)}$ can be successfully used to obtain a constant factor approximation for MAX POS ONE-IN-k-SAT.

Corollary 5 *There exists a polynomial-time* $\frac{4(k-1)}{k}$*-approximation for* MAX POS ONE-IN-k-SAT.

Sketch of the proof. Given a set $\mathcal{C}$ of clauses with each clause containing exactly k positive literals, for any clause $C = \{x_{i(1)}, x_{i(2)}, \ldots, x_{i(k)}\}$ in $\mathcal{C}$, we consider the subset $F(C) = \{i(1), i(2), \ldots, i(k)\} \subseteq [n]$. Then, we apply Algorithm $\texttt{greedy}_{\texttt{MSF}(\Delta)}$ on the instance $\mathcal{F}(\mathcal{C}) = \{F(C) \ : \ C \in \mathcal{C}\}$ (notice that $\Delta = k$). The same probabilistic argument adopted in the proof of Theorem 1 guarantees that the first selector S_1, computed by the algorithm, satisfies at least $\frac{k}{4(k-1)}$ clauses of $\mathcal{C}$. The corollary, hence, follows. □

4 Hardness Results

By adopting the definitions in the proof of Corollary 5, and from the fact that MAX POS ONE-IN-k-SAT is NP-hard (see [17,13]), it easily follows

Theorem 6 *It is* NP*-hard to approximate* MIN SELECTIVE FAMILY *within a factor smaller than 2.*

The following result reverts the connection between selective families and broadcast protocols. Indeed, it shows that any non-approximability result for the MIN SELECTIVE FAMILY directly translates into an equivalent negative result for MIN BROADCAST, when restricted to networks of constant eccentricity.
Let MIN D-BROADCAST denote the restriction of MIN BROADCAST to networks of eccentricity D.

Theorem 7 *For any fixed positive integer $D \geq 2$, if* MIN D-BROADCAST *is r-approximable, then* MIN SELECTIVE FAMILY *is $\frac{rD-1}{D-1}$-approximable.*

Sketch of the proof. Let $\mathcal{F}$ be an instance of MIN SELECTIVE FAMILY, where $\mathcal{F} = \{F_1, \ldots, F_m\}$ is a collection of subsets of $[n]$. We construct (in polynomial time) an instance $\langle G_D^{\mathcal{F}}, s\rangle$ of MIN D-BROADCAST such that $\mathcal{F}$ has a selective family of size k if and only if $\langle G_D^{\mathcal{F}}, s\rangle$ has a broadcast protocol with completion time equal to $1 + k(D-1)$. The network $G_D^{\mathcal{F}}$ is a $D+1$ layered graph with layers $L_0, \ldots, L_D$, with $L_0 = \{s\}$ and the number of nodes in $G_D^{\mathcal{F}}$ is at most $n|\mathcal{F}|^D$. The graph $G_D^{\mathcal{F}}$ is defined by induction on D:
Base Step ($D = 2$). The network $G_2^{\mathcal{F}}$ consists of three levels: $L_0 = \{s\}$, $L_1 = \{x_1, \ldots, x_n\}$ and $L_2 = \{y_1, \ldots, y_m\}$, where s is connected to every $x_i \in L_1$, and the edge (x_i, y_j) exists iff $x_i \in F_j$.
Inductive Step. The graph $G_{D+1}^{\mathcal{F}}$ can be obtained from $G_D^{\mathcal{F}}$ as follows: The layer L_{D+1} of $G_{D+1}^{\mathcal{F}}$ is obtained by replacing every node in the layer L_D of $G_D^{\mathcal{F}}$ by a copy of the graph $G_2^{\mathcal{F}} \setminus \{s\}$. More formally,

1. Replace every $z_i \in L_D$ by the set $X_{D+1}^i = \{x_1^i(D+1), \ldots, x_n^i(D+1)\}$; each of such new vertices has the same in-neighborhood of z_i.
2. Add a set $Y_{D+1}^i = \{y_1^i(D+1), \ldots, y_m^i(D+1)\}$ of m new nodes. Then, add the edge $(x_k^i(D+1), y_l^i(D+1))$ if and only if (x_k, y_l) is an edge in $G_2^{\mathcal{F}}$.

So, the layer L_D of $G_{D+1}^{\mathcal{F}}$ is the union of all X_{D+1}^i's determined by the last level of $G_D^{\mathcal{F}}$ and the layer L_{D+1} of $G_{D+1}^{\mathcal{F}}$ is the union of all Y_{D+1}^i's.

Claim 3 *$\mathcal{F}$ has a selective family of size k if and only if $\langle G_D^{\mathcal{F}}, s\rangle$ has a broadcast protocol with completion time equal to $1 + k(D-1)$.*

Sketch of the proof. The proof is by induction on D.
Base Step ($D = 2$). Consider the family $\mathcal{F}^{L_2}$ of in-neighborhoods of the nodes in L_2. Then, $G_2^{\mathcal{F}}$ admits a broadcast protocol of completion time $k+1$ iff $\mathcal{F}^{L_2}$ has a selective family of size k. Since $\mathcal{F} = \mathcal{F}^{L_2}$, then the theorem follows.

Inductive Step. ($\Rightarrow$). It is easy to show that by a suitable iteration of the broadcast protocol for $G_2^{\mathcal{F}}$ (yielded by the selective family $\mathcal{S}$ for $\mathcal{F}$) on $G_D^{\mathcal{F}}$, we obtain a completion time $1 + k \cdot (D - 1)$, for any D.
($\Leftarrow$). Consider any broadcast protocol P for $\langle G_{D+1}^{\mathcal{F}}, s\rangle$ with completion time $1 + kD$. Also, let t be the number of time slots required by P to inform all the nodes in the second-last layer L_D of $G_{D+1}^{\mathcal{F}}$. It is easy to see that P completes broadcasting on $\langle G_D^{\mathcal{F}}, s\rangle$ within time-slot t.

If $t \leq 1 + k(D - 1)$, then by inductive hypothesis, $\mathcal{F}$ has a selective family of size k. Otherwise, we first observe that for any i, all the nodes in X_{D+1}^i have the same in-neighborhood, thus implying that they are informed at the same time slot. Hence, there must exist a set X_{D+1}^{last} that is informed (according to P) at time slot t. Let $t' = t + \Delta t$ be the number of time slots necessary to P to inform Y_{D+1}^{last}, that is, the set of out-neighbors of X_{D+1}^{last}. From the fact that $t > 1 + k(D - 1)$ and $t' = t + \Delta t \leq 1 + kD$, we obtain $\Delta t < k$. By construction, the subgraph induced by $X_{D+1}^{last} \cup Y_{D+1}^{last}$ is isomorphic to $G_2^{\mathcal{F}} \setminus \{s\}$. Hence, there exists a protocol for $\langle G_2^{\mathcal{F}}, s\rangle$ with broadcasting time $1 + \Delta t < 1 + k$. By inductive hypothesis, $\mathcal{F}$ has a selective family of size k. □

The proof of Claim 3 easily implies the following

Claim 4 *Given any broadcast protocol P with completion time on $G_D^{\mathcal{F}}$ equal to t, it is possible to construct (in time polynomial in $|G_D^{\mathcal{F}}|$) a protocol P' with completion time on $G_D^{\mathcal{F}}$ equal to $t' = 1 + k(D - 1) \leq t$, for some integer $k \geq 1$.*

Consider any r-approximation algorithm for Min D-Broadcast. From Claim 4, we can assume that such an algorithm returns a broadcast protocol for $\langle G_D^{\mathcal{F}}, s\rangle$ of completion time $APX(G_D^{\mathcal{F}}) = 1 + k \cdot (D - 1)$, for some $k \geq 1$. By hypothesis, it holds that

$$\frac{APX(G_D^{\mathcal{F}})}{OPT(G_D^{\mathcal{F}})} = \frac{1 + k \cdot (D - 1)}{1 + OPT(\mathcal{F}) \cdot (D - 1)} \leq r, \tag{2}$$

which implies that

$$\frac{k}{OPT(\mathcal{F})} \leq \frac{r[1 + OPT(\mathcal{F}) \cdot (D - 1)] - 1}{OPT(\mathcal{F}) \cdot (D - 1)} \leq \frac{rD - 1}{D - 1}.$$

Finally, by applying Claim 3, we can construct (in polynomial time) a selective family for $\mathcal{F}$ of size at most $k \leq OPT(\mathcal{F})\frac{rD-1}{D-1}$. Hence the theorem follows. □

By making use of Theorem 6 and Theorem 7, we can easily obtain the following result (whose proof is here omitted).

Corollary 8 *For any constant $D \geq 2$, it is* NP*-hard to approximate* Min D-Broadcast *within a factor less than $2 - 1/D$. Moreover, for any positive integer $c \geq 1$,* Min $(\log^{c/(c+1)} n)$-Broadcast *cannot be approximated by a factor less than 2 (unless* NP $\subseteq$ DTIME$[n^{\log^c n}]$*).*

Remark 9 *The $O(D + \log^5 n)$ broadcasting protocol of [11] implies that* Min $D(n)$-Broadcast *is in* APX *for any $D(n) \in \Omega(\log^5 n)$.*

5 Open Problems

The main open problem which is related to this paper consists of determining whether the de-randomization techniques can also be applied to the probabilistic construction of selective families given in [9]. We suspect that this is not true and, hence, that, in order to constructively achieve the upper bound of [9], alternative techniques have to be used.

References

1. N. Alon,, A. Bar-Noy, N. Linial, and D. Peleg (1991), A lower bound for radio broadcast, *J. Comput. System Sci.*, 43, 290-298.
2. N. Alon and J. Spencer (1992), *The probabilistic method*, Wiley.
3. R. Bar-Yehuda, O. Goldreich, and A. Itai (1992), On the time-complexity of broadcast operations in multi-hop radio networks: an exponential gap between determinism and randomization, *J. Comput. System Sci.*, 45, 104–126.
4. I. Chlamtac and S. Kutten (1985), On Broadcasting in Radio Networks - Problem Analysis and Protocol Design, *IEEE Trans. on Communications*, 33, 1240–1246.
5. I. Chlamtac, O. Weinstein (1991), The Wave Expansion Approach to Broadcasting in Multihop Radio Networks, *IEEE Trans. on Communications*, 39, 426–433.
6. B.S. Chlebus, L. Gąsieniec, A.M. Gibbons, A. Pelc, and W. Rytter (2000), Deterministic broadcasting in unknown radio networks, *Proc. of 11th ACM-SIAM SODA*, 861–870.
7. B. S. Chlebus, L. Gąsieniec, A. Ostlin, and J. M. Robson (2000), Deterministic radio broadcasting, *Proc. of 27th ICALP*, LNCS, 1853, 717–728.
8. M. Chrobak, L. Gąsieniec, W. Rytter (2000), Fast broadcasting and gossiping in radio networks, *Proc. of 41st IEEE FOCS*, 575–581.
9. A.E.F. Clementi, A. Monti, and R. Silvestri (2001), Selective Families, Superimposed Codes, and Broadcasting on Unknown Networks, *Proc. 12th ACM-SIAM SODA*, 709–718.
10. R. Gallager (1985), A perspective on multiaccess channels, *IEEE Trans. Inform. Theory*, 31, 124–142.
11. I. Gaber and Y. Mansour (1995), Broadcast in Radio Networks, *Proc. 6th ACM-SIAM SODA*, 577–585.
12. R. Motwani and P. Raghavan (1995), *Randomized Algorithms*, Cambridge University Press.
13. C.H. Papadimitriou (1994), *Computational Complexity*, Addison Wesley.
14. C.H. Papadimitriou and M. Yannakakis (1991), Optimization, approximation, and complexity classes, *J. Comput. System Sci.*, 43, 425–440.
15. A. Pelc (2000), Broadcasting in Radio Networks, unpublished manuscript.
16. L.G. Roberts (1972), Aloha Packet System with and without Slots and Capture, ASS Notes 8, Advanced Research Projects Agency, Network Information Center, Stanford Research Institute.
17. T.J. Schaefer (1979), The complexity of satisfiability problems, *Proc. 10th ACM STOC*, 216–226.

L Infinity Embeddings

Don Coppersmith

IBM TJ Watson Research Center, Yorktown Heights NY 10598, USA
dcopper@us.ibm.com
http://www.research.ibm.com/people/c/copper/

Abstract. Given ϵ, for N sufficiently large, we give a metric on N points which cannot be isometrically embedded in ℓ_∞^b for $b < N - N^\epsilon$.

1 Introduction

Consider an undirected graph G on N vertices, and a collection $H = \{h_1, \ldots, h_b\}$ of complete bipartite graphs on subsets of these N vertices, such that the edge set of G is the union of the edge sets of h_i. Let $f(G, N)$ be the least integer b allowing such a representation, and let $f(N)$ be the maximum of $f(G, N)$ over all undirected graphs G on N vertices.

We will show that for each $\epsilon > 0$ there is an N_ϵ such that for all $N > N_\epsilon$ we have $f(N) \geq N - N^\epsilon$.

As a corollary, we will exhibit a metric T on N points, obeying the triangle inequality, such that any isometric embedding of T into a space ℓ_∞^b requires $b \geq f(N)$. That is, there is an N-point set with a metric, which cannot be isometrically embedded into ℓ_∞^b unless $b \geq N - N^\epsilon$.

The construction uses random graphs.

2 Notation

For positive integers r, s, let $K_{r,s}$ be the complete bipartite graph on r and s vertices. When there are $r+s$ vertices $x_1, x_2, \ldots, x_r, y_1, y_2, \ldots, y_s$ in G such that x_i and y_j are joined by an edge in G for each i and j, we say that G *contains* $K_{r,s}$. This definition disregards edges (x_i, x_j) or (y_i, y_j); $K_{r,s}$ is not necessarily an *induced* subgraph of G.

A *cover* $H = \{h_1, h_2, \ldots, h_b\}$ of the undirected graph G is a collection of complete bipartite graphs h_i on subsets of the vertex set of G, the union of whose edge sets gives the edge set of G.

For each bipartite graph $h_i = K_{r,s}$ in the cover, with $r \leq s$, say that the r nodes *share a base* of h_i; further, if $r = 1$ that node is a *singleton*. If $r = s$, select one side arbitrarily to be the base.

3 Outline of Construction

Select an integer $L > 2/\epsilon$. Set $K = N^\epsilon$. We will generate a random graph G, with edge probability sufficiently small that the only complete bipartite graphs

M. Goemans et al. (Eds.): APPROX-RANDOM 2001, LNCS 2129, pp. 223–228, 2001.

$K_{r,s}, (r \leq s)$ contained in G are those with $r < L$. Consider G and a cover $H = \{h_1, \ldots, h_b\}$ with $b < N - K$. We will assign a positive "weight" $w(h, x)$ to each pair (h, x) where $h \in H$ and where the vertex x "shares in the base" of h. We use these weights and a second random construction, to identify an "assignment" J, namely two disjoint sets of vertices X, Y and a mapping $g : Y \to X$ such that each $x \in X$ shares in the bases of fewer than L graphs h, and for each such graph some vertex $y \in g^{-1}(x)$ also shares in the basis. This setup J will determine conditions on edges among the vertices in X and Y. We bound the probability that a given G satisfies such conditions. Putting them together, we find that, with high probability, a random G is compatible with no such J and thus admits no such cover.

4 Construction

Theorem 1. *Given $\epsilon > 0$ there is an N_ϵ such that for each $N > N_\epsilon$ there is a graph G on N vertices such that any cover $H = \{h_1, h_2, \ldots, h_b\}$ of G must have $b \geq N - N^\epsilon$.*

Proof. Select an integer L and real δ satisfying $L > 2/\delta > 2/\epsilon$. Select N_ϵ large enough that for $N > N_\epsilon$ we have $N^{\epsilon-\delta} > 24L^4 \log N$. Given $N > N_\epsilon$, construct a random undirected graph G on N vertices, with independent edge probability $p = N^{-\delta}$.

Set $K = N^\epsilon$, and suppose $b < N - K$.

The expected number of copies of the complete bipartite graph $K_{L,L}$ in G is bounded by $N^{2L}p^{L^2}/(L!)^2 = o(1)$. So with good probability, the only complete bipartite graphs $K_{r,s}$ $(r \leq s)$ that G will contain will have $r < L$. Hereafter we assume that this is the case, and revisit this assumption later.

Consider a cover $H = \{h_1, h_2, \ldots, h_b\}$ of G.

Without loss of generality, we will assume that a singleton vertex x will not share in the bases for other dimensions, since each singleton x can be removed from all other bipartite graphs h_i without increasing the size b of the cover.

We assign positive weights $w(h, x)$ to each pair (h, x) of a bipartite graph $h \in H$ and a vertex x which shares in the base of h. If x does not share in the base of h we set $w(h, x) = 0$. If x shares in the bases of $M \geq L$ bipartite graphs, we assign $w(h, x) = 1/M$ to each such instance. (For any graph h all of whose weights have been assigned, we will have $\sum_x w(h, x) < 1$.) For each graph h with some weights unassigned, assign arbitrary positive weights $w(h, x)$ to these unassigned positions x, subject to $\sum_x w(h, x) = 1$. (For example, for each h the unassigned weights could be made equal.) So for each h we get $\sum_x w(h, x) \leq 1$.

Each vertex x has total weight $W(x) = \sum_h w(h, x)$ between 0 and $L - 1$. A singleton has total weight 1. If $W(x) < 1$ we know that x shares in fewer than L bases.

A standard counting argument shows that

$$\sum_x (1 - W(x)) \geq N - b > K.$$

Now we define a random set X'. Set a parameter $\alpha = 1/(2L^3)$. For each x independently, include x in X' with probability $\max(0, \alpha(1 - W(x)))$. By our counting argument, $E(|X'|) > \alpha K$.

Each $x \in X'$ shares in fewer than L bases. For each $h \in H$ for which x shares in the base, we randomly select another vertex y which shares in this base, with the conditional probability of selecting y being given by

$$\frac{w(h,y)}{\sum_{z \neq x} w(h,z)}.$$

If x, y share in several bases and y gets selected several times in association with x, we pretend it was only selected once. We let Z' be the set of triples (x, y, h) selected in this process.

For given x, h, the probability that y is selected, given that $x \in X'$, is bounded by

$$\frac{w(h,y)}{\sum_{z \neq x} w(h,z)} = \frac{w(h,y)}{1 - w(h,x)} \leq \frac{w(h,y)}{1 - W(x)},$$

since h is one of the graphs for which $\sum_z w(h,z) = 1$. Factoring in the probability $\alpha(1 - W(x))$ that $x \in X'$, we see that the probability of the triple (x, y, h) is bounded by $\alpha w(h, y)$. Thus for a given y, the expected number of triples (x, y, h) is at most

$$L\alpha \sum_h w(h,y) \leq \alpha L W(y) \leq \alpha L^2.$$

(The first factor L recognizes the fact that several x might share the same h.) Add to this the probability that the node y was also selected to be in X', and the expected number of occurrences of y is bounded by

$$\alpha L W(y) + \alpha \max(0, 1 - W(y)),$$

which is αL^2 (if $W(y) > 1$) or at most αL (if $W(y) \leq 1$). By our choice of α, either bound is less than $1/(2L)$.

By independence, if we condition on the event that the triple $(x, y, h) \in Z'$, the expected number of *other* occurrences of y in Z' (that is, in triples with other $x' \neq x$) is also less than $1/(2L)$.

Now we refine X' by throwing out duplicates. Let X be the set of $x \in X'$ for which neither x nor the associated y appear elsewhere in Z'. By the arguments above, each x will be deleted with probability at most $L \times (1/(2L)) = 1/2$, so that the expected size of X is at still least $\alpha K/2$.

Repeat the random selection until we find an X at least that large, and delete vertices to get $|X| = q = \lceil \alpha K/2 \rceil$. Let Y be the set of associated vertices y, so that $|Y| \leq (L-1)|X|$, and let $Z \subseteq Z'$ be the result of deleting triples there. Define $g : Y \to X$ as the (unique) association $g(y) = x$ if $(x, y, h) \in Z$. Together, (X, Y, g) form our assignment J.

$J = (X, Y, g)$ enjoys the following properties:

1. $|X| = q$.

2. $|Y| \leq (L-1)|X|$; in fact $|g^{-1}(x)| \leq L-1$.
3. X, Y are disjoint.
4. Each $x \in X$ shares in fewer than L bases (possibly 0).
5. Each $x \in X$ is not a singleton.
6. For each graph h with x sharing the base, there is $y \in Y$ with $g(y) = x$ also sharing the base.

From these properties we deduce the following:

Lemma 1. *Given $x \neq x' \in X$, if the edge (x, x') is in G then either there is $y \in g^{-1}(x)$ with edge (y, x') in G, or there is $y' \in g^{-1}(x')$ with edge (x, y') in G.*

Proof. The edge (x, x') is either in one of the graphs h in which x shares a base, or in one of the graphs h' in which x' shares a base; in that graph, the associated y has an edge to x' (or y' to x), since it is a *complete* bipartite graph.

Lemma 2. *Given an assignment J, the probability that a random graph G satisfies the conditions of Lemma 1 is bounded by*

$$P = \left(1 - p(1-p)^{2(L-1)}\right)^{q(q-1)/2} < e^{-pq^2/3}.$$

Proof. For each of $q(q-1)/2$ unordered pairs of points (x, x') in X, we outlaw the event that (x, x') is an edge in G while each of (x, y') (for $y' \in g^{-1}(x')$) and (x', y) (for $y \in g^{-1}(x)$) is not an edge in G. This event has probability at least $p(1-p)^{2(L-1)}$. Further, each possible edge is examined only once (since each y is associated with a single $x = g(y)$), so the probabilities are independent. The bound $P < e^{-pq^2/3}$ is assured for N sufficiently large.

Lemma 3. *The number of assignments J is bounded by N^{qL}.*

Proof. Counting.

So the expected number of assignments J compatible with a random G is bounded by

$$N^{qL} e^{-pq^2/3} = e^{q(L \log N - pq/3)}.$$

Noticing that

$$\frac{pq}{3} \geq \frac{N^{-\delta} \alpha K}{6} = \frac{N^{-\delta} N^{\epsilon}}{12 L^3} > 2L \log N,$$

we find that the $-pq/3$ term dominates the exponent, so that the expected number of valid assignments J is $o(1)$.

Now we can revisit the assumption that G contains no large bipartite graphs. Let E denote the event that G contains a subgraph $K_{L,L}$. Let D denote the event that G has a cover H with $b < N - K$. Then

$$Prob(D) \leq Prob(E) + Prob(D \& \neg E) \leq o(1) + o(1) < 1.$$

So there exist graphs G on N vertices for which D fails. Any such graph validates the theorem.

5 Relation to Embeddings

Given a graph G on N vertices, which cannot be covered with b complete bipartite graphs, one immediately constructs a finite metric space on N points that cannot be isometrically embedded into ℓ^b_∞.

Let T be the metric defined on N points, where $d(x,y) = 2$ if (x,y) is an edge of G, and $d(x,y) = 1$ if (x,y) is not an edge of G. Consider an isometric embedding of T into ℓ^b_∞. For each dimension i, we perform the following normalizations in turn. It is a simple matter to check that the embedding remains valid at each step.

1. Round each coordinate down to the nearest integer.
2. If k is the largest coordinate, subtract $k-2$ from each coordinate, so that the largest coordinate is 2.
3. (Now all coordinates are 0, 1 or 2.) If more points have coordinate 0 than 2, interchange coordinates 0 and 2; this is the linear map $x \to 2-x$.

Now define complete bipartite graphs $h_i, i = 1, \ldots, b$: If in dimension i the points $x_1, \ldots, x_r$ have coordinate 0 and the points $y_1, \ldots, y_s$ have coordinate 2, then $h_i = K_{r,s}$ is the complete bipartite graph on $(x_1, \ldots, x_r; y_1, \ldots, y_s)$.

h_i can be empty if no points have coordinate 0 in dimension i.

It is clear that the bipartite graphs h_i form a cover of G, and that there are at most b graphs in this cover. This shows:

Theorem 2. *Given a graph G on N vertices with no cover $\{h_1, \ldots, h_b\}$ of a given size b, there is a finite metric T on N points which cannot be isometrically embedded into ℓ^b_∞.*

Corollary 1. *Given $\epsilon > 0$ there is an N_ϵ such that for all $N > N_\epsilon$ there is a metric T on N points which cannot be isometrically embedded into ℓ^b_∞ with $b < N - N^\epsilon$.*

Remark:

For the isometric embedding problem, a trivial lower bound is $b \geq N/2$, as seen by considering a metric with $N(N-1)/2$ distances linearly independent over $\mathbf{Q}$, and noting that each dimension can satisfy at most $N-1$ of these distances.

The upper bound $N-1$ comes from the "Frobenius embedding": in dimension $i, 1 \leq i \leq N-1$ let point x_j have coordinate $d(x_i, x_j)$.

A better lower bound of $\lfloor 2N/3 \rfloor$ is given by Holsztysnki [2]. A better upper bound of $N-2$ is given by Wolfe [3]. Both can be found in the book of Deza and Laurent [1].

Acknowledgments

Alan Hoffman and John Lew proposed the problem to me. Alan then encouraged me to seek publication, and even made it possible by unearthing a copy of an earlier version of my manuscript after twenty years.

References

1. M.M. Deza and M. Laurent, *Geometry of Cuts and Metrics*, Springer 1997, series *Algorithms and Combinatorics* **15**, page 156.
2. W.Holsztysnki, "$\mathbf{R}^n$ as universal metric space," *Notices of the AMS*, **25:A-367**, 1978.
3. D.Wolfe, "Imbedding of a finite metric set in an n-dimensional Minkowskispace," *Proc. Konin. Neder. Akad. Wetenschappen, Series A, Math. Sciences,* **70**: 136-140, 1967.

On Euclidean Embeddings and Bandwidth Minimization*

John Dunagan and Santosh Vempala

Department of Mathematics, MIT, Cambridge MA, 02139
{jdunagan, vempala}@math.mit.edu

Abstract. We study Euclidean embeddings of Euclidean metrics and present the following four results: (1) an $O(\log^3 n\sqrt{\log\log n})$ approximation for minimum bandwidth in conjunction with a semi-definite relaxation, (2) an $O(\log^3 n)$ approximation in $O(n^{\log n})$ time using a new constraint set, (3) a lower bound of $\Theta(\sqrt{\log n})$ on the least possible volume distortion for Euclidean metrics, (4) a new embedding with $O(\sqrt{\log n})$ distortion of point-to-subset distances.

1 Introduction

The minimum bandwidth problem asks for a permutation of the vertices of an undirected graph that minimizes the maximum difference between the endpoints of its edges. This maximum difference is called the bandwidth. Minimizing the bandwidth is NP-hard [5].

The question of finding good *approximations* to the minimum bandwidth has led to two different Euclidean embeddings of graphs. One of them is obtained as a solution to a semi-definite relaxation of the problem [1]. The other is an embedding that preserves *tree volumes* of subsets of a given metric [3,4]. The tree volume of a subset is the product of the edge lengths of a minimum spanning tree of the subset. These two embeddings were used in separate approximation algorithms.

In this paper we combine the two embeddings to obtain an improved approximation. The quality of the approximation is $O(\rho\log^{2.5} n)$ where ρ is the best possible volume distortion (defined in section 2) of a Euclidean metric. Using Rao's upper bound [6] of $O(\sqrt{\log n\log\log n})$ on ρ we obtain an approximation guarantee of $O(\log^3 n\sqrt{\log\log n})$ which improves on [4] by a factor of $\Theta(\sqrt{\log n})$.

Our approach immediately leads to the question of whether a better upper bound is possible. In section 5, we show a lower bound of $\Omega(\sqrt{\log n})$ on the volume distortion even for the path graph. Thus further improvements to bandwidth approximation will have to come from other avenues.

We then turn to the general question of embedding metrics in Euclidean space. Finding an embedding of a metric that "preserves" properties of the original metric is a classical problem. A natural property to consider in this regard is the original distance function itself. J. Bourgain [2] gave an embedding that

* Supported in part by NSF Career Award CCR-9875024.

M. Goemans et al. (Eds.): APPROX-RANDOM 2001, LNCS 2129, pp. 229–240, 2001.

achieves a *distortion* of $O(\log n)$ for any metric on n points, i.e. the distance between points in the embedding is within a factor of $O(\log n)$ of their distance in the metric. In other words, it "preserves" distances any point and any another point. A natural generalization would be an embedding that preserves the distance between any point and any subset of points. The distance of a point u to a subset of points S, is simply the distance of u to the closest point in S. For a Euclidean embedding, the distance of a point u to a subset S can be defined as the Euclidean distance from u to the *convex hull* of S. Thus point-to-subset distance is a direct generalization of point-to-point distance for metrics as well as for points in Euclidean space. In section 6, we give an embedding whose point-to-subset distortion is $O(\sqrt{\log n})$ for any Euclidean metric where the shortest distance is within a $poly(n)$ factor of the longest distance.

Replacing "convex" in the definition above by "affine" leads to another interesting property. In section 6 we observe that for any Euclidean embedding, the distortion of affine point-to-subset distances is also an upper bound on its volume distortion. In section 7, we formulate a new system of constraints that are separable in $O(n^{\log n})$ time, and which result in an $O(\log^3 n)$ approximation to the minimum bandwidth using the results of section 6. We conclude with the conjecture that our embedding (section 6.1) achieves the optimal volume distortion for Euclidean metrics.

2 Euclidean Embeddings of Metrics

Let $G = (V, E)$ be a finite metric with distance function $d(u, v)$. We restrict our attention throughout the paper to Euclidean embeddings ϕ of G that are contractions, i.e. the distances between embedded points are at most the original distances. As mentioned in the introduction, the *distortion* of a contraction embedding, $\phi(G)$, is

$$\max_{u,v \in V} \frac{d(u,v)}{|\phi(u) - \phi(v)|}$$

where $|\cdot|$ is the Euclidean distance (L_2 norm). A Euclidean metric on n points is a metric that is exactly realizable as the distances between n points in Euclidean space.

The *Tree Volume (Tvol)* of a metric is the product of the edge lengths of the minimum spanning tree. A subset S of a metric also induces a metric, and its tree volume, $Tvol(S)$ is the product of the edges of the minimum spanning tree of the metric induced by S.

The *Euclidean Volume (Evol)* of a subset of points $\{x_1, \ldots, x_k\}$ in some Euclidean space is the volume of the $(k-1)$-dimensional simplex spanned by the points.

Definition 1. *The* k**-volume distortion** *of a contraction embedding* ϕ *is defined as*

$$\max_{S \subseteq V, |S|=k} \left(\frac{Tvol(S)}{(k-1)! Evol(\phi(S))} \right)^{\frac{1}{k-1}}$$

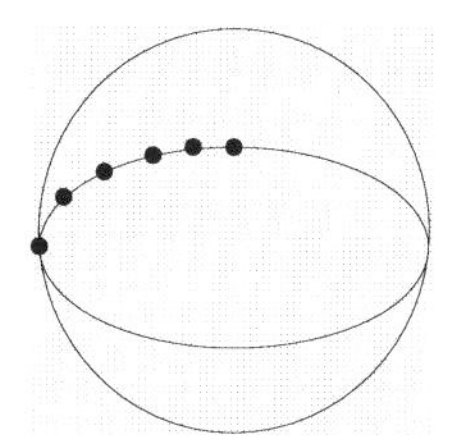

Fig. 1. Not quite points on a line, but close.

Remark. The factor $(k-1)!$ in the denominator is a normalization that connects volume of a simplex to volume of a parallelepiped. The distortion as defined above is within a factor of 2 of the distortion as defined in [3,4]. We find it unnecessary to go through the notion of "best possible volume" (Vol) used there.

The following theorem, due to Rao [6], connects the tree volume with Euclidean volume for the special case of Euclidean metrics.

Theorem 1. *For any Euclidean metric G, there exists a Euclidean embedding $\phi(G)$ whose k-volume distortion is $O(\sqrt{\log n \log\log n})$ for all k up to $\log n$.*

3 A Semi-Definite Relaxation

To arrive at the semi-definite relaxation of [1], we can start by imagining that the points of the graph are arranged along a great circle of the sphere at regular intervals spanning an arc of 90 degrees as in Figure 1. This is our approximation of laying out all the points on a line, and it is good to within a factor of 2. We relax this to allow the points to wander around the sphere, but maintaining that no two lie more than 90 degrees apart, and that they satisfy the "spreading" constraints. The objective function for our relaxation is to minimize the maximum distance between any pair of points connected by an edge in the original graph. We now give the SDP explicitly, where $G = (V, E)$ is our original graph. Note that G does not neccesarily induce a Euclidean metric, but the solution to the SDP below, where the vectors correspond to vertices of G, does induce a Euclidean metric. It is shown in [1] that this is a relaxation and that it can be solved in polytime.

$$\begin{aligned} \min\ & b \\ u_i \cdot u_j \geq 0 \quad & \forall i, j \in V \\ |u_i| = n \quad & \forall i \in V \\ |u_i - u_j| \leq b \quad & \forall (i,j) \in E \\ \sum_{j \in S} (u_i - u_j)^2 \geq \frac{1}{12}|S|^3 \quad & \forall S \subseteq V, \forall i \in V \end{aligned}$$

4 A Rounding Algorithm

Let the Euclidean embedding obtained by solving the relaxation be $U = \{u_1, \ldots, u_n\}$. The algorithm below rounds this solution to an ordering of the vertices of G.

1. Find a volume respecting embedding of U, $\phi(U) = \{v_1, \ldots, v_n\}$, using Rao's algorithm [6] with $k = \log n$.
2. Pick a random line ℓ passing through the origin.
3. Project the points of the embedding $\phi(U)$ to ℓ and output the ordering obtained.

Denote the dimension of the embedding $\phi(U)$ by d. Upon random projection, edge lengths shrink by a factor of $\frac{1}{\sqrt{d}}$ in expectation. To analyze the quality of the approximation we obtain, we show that every edge shrinks by at least a factor of $\frac{\sqrt{\log n}}{\sqrt{d}}$, and that not too many points fall in any interval of length $\frac{1}{\sqrt{d}}$. To show that no more than m points fall in an interval, we show that no more than $\binom{m}{k}$ sets of k points fall in the interval.

We will use the following lemmas. Lemma 1 is from [4] and lemmas 2 and 3 are from [7].

Lemma 1.

$$\sum_{S \subset U, |S|=k} \frac{1}{Tvol(S)} \leq n(\log n)^{k-1}$$

Lemma 2. *Let $v \in \mathbb{R}^d$. For a random unit vector ℓ,*

$$\Pr\left[|v \cdot \ell| \leq \frac{c}{\sqrt{d}}|v|\right] \geq 1 - e^{-c^2/4}.$$

Lemma 3. *Let S be a set of vectors $v_1, \ldots, v_k \in \mathbb{R}^d$. For a random unit vector ℓ*

$$\Pr\left[max_i\{v_i \cdot \ell\} - min_i\{v_i \cdot \ell\} \leq W\right] = O\left(\frac{W^{k-1} d^{\frac{k-1}{2}}}{(k-1)! Evol(S)}\right)$$

Lemma 4. *After random projection, the number of vertices that fall in any interval of length $\frac{1}{\sqrt{d}}$ is $O(\rho \log^2 n)$, where ρ is the k-volume distortion of the embedding.*

Proof. Consider an interval of length $W = \frac{1}{\sqrt{d}}$. For a subset S of V, let X_S be a random variable that is 1 if all the vectors in S fall in the interval. Let us estimate the total number of sets S of size k that fall in the interval.

$$\mathbf{E}\left[\sum_{|S|=k} X_S\right] = \sum_{|S|=k} \mathbf{E}[X_S] \tag{1}$$

$$= \sum_{|S|=k} \Pr(X_S = 1)$$

$$\leq \sum_{|S|=k} \frac{W^k d^{\frac{k}{2}}}{(k-1)! Evol(S)} \tag{2}$$

$$= \sum_{|S|=k} \frac{1}{(k-1)! Evol(S)}$$

$$\leq \sum_{|S|=k} \frac{(\rho)^{k-1}}{Tvol(S)} \tag{3}$$

$$\leq (\rho)^k \sum_{|S|=k} \frac{1}{Tvol(S)}$$

$$\leq (\rho)^k n (\log n)^k \tag{4}$$

$$\leq (2\rho \log n)^k \tag{5}$$

Step 2 is from lemma 3, step 3 is an application of theorem 1, step 4 is from lemma 1, and step 5 follows from $k = \log n$.

We need to consider only $O(n\sqrt{\log n})$ intervals of length $\frac{1}{\sqrt{d}}$ (the longest distance is $O(n)$ originally, and by lemma 2, it maps to a distance of at most $O(n\sqrt{\log n})$ with high probability). Now by Markov's inequality, with high probability, the number of k subsets that fall in any interval of length $\frac{1}{\sqrt{d}}$ is at most $n^2(2\rho \log n)^k \leq (8\rho \log n)^k$. Thus if the number of points in such an interval is m then $\binom{m}{k} \leq (8\rho \log n)^k$ which implies that $m = O(\rho \log^2 n)$ (using $k = \log n$). □

Theorem 2. *The algorithm finds an* $O(\rho \log^{2.5} n) = O(\log^3 n \sqrt{\log \log n})$ *approximation with high probability.*

Proof. Consider an edge (i, j) in the original graph that is mapped to vectors v_i and v_j after the volume-preserving embedding. Then $\max_{(i,j)\in E} |v_i - v_j|$ is a lower bound on the bandwidth of the graph (the distance between the solution vectors of the SDP is a lower bound and this distance is only contracted during the volume-preserving embedding).

After the last step, with high probability the distance between the projections of v_i and v_j is at most $O(\frac{\sqrt{\log n}}{\sqrt{d}}|v_i - v_j|)$ for *every* pair (i, j) (lemma 2).

Thus the maximum number of intervals of length $\frac{1}{\sqrt{d}}$ any edge (i, j) can span along the random line is $O(\sqrt{\log n} \cdot |v_i - v_j|)$. Along with lemma 4 this implies that the bandwidth of the final ordering is $O(\rho \log^{2.5} n)$ times the optimum with high probability. □

5 A Lower Bound on Volume Distortion

Our bandwidth algorithm and its analysis motivate the question of whether there are embeddings with better volume distortion. In this section we show that even for a path on n vertices, the best possible volume distortion is $\Omega(\sqrt{\log n})$. Thus a further improvement in bandwidth approximation will have to come from other sources.

Theorem 3. *Let G be a path on n vertices. Then for any Euclidean embedding of G, the distortion for subsets of size up to k is $\Omega((\log n)^{1/2-1/k})$. For $k = \Omega(\log \log n)$, the distortion is $\Omega((\log n)^{1/2})$.*

We begin by proving that the distortion is $\Omega((\frac{\log n}{\log \log n})^{1/4})$ for subsets of size 3.

Proof of weaker bound. Let our embedding map $\{u_1, ...u_n\}$ to $\{\phi(u_1), ..., \phi(u_n)\}$ and let $P_1 = \phi(u_1), P_2 = \phi(u_2), P_3 = \phi(u_3)$. We will show a tradeoff between the area of $\{P_1, P_2, P_3\}$ and the length of P_1P_3. Applying this recursively will yield the claimed bound.

Let $|P_1P_2| = |P_2P_3| = 1$ and let the perpendicular distance from P_2 to P_1P_3 be d. Let $|P_1P_3| = 2c$. The area of the triangle is dc and the Pythagorean identity yields $1 = d^2 + c^2$. Assume that β is an upper bound on the 3-volume distortion of any subset of three points in our embedding. Then

$$\left(\frac{1}{2dc}\right)^{1/2} = \left(\frac{Tvol(S)}{(k-1)!Evol(\phi(S))}\right)^{\frac{1}{k-1}} \leq \beta$$

and since $c \leq 1$, we find $d \geq \frac{1}{2\beta^2}$.

Using the Pythagorean identity, this implies $c = \sqrt{1-d^2} \approx 1 - d^2/2 \leq 1 - \frac{1}{8\beta^4}$. Thus every distance between two points $\phi(u_i), \phi(u_{i+2})$ is at most $2 \cdot (1 - \frac{1}{8\beta^4})$. Now we apply the same argument to subsets of three points at distance 2 apart, $\{\phi(u_i), \phi(u_{i+2}), \phi(u_{i+4})\}$. We obtain that the distance between $\phi(u_i)$ and $\phi(u_{i+4})$ is at most $4 \cdot (1 - \frac{1}{8\beta^4})^2$. Continuing this analysis, we find that the distance $|\phi(u_1) - \phi(u_n)|$ is at most $n \cdot (1 - \frac{1}{8\beta^4})^{\log n}$.

However, our assumption that we have distortion at most β implies $|\phi(u_1) - \phi(u_n)| \geq n/\beta$. Thus we have

$$n \cdot (1 - \frac{1}{8\beta^4})^{\log n} \geq n/\beta$$

implying $\beta \geq \left(\frac{\log n}{\log \log n}\right)^{1/4}$. □

Proof of Stronger Bound. Consider the volume of $\{P_1, ...P_k\}$, and assume without loss of generality that $\forall i, |P_iP_{i+1}| = 1$. Now let $c_i = \frac{1}{2}|P_iP_{i+2}|$, and $d_i =$ orthogonal distance from P_{i+1} to P_iP_{i+2}. We first claim that $Evol(P_1, ...P_k) \leq \frac{\prod_{i=1}^{k-2}(2d_i)}{(k-1)!}$. The proof is by induction. Our base case is $Evol(P_1, P_2) \leq 1$, which is clear. Assume that $Evol(P_1, ...P_j) \leq \prod_{i=1}^{j-2}(2d_i)/(j-1)!$ and consider P_{j+1}. We have that the midpoint of $P_{j-1}P_{j+1}$ is d_{j-1} away from P_j. This implies that P_{j+1} is no more than $2d_{j-1}$ away from the subspace spanned by $\{P_1, ...P_j\}$. The claim follows. Our new bound on the $\{d_i\}$ follows from

$$\left(\frac{1}{\prod_{i=1}^{k-2} d_i}\right)^{\frac{1}{k-1}} \leq \left(\frac{Tvol(S)}{(k-1)!Evol(\phi(S))}\right)^{\frac{1}{k-1}} \leq \beta$$

and the bound is $\frac{\sum_{i=1}^{k-2} 2d_i^2}{k-2} \geq (\prod_{i=1}^{k-2} 2d_i^2)^{\frac{1}{k-2}} \geq \frac{1}{2}\beta^{-2(\frac{k-1}{k-2})}$ where the first inequality follows from the arithmetic mean-geometric mean inequality. As before, we have $c_i \leq 1 - d_i^2/2$. Since $|P_1P_k| \leq 2(c_1 + c_3 + ...c_{k-3} + 1)$ and $|P_1P_k| \leq 2(1 + c_2 + c_4 + ...c_{k-2})$, we find that $|P_1P_k| \leq 2 + \sum_{i=1}^{k-2} c_i = (k-1)(\frac{k}{k-1} - \frac{k-2}{k-1}\frac{\sum_{i=1}^{k-2} d_i^2}{2(k-2)}) \approx (k-1)(1 - \frac{\sum_{i=1}^{k-2} d_i^2}{2(k-2)})$. Our bound on the length of P_1P_k becomes $|P_1P_k| \leq (k-1)(\frac{k}{k-1} - \frac{k-2}{k-1}\frac{1}{8}\beta^{-2\frac{k-1}{k-2}})$. Now we apply our recursive construction again, this time on sets of size k at a time. Since we are no longer just doubling each time, we can apply our analysis only $\log_k n$ times. Plugging this in yields the bound

$$\left(\frac{k}{k-1} - \frac{k-2}{k-1}\frac{1}{8\beta^{2(\frac{k-1}{k-2})}}\right)^{\frac{\log n}{\log k}} \geq \frac{1}{\beta}$$

which simplifies to $\log n \leq 16(\log k)\beta^{2(1+\frac{1}{k-2})}\log\beta$, implying $\beta \geq (\log n)^{(1/2-1/k)}$.

□

6 Embeddings Preserving Point-to-Subset Distances

The distance of a point (or vertex) u of G to a subset of points S is simply $d(u, S) = \min_{v \in S} d(u, v)$. For points in Euclidean space, let us define the distance of a point u to a set of points S as the minimum distance from u to the *convex hull* of S, which we denote with the natural extension of $|\cdot|$. We denote the convex hull of a set of points S by conv(S), and the affine hull by aff(S).

Definition 2. *The* **point-to-subset distortion** *of an embedding* $\phi(G)$ *is*

$$\max_{u \in V, S \subset V} \frac{d(u, S)}{|\phi(u) - \text{conv}(\phi(S))|}$$

In this section we investigate the question of the best possible point-to-subset distortion of a Euclidean metric. Besides its geometric appeal, the question has the following motivation. Suppose we replaced "convex" in the definition above by "affine" and called the related distortion the *affine point-to-subset* distortion. Then we would have the following connection with volume distortion.

Lemma 5. *Let* $\phi(G)$ *be a contraction embedding of a metric* G. *Then the affine point-to-subset distortion is an upper bound on the* k*-volume distortion, for all* $2 \leq k \leq n$.

Proof of lemma 5. Consider a set S of vertices in G, and a mimimum spanning tree T of S. Consider any leaf u of T. If the point-to-subset distortion of our embedding ϕ is β, then

$$d(u, S \setminus \{u\}) \leq \beta|\phi(u) - \text{aff}(\phi(S \setminus \{u\}))|$$

Proceeding inductively, we find that

$$\begin{aligned} Tvol(S) &\leq \beta^{k-1}(\text{volume of parallelepiped defined by } \phi(S)) \\ &\leq \beta^{k-1}(k-1)! Evol(\phi(S)) \end{aligned}$$

□

We now state our main theorem on point-to-subset distortion. In the next two subsections, we define the embedding, and then prove that the embedding satisfies the theorem.

Theorem 4. *For any Euclidean metric G where the shortest distance is within a $poly(n)$ factor of the longest distance, there exists a Euclidean embedding $\phi(G)$ whose point-to-subset distortion is $O(\sqrt{\log n})$.*

6.1 The Embedding

Let $G = (V, E)$ be a Euclidean metric with distances (edge lengths) $d(u, v)$ for all pairs of vertices $u, v \in V$. We assume that all the distances lie between 8 and $8n$. (Any polynomial upper bound on the ratio of the shortest to the longest distance would suffice). Since G is Euclidean, we can assume without loss of generality that the vertices are points in some d-dimensional Euclidean space. Given only the distances, it is trivial to find points realizing the distances by solving an SDP. The embedding we now describe was inspired by the work of Rao [6].

Before defining the embedding in general, let us consider the following illustrative example. Suppose that $d = 1$, i.e., all the points lie on a line. In this case, we could proceed by generating coordinates according to the following random process: for each R in the set $\{1, 2, 2^2, \ldots, 2^{\lfloor \log n \rfloor}\}$, we repeat the following procedure N times: choose each point from the subset $\{1, \ldots, 8n\}$ with probability $1/R$ for inclusion in a set S, and come up with a coordinate $\phi_S(v)$ for every $v \in V$. The coordinate $\phi_S(v)$ is defined to be $\min_{w \in S} |v - w|$, and then $\phi(v)$ is the vector given by the set of coordinates for v. This will yield $N \log n$ coordinates.

We now explain why this yields a $\sqrt{\log n}$ affine point-to-subset distortion. This is a stronger property than $\sqrt{\log n}$ point-to-subset distortion, and it will only be proved for $d = 1$. Consider a set $U \subset V$ and a point u, with distance on the line $d(u, U)$. For every S, we have that $|\phi_S(u) - \text{aff}(\phi_S(U))| \leq d(u, U)$, and it is a simple application of Cauchy-Shwarz to get that $|\phi(u) - \text{aff}(\phi(U))| \leq d(u, U)\sqrt{N \log n}$. To obtain a lower bound of $\Omega(d(u, U)\sqrt{N})$, we consider the largest R such that $R \leq d(u, U)$. Denote this value of R by r; we now show that with constant probability, a set S chosen by including points in S with probability $\frac{1}{r}$ yields $|\phi_S(u) - \text{aff}(\phi_S(U))| = \Omega(d(u, U))$. We get this from the following view of the random process: fix some particular affine combination aff_0, pick points for inclusion in S at distances in $(d(u, U)/2, d(u, U))$ to the left and to the right of u, pick the rest of the points not near u with probability $\frac{1}{r}$, and we still have constant probabilty of picking another point within the two points bracketing u; the variation in $|\phi_S(u) - \text{aff}_0(\phi_S(U))|$ due to the distinct choices for this last point included in S is $\Omega(d(u, U))$ with constant probability. Feige proves [4] that the number of "distinct" affine combinations is not too great, and thus taking N sufficiently large, but still polynomial, yields that this occurs with high probability simultaneously for all "distinct" affine combinations, and thus over the uncountable set of all affine combinations. Taking N a little bit larger still (but still polynomial) then yields that this is simultaneously true for all point-subset pairs.

For the general case ($d \geq 1$), our algorithm chooses a random line, projects all the points to this random line, and then computes the coordinates as above. In detail, for each R in the set $\{1, 2, 2^2, \ldots, 2^{\lfloor \log n \rfloor}\}$, we repeat the following procedure N times:

1. Pick a random line ℓ through the origin.
2. Project all the points to ℓ, and scale up by a factor of $\sqrt{d}$. Let the projection of u be u^ℓ.
3. Place points along ℓ at unit intervals. Pick a random subset S of these points, by choosing each point with probability $\frac{1}{R}$, independently.
4. The coordinate for each vertex u along the axis corresponding to the S and ℓ pair is $\phi_S(u) = d(u^\ell, S) = \min_{w \in S} |u^\ell - w|$.

Thus the total number of dimensions is $O(N \log n)$. For the same reasons as cited above, $N = poly(n)$ will suffice. Thus the dimension of the final embedding is polynomial in n.

6.2 The Proof

We first upper bound the point-to-subset distances in our final embedding. Consider any point u and subset U. It is enough to consider the point $v \in U$ minimizing $d(u, v)$ by the following lemma.

Lemma 6. *For every pair $u, v \in V$,*

$$|\phi(u) - \phi(v)| \leq 2d(u, v)\sqrt{N \log n}$$

Proof. After scaling up by a factor of $\sqrt{d}$, we have that for any pair $u, v \in V$,

$$\begin{aligned} |\phi(u) - \phi(v)|^2 &= \sum_{(S,\ell)} |\phi_S(u) - \phi_S(v)|^2 \\ &= \sum_{(S,\ell)} |d(u^\ell, S) - d(v^\ell, S)|^2 \end{aligned}$$

$$\leq \sum_{(S,\ell)} d(u^\ell, v^\ell)^2$$

$$\leq \sum_{(S,\ell)} 2d(u,v)^2 \tag{6}$$

$$= d(u,v)^2 N \log n$$

where step 6 is true with constant probability for a single random line, and with very high probability when summing over all the random lines. □

Since $|\phi(u) - \text{conv}(\phi(U))| \leq |\phi(u) - \phi(v)|, v \in U$, we have our upper bound.

Now we lower bound the point-to-subset distances. Consider again a particular point u and subset $U = \{u_i\}$, and some fixed convex combination $\{\lambda_i\}$ such that $\sum \lambda_i = 1$ and $\forall i, \lambda_i \geq 0$. Let r be the highest power of 2 less than $d(u, U)$. For any coordinate corresponding to a subset S generated using $R = r$, we show that $|\phi(u) - \sum_i \lambda_i \phi(u_i)| = \Omega(d(u, U))$ with constant probability.

Towards this goal, we claim there is a constant probability that the following two events both happen.

(i) $\sum_i \lambda_i \phi_S(u_i) \geq r/16$
(ii) $\phi_S(u) \leq r/32$

First we condition on some point within $r/32$ of u^ℓ being chosen for inclusion in S. This happens with constant probability (over choice of S). We have at least a constant probability of the λ_i's corresponding to u_i's at least $r/4$ away from u adding up to at least $2/3$ (over choice of ℓ). Condition on this as well. Then we lower bound the expected value of $\sum_i \lambda_i \phi_S(u_i)$ by

$$E[\sum_i \lambda_i \phi_S(u_i)] \geq \sum_{i:|u_i^\ell - u_i| \geq r/4} \lambda_i E[\phi_S(u_i)]$$

Since $E[\phi_S(u_i)] \geq r/8$, we have that the expectation is at least $(2/3)(r/8) = r/12$. By Markov's inequality, the value of $\sum_i \lambda_i \phi_S(u_i)$ is at least $r/16$ with constant probability.

Since this happens for all coordinates with $R = r$, i.e. N of the coordinates, we obtain the lower bound. As before, a polynomially large N suffices to make the statement true with high probability for every point, every subset, and every convex combination simultaneously.

7 Convexity of kth Moments

In section 6.1, we proved that our embedding did preserve all *affine* point-to-subset distances to within $O(\sqrt{\log n})$ for the case $d = 1$. Since the optimal solution to the bandwidth problem is an arrangement of points on a line, it is the case that an embedding realizing this distortion of the optimal solution exists. This implies that the constraint

$$\sum_{|S|=k} \frac{1}{Evol(S)} \leq \sum_{|S|=k} \frac{(k-1)!\rho^k}{Tvol(S)} \leq (2\rho k \log n)^k$$

with $\rho = \sqrt{\log n}$ is satisfied by a Euclidean embedding of the optimal solution. We call this constraint the kth *moment constraint*. We show in this section that the above constraint is convex, and thus we can impose it explicitly in our SDP (replacing the spreading constraint), separate over it, and then apply the machinery of section 4 with $\rho = \sqrt{\log n}$ to obtain an $O(\log^3 n)$ approximation to the optimal bandwidth. The only caveat is that the constraint has $\binom{n}{k} = O(n^{\log n})$ terms, so this is not quite a polynomial time algorithm. We proceed with

Lemma 7. *Let c be fixed. The following is a convex constraint over the set of Postive Semi-Definite (PSD) matrices X.*

$$\sum_{|S|=k} \frac{1}{Evol(X_S)} \leq c$$

Proof. We analyse the constraint given in the lemma on a term by term basis. Suppose X and Y are PSD matrices, and $(X+Y)/2$ is their convex combination. Then it suffices to show that

$$\frac{1}{Evol((X+Y)/2)} \leq \frac{1}{2}\left(\frac{1}{Evol(X)} + \frac{1}{Evol(Y)}\right)$$

because the constraint in the lemma statement is just a sum over many submatrices. We actually prove the stronger statement that

$$\frac{1}{Evol((X+Y)/2)} \leq \sqrt{\frac{1}{Evol(X)}\frac{1}{Evol(Y)}}$$

which implies the former statement by the arithmetic mean-geometric mean inequality (GM $\leq$ AM). This last statement is equivalent to (clearing denominators and squaring twice)

$$Det(XY) \leq Det^2((X+Y)/2)$$

which is equivalent to

$$\begin{aligned}
1 \leq & \frac{Det^2((X+Y)/2)}{Det(XY)} \\
&= Det(\frac{1}{4}(X+Y))Det(X^{-1})Det(X+Y)Det(Y^{-1}) \\
&= Det(\frac{1}{4}(X+Y)(X^{-1})(X+Y)(Y^{-1})) \\
&= Det(\frac{1}{4}(I+YX^{-1})(XY^{-1}+I)) \\
&= Det(\frac{1}{4}(YX^{-1}+2I+XY^{-1})) \\
&= Det(\frac{1}{4}(A+2I+A^{-1}))
\end{aligned}$$

where we let $A = YX^{-1}$ at the very end. Also let $B = \frac{A+2I+A^{-1}}{4}$. We have reduced our original claim to showing that $Det(B) \geq 1$. We will show the stronger property that every eigenvalue of B is at least 1. Consider an arbitrary (eigenvector, eigenvalue)-pair of A , given by (e, λ). Then

$$Be = \frac{1}{4}(\lambda + 2 + \frac{1}{\lambda})e$$

Since $\frac{1}{4}(\lambda + 2 + \frac{1}{\lambda}) \geq 1$, we have that e is an eigenvector of eigenvalue at least 1 for B (this used that $\lambda \geq 0$, which is true since A is PSD). Since the eigenvectors of A form an orthonormal basis of the whole space, all of B's eigenvectors are also eigenvectors of A. □

8 Conclusion

We conjecture that the embedding described in section 6.1 has $O(\sqrt{\log n})$ *affine* point-to-subset distortion as well. This would directly imply that our algorithm achieves an $O(\log^3 n)$ approximation for the minimum bandwidth in polynomial time.

References

1. A. Blum, G. Konjevod, R. Ravi, and S. Vempala, "Semi-Definite Relaxations for Minimum Bandwidth and other Vertex-Ordering Problems," Proc. 30th ACM Symposium on the Theory of Computing, 1998.
2. J. Bourgain, "On Lipshitz embedding of finite metric spaces in Hilbert space," Israel J. Math. 52 (1985) 46-52.
3. U. Feige, "Approximating the bandwidth via volume respecting embeddings," in Proc. 30th ACM Symposium on the Theory of Computing, 1998.
4. U. Feige, "Improved analysis of the volume distortion of the random subsets embedding," Manuscript.
5. C. H. Papadimitriou, The NP-completeness of the bandwidth minimization problem, *Computing*, 16: 263-270, 1976.
6. S. Rao, "Small distortion and volume preserving embeddings for planar and Euclidean metrics," Proc. of Symposium on Computational Geometry, 1999.
7. S. Vempala, "Random Projection: A new approach to VLSI layout," Proc. of FOCS 1998.

The Non-approximability of Non-Boolean Predicates

Lars Engebretsen

MIT Laboratory for Computer Science
200 Technology Square, NE43-367
Cambridge, Massachusetts 02139-3594
enge@mit.edu

Abstract. Constraint satisfaction programs where each constraint depends on a constant number of variables have the following property: The randomized algorithm that guesses an assignment uniformly at random satisfies an expected constant fraction of the constraints. By combining constructions from interactive proof systems with harmonic analysis over finite groups, Håstad showed that for several constraint satisfaction programs this naive algorithm is essentially the best possible unless **P** = **NP**. While most of the predicates analyzed by Håstad depend on a small number of variables, Samorodnitsky and Trevisan recently extended Håstad's result to predicates depending on an arbitrarily large, but still constant, number of Boolean variables.
We combine ideas from these two constructions and prove that there exists a large class of predicates on finite non-Boolean domains such that for predicates in the class, the naive randomized algorithm that guesses a solution uniformly is essentially the best possible unless **P** = **NP**. As a corollary, we show that the k-CSP problem over domains with size D cannot be approximated within $D^{k-O(\sqrt{k})} - \epsilon$, for any constant $\epsilon > 0$, unless **P** = **NP**. This lower bound matches well with the best known upper bound, D^{k-1}, of Serna, Trevisan and Xhafa.

1 Introduction

In a breakthrough paper, Håstad [7] studied the problem of giving approximate solutions to maximization versions of several constraint satisfaction problems. An instance of a such a problem is given as a collection of constraints, i.e., functions from some domain to $\{0, 1\}$, and the objective is to satisfy as many constraints as possible. An approximate solution of a constraint satisfaction program is simply an assignment that satisfies roughly as many constraints as possible. In this setting, we are interested in proving either that there exists a polynomial time algorithm producing approximate solutions some constant fraction from the optimum or that no such algorithms exist.

Typically, each individual constraint depends on a fixed number k of the variables and the size of the instance is given as the total number of variables that appear in the constraints. In this case, which is usually called the Max k-CSP

M. Goemans et al. (Eds.): APPROX-RANDOM 2001, LNCS 2129, pp. 241–249, 2001.

problem, there exists a very naive algorithm that approximates the optimum within a constant factor: The algorithm that just guesses a solution at random. In his paper, Håstad [7] proved the very surprising fact that this algorithm is essentially the best possible efficient algorithm for several constraint satisfaction problems, unless **P** = **NP**. The proofs unify constructions from interactive proof systems with harmonic analysis over finite groups and give a general framework for proving strong impossibility results regarding the approximation of constraint satisfaction programs. Håstad [7] suggests that predicates with the property that the naive randomized algorithm is the best possible polynomial time approximation algorithm should be called *non-approximable beyond the random assignment threshold.*

Definition 1. *A Max k-CSP on k variables is* non-approximable beyond the random assignment threshold *if, for any constant $\epsilon > 0$, it is **NP**-hard to approximate the optimum of the CSP within a factor $w - \epsilon$, where $1/w$ is the expected fraction of constraints satisfied by a solution guessed uniformly at random.*

Håstad's paper [7] deals mainly with constraint satisfaction programs involving a small number, typically three or four, variables. In most of the cases, the variables are Boolean, but Håstad also treats the case of linear equations over Abelian groups. In the Boolean case, Håstad's techniques have been extended by Trevisan [13], Sudan and Trevisan [11], and Samorodnitsky and Trevisan [9] to some predicates involving a large, but still constant, number of Boolean variables. In this paper, we prove that those extensions can be adapted also to the non-Boolean case—a fact that is not immediately obvious from the proof for the Boolean case. This establishes non-approximability beyond the random assignment threshold for a large class of non-Boolean predicates. Our proofs use Fourier analysis of functions from finite Abelian groups to the complex numbers combined with what has now become standard constructions from the world of interactive proof systems. As a technical tool, Sudan and Trevisan [11] developed a certain composition lemma. In this paper, we extend this lemma to the non-Boolean setting. By using the lemma as an integrated part of the construction rather than a black box, we are also able to improve some of the constants involved.

A consequence of our result is that it is impossible to approximate Max k-CSP over domains of size D within $D^{k-O(\sqrt{k})} - \epsilon$, for any constant $\epsilon > 0$, in polynomial time unless **P** = **NP**. This lower bound matches well with the best known upper bound, D^{k-1}, following from a linear relaxation combined with randomized rounding [10,12].

The paper is outlined as follows: We give the general ideas behind our construction in Sec. 2. Then we give the construction of our PCP in Sec. 3 and the connection to Max k-CSP and the non-approximability beyond the random assignment threshold of several non-Boolean predicates in Sec. 4. Finally, we conclude with some directions for future research. The proof of the non-approximability result is somewhat technical and we omit it from this extended abstract. A more complete version of our results is available as an ECCC technical report [3].

2 Outline of the Construction

The underlying idea in our construction is the same as in Håstad's [7]. We start with an instance of μ-gap E3-Sat(5).

Definition 2. μ-gap E3-Sat(5) *is the following decision problem: We are given a Boolean formula ϕ in conjunctive normal form, where each clause contains exactly three literals and each literal occurs exactly five times. We know that either ϕ is satisfiable or at most a fraction $\mu < 1$ of the clauses in ϕ are satisfiable and are supposed to decide if the formula is satisfiable.*

It is known [2,4] that μ-gap E3-Sat(5) is **NP**-hard.

There is a well-known two-prover one-round (2P1R) interactive proof system that can be applied to μ-gap E3-Sat(5). It consists of two provers, P_1 and P_2, and one verifier. Given an instance, i.e., an E3-Sat formula ϕ, the verifier picks a clause C and variable x in C uniformly at random from the instance and sends x to P_1 and C to P_2. It then receives an assignment to x from P_1 and an assignment to the variables in C from P_2, and accepts if these assignments are consistent and satisfy C. If the provers are honest, the verifier always accepts with probability 1 when ϕ is satisfiable, i.e., the proof system has *completeness* 1. It can be shown that the provers can fool the verifier with probability at most $(2 + \mu)/3$ when ϕ is not satisfiable, i.e., that the above proof system has *soundness* $(2 + \mu)/3$.

The soundness can be lowered to $((2 + \mu)/3)^u$ by repeating the protocol u times independently, but it is also possible to construct a one-round proof system with lower soundness by repeating u times in parallel as follows: The verifier picks u clauses $\{C_1, \ldots, C_u\}$ uniformly at random from the instance. For each C_i, it also picks a variable x_i from C_i uniformly at random. The verifier then sends $\{x_1, \ldots, x_u\}$ to P_1 and the clauses $\{C_1, \ldots, C_u\}$ to P_2. It receives an assignment to $\{x_1, \ldots, x_u\}$ from P_1 and an assignment to the variables in $\{C_1, \ldots, C_u\}$ from P_2, and accepts if these assignments are consistent and satisfy $C_1 \wedge \cdots \wedge C_u$. As above, the completeness of this proof system is 1, and it can be shown [8] that the soundness is at most c_μ^u, where $c_\mu < 1$ is some constant depending on μ but not on u or the size of the instance.

In the above setting, the proof is simply an assignment to all the variables. In that case, the verifier can just compare the assignments it receives from the provers and check if they are consistent and satisfying. The construction we use to prove that several non-Boolean constraint satisfaction programs are non-approximable beyond the random assignment threshold can be viewed as a simulation of the u-parallel repetition of the above 2P1R interactive proof system for μ-gap E3-Sat(5). We use a probabilistically checkable proof system (PCP) with a verifier closely related to the particular constraint we want to analyze. To find predicates that depend on variables from some domain of size D and are non-approximable beyond the random assignment threshold, we work with an Abelian group G of size D. The predicates we study are ANDs of linear equations involving three variables in G. The proof is what Håstad [7] calls a Standard Written G-Proof with parameter u. It is supposed to be a very redundant encoding of a string of length n, which when ϕ is a satisfiable formula should be a satisfying assignment.

Definition 3. *If U is some set of variables taking values in $\{-1, 1\}$, we denote by $\{-1, 1\}^U$ the set of every possible assignment to those variables. The* Long G-Code *of some string x of length $|U|$ is the value of all functions from $\{-1, 1\}^U$ to G evaluated on the string x; $A_{U,x}(f) = f(x)$.*

Since there are $|G|^{2^{|U|}}$ functions from $\{-1, 1\}^U$ to G, the *Long G-Code* of a string of length u has length $|G|^{2^u}$. The proof introduced by Håstad [7] contains the Long G-Code of several subsets containing a constant number of variables. Each such subset is supposed to represent either an assignment to the variables sent to P_1 or the clauses sent to P_2 in the 2P1R interactive proof system for μ-gap E3-Sat(5).

Definition 4. *A* Standard Written G-Proof with parameter u *contains for each set $U \subseteq [n]$ of size at most u a string of length $|G|^{2^{|U|}}$, which we interpret as the table of a function $A_U: \mathcal{F}_U^G \to G$. It also contains for each set W constructed as the set of variables in u clauses a function $A_W: \mathcal{F}_W^G \to G$.*

Definition 5. *A Standard Written G-Proof with parameter u is a* correct proof *for a formula ϕ of n variables if there is an assignment x, satisfying ϕ, such that A_V is the Long G-Code of $x|_V$ for any V of size at most u or any V constructed as the set of variables of u clauses.*

To check that the proof is a correct proof, we—following the construction of Samorodnitsky and Trevisan [9]—first query $2k$ positions from the proof and then, as a checking procedure, construct k^2 linear equations, each of them involving two of the first $2k$ queried positions and one extra variable. To give a more illustrative picture of the procedure, we let the first $2k$ queries correspond to the vertices of a complete $k \times k$ bipartite graph. The k^2 linear equations that we check then correspond to the edges of this graph.

As for the non-approximability beyond the random assignment threshold, a random assignment to the variables satisfy all k^2 linear equations simultaneously with probability $|G|^{-k^2}$—the aim of our analysis is to prove that this is essentially the best possible any polynomial time algorithm can accomplish. This follows from the connection between our PCP and the 2P1R interactive proof system for μ-gap E3-Sat(5): We assume that it is possible to satisfy a fraction $|G|^{-k^2} + \epsilon$ for some constant $\epsilon > 0$ and prove that this implies that there is a correlation between the tables queried by the verifier in our PCP. We can then use this correlation to explicitly construct strategies for the provers in the 2P1R proof system for μ-gap E3-Sat(5) such that the verifier in that proof system accepts with probability larger than c_μ^u. The final link in the chain is the observation that since our verifier uses only logarithmic randomness, we can form a CSP with polynomial size by enumerating the checked constraints for every possible outcome of the random bits. If the resulting constraint satisfaction program is approximable beyond the random assignment threshold, we can use it to decide the **NP**-hard language μ-gap E3-Sat(5) in polynomial time.

We remark, that by checking the equations corresponding to some subset E of the edges in the complete bipartite graph we also get a predicate which is non-approximable beyond the random assignment threshold: It is satisfied with probability $|G|^{-|E|}$ by a random assignment and our proof methodology works also for this case.

3 The PCP

The proof is a Standard Written G-Proof with parameter u. It is supposed to represent a string of length n. When ϕ is a satisfiable formula this string should be a satisfying assignment.

The verifier is parameterized by the integers ℓ and m, a set $E \subseteq [\ell] \times [m]$, and a constant $\delta_1 > 0$; and it should accept with high probability if the proof is a correct Standard Written G-Proof for a given formula ϕ.

1. Select uniformly at random u variables $x_1, \dots, x_u$. Let U be the set of those variables.
2. For $j = 1, \dots, m$, select uniformly at random u clauses $C_{j,1}, \dots, C_{j,u}$ such that clause $C_{j,i}$ contains variable x_i. Let Φ_j be the Boolean formula $C_{j,1} \wedge \dots \wedge C_{j,u}$. Let W_j be the set of variables in the clauses $C_{j,1}, \dots, C_{j,u}$.
3. For $i = 1, \dots, \ell$, select uniformly at random $f_i \in \mathcal{F}_U^G$.
4. For $j = 1, \dots, m$, select uniformly at random $g_j \in \mathcal{F}_{W_j}^G$.
5. For all $(i, j) \in E$, choose $e_{ij} \in \mathcal{F}_{W_j}^G$ such that, independently for all $y \in W$,
   (a) With probability $1 - \delta_1$, $e_{ij}(y) = 1_G$.
   (b) With probability δ_1, $e_{ij}(y)$ is selected uniformly at random from G.
6. Define h_{ij} such that $h_{ij}(y) = \left(f_i(y|_U)g_j(y)e_{ij}(y)\right)^{-1}$.
7. If for all $(i, j) \in E$, $A_U(f_i)A_{W_j}(g_j \wedge \Phi_j)A_{W_j}(h_{ij} \wedge \Phi_j) = 1$, then accept, else reject.

Lemma 1. *The completeness of the above test is at least* $(1 - \delta_1)^{|E|}$.

Proof. Given a correct proof, the verifier can only reject if one of the error functions e_{ij} are not 1_G for the particular string encoded in the proof. Since the error functions are chosen pointwise uniformly at random, the probability that they all evaluate to 1_G for the string encoded in the proof is $(1 - \delta_1)^{|E|}$. Thus, the verifier accepts a correct proof with probability at least $(1 - \delta_1)^{|E|}$.

Lemma 2. *For every constant $\delta_2 > 0$, it is possible to select a constant u such that the soundness of the above PCP is at most $1/|G|^{|E|} + \delta_2$.*

Proof (sketch). The proof of this lemma uses the Fourier expansion of the indicator for the event that the verifier in the PCP accepts to prove that if the verifier accepts with probability $1/|G|^{|E|} + \delta_2$ the tables U and $W_1, \dots, W_k$ are correlated. This correlation is then used to provide a strategy for the provers in the 2P1R proof system for μ-gap E3-Sat(5). This strategy makes the verifier in the 2P1R proof system accept with probability at least $\delta_1\delta_2^2/(|G|-1)^{|E|}$. Finally, by selecting the constant u such that $\delta_1\delta_2^2/(|G|-1)^{|E|} > c_\mu^u$, we conclude that the soundness of the above PCP is at most $1/|G|^{|E|} + \delta_2$.

For a more detailed proof of the above lemma, see the full version of this paper [3].

4 The Reduction to Non-Boolean CSPs

We now show how the above PCP can be connected with CSPs to prove that the corresponding CSPs are non-approximable beyond the random assignment threshold. As for the completeness c and the soundness s of the PCP from the previous section, we have shown that $c \geq (1-\delta_1)^{|E|}$ and $s \leq |G|^{-|E|} + \delta_2$, for arbitrarily small constants $\delta_1, \delta_2 > 0$.

Theorem 1. *Let G be any finite Abelian group, ℓ and m be arbitrary positive integers, and $E \subseteq [\ell] \times [m]$. Then the predicate*

$$\bigwedge_{i,j:(i,j)\in E} (x_i x_j x_{i,j} = a_{i,j}),$$

where $a_{i,j} \in G$ and x_i, x_j, and $x_{i,j}$ assume values in G, is non-approximable beyond the random assignment threshold.

Before proving the theorem, we restate it in slightly different words.

Definition 6. Max k-CSP-G *is the following maximization problem: Given a number of functions from G^k, where G is a finite Abelian group, to $\mathbf{Z}_2$, find the assignment maximizing the number of functions evaluating to 1. The total number of variables in the instance is denoted by n.*

Theorem 2. *Let G be any finite Abelian group, ℓ and m be arbitrary positive integers, $E \subseteq [\ell] \times [m]$, and $k = |E| + \ell + m$. Then it is* **NP***-hard to approximate Max k-CSP-G within $|G|^{|E|} - \epsilon$ for any constant $\epsilon > 0$.*

Proof. Select the constants $\delta_1 > 0$ and $\delta_2 > 0$ such that

$$\frac{(1-\delta_1)^{|E|}}{|G|^{-|E|} + \delta_2} \geq |G|^{|E|} - \epsilon.$$

Then select the constant u such that $\delta_1\delta_2^2/(|G|-1)^{|E|} > c_\mu^u$. Now consider applying the PCP from Sec. 3 to an instance of the **NP**-hard problem μ-gap E3-Sat(5).

Construct an instance of Max k-CSP-G as follows: Introduce variables $x_{U,f}$ and $y_{\Phi_j,g}$ for every $A(f)$ and $B_j(g)$, respectively. For all possible combinations of a set U, clauses $\Phi_1,\dots,\Phi_m$, and functions $f_1,\dots,f_\ell$, $g_1,\dots,g_m$, $h_{1,1},\dots,h_{\ell,m}$, introduce a constraint that is one if $x_{U,f_i}y_{\Phi_j,g_j} = y_{\Phi_j,h_{ij}}$ for all $(i,j)\in E$. Set the weight of this constraint to the probability of the event that the set U, the clauses $\Phi_1,\dots,\Phi_m$, and the functions $f_1,\dots,f_\ell$, $g_1,\dots,g_m$, and $h_{1,1},\dots,h_{\ell,m}$ are chosen by the verifier in the PCP. Each constraint is a function of at most $|E|+\ell+m$ variables. The total number of constraints is at most

$$n^u 5^{mu} |G|^{\ell 2^u + m2^{3u} + \ell m 2^{3u}},$$

which is polynomial in n if ℓ, m, $|G|$, and u are constants. The weight of the satisfied equations for a given assignment to the variables is equal to the probability that the PCP from Sec. 3 accepts the proof corresponding to this assignment. Thus, any algorithm approximating the optimum of the above instance within

$$\frac{(1-\delta_1)^{|E|}}{|G|^{-|E|}+\delta_2} \geq |G|^{|E|} - \epsilon$$

decides the **NP**-hard problem μ-gap E3-Sat(5).

Corollary 1. *For any integer $k \geq 3$ and any constant $\epsilon > 0$, it is* ***NP****-hard to approximate Max k-CSP-G within $|G|^{k-2\sqrt{k+1}+1} - \epsilon$.*

We omit the proof of Corollary 1 from this extended abstract—it is present in the full version of this paper [3]. From the details of the proof, it is possible to see that we can rephrase the result in the following slightly stronger form.

Corollary 2. *For any integer $s \geq 2$ and any constant $\epsilon > 0$, it is* ***NP****-hard to approximate Max s^2-CSP-G within $|G|^{(s-1)^2} - \epsilon$. For any integer $k \geq 3$ that is not a square and any constant $\epsilon > 0$, it is* ***NP****-hard to approximate Max k-CSP-G within $|G|^{k-2\sqrt{k+1}+2} - \epsilon$.*

5 Conclusions

We have shown that it is possible to combine the harmonic analysis introduced by Håstad [7] with the recycling techniques used by Samorodnitsky and Trevisan [9] to obtain a lower bound on the approximability of Max k-CSP-G. The proof of results of this type typically study some predicate on a constant number of variables such that a random assignment to the variables satisfies the predicate with probability $1/w$. Starting from the 2P1R interactive proof system for μ-gap E3-Sat(5), instances such that it is **NP**-hard to approximate the number of satisfied constraints within $w-\epsilon$, for any constant $\epsilon > 0$, are constructed. Our proof is no exception to this rule.

The current state of the art regarding the (non-)approximability of predicates is that there are a number of predicates—such as linear equations mod p with three unknowns in every equation, E3-satisfiability, and the predicates of this paper—that are non-approximable beyond the random assignment threshold [7,9]. There also exists some predicates—such as linear equations mod p with two unknowns in every equation and E2-satisfiability—where there are polynomial time algorithms beating the bound obtained from a random assignment [1,5,6].

A very interesting direction for future research is to try to determine criteria identifying predicates that are non-approximable beyond the random assignment threshold. Some such attempts have been made for special cases. For predicates of three Boolean variables, it is known that the predicates that are non-approximable beyond the random assignment threshold are precisely those that are implied by parity [7,14]. However, the general question remains completely open.

Acknowledgments

The author thanks Johan Håstad for many clarifying discussions on the subject of this paper.

References

1. G. Andersson, L. Engebretsen, and J. Håstad. A new way of using semidefinite programming with applications to linear equations mod p. *J. Alg.*, 39(2):162–204, May 2001.
2. S. Arora, C. Lund, R. Motwani, M. Sudan, and M. Szegedy. Proof verification and the hardness of approximation problems. *J. ACM*, 45(3):501–555, May 1998.
3. L. Engebretsen. Lower bounds for non-Boolean constraint satisfaction. Technical Report TR00-042, ECCC, June 2000.
4. U. Feige. A threshold of $\ln n$ for approximating set cover. *J. ACM*, 45(4):634–652, July 1998.
5. U. Feige and M. X. Goemans. Approximating the value of two prover proof systems, with applications to MAX 2SAT and MAX DICUT. In *Proc. 3rd ISTCS*, pages 182–189, 1995.
6. M. X. Goemans and D. P. Williamson. Improved approximation algorithms for maximum cut and satisfiability problems using semidefinite programming. *J. ACM*, 42(6):1115–1145, Nov. 1995.
7. J. Håstad. Some optimal inapproximability results. In *Proc. 29th STOC*, pages 1–10, 1997. Accepted for publication in *J. ACM*.
8. R. Raz. A parallel repetition theorem. *SIAM J. Comput.*, 27(3):763–803, June 1998.
9. A. Samorodnitsky and L. Trevisan. A PCP characterization of NP with optimal amortized query complexity. In *Proc. 32nd STOC*, pages 191–199, 2000.
10. M. Serna, L. Trevisan, and F. Xhafa. The (parallel) approximability of non-Boolean satisfiability problems and restricted integer programming. In *Proc. 15th STACS*, vol. 1373 of *LNCS*, pages 488–498, 1998. Springer-Verlag.
11. M. Sudan and L. Trevisan. Probabilistically checkable proofs with low amortized query complexity. In *Proc. 39th FOCS*, pages 18–27, 1998. IEEE.

12. L. Trevisan. Parallel approximation algorithms by positive linear programming. *Algorithmica*, 21(1):72–88, May 1998.
13. L. Trevisan. Recycling queries in PCPs and in linearity tests. In *Proc. 30th STOC*, pages 299–308, 1998.
14. U. Zwick. Approximation algorithms for constraint satisfaction programs involving at most three variables per constraint. In *Proc. 9th SODA*, pages 201–210, 1998.

On the Derandomization of Constant Depth Circuits

Adam R. Klivans*

Laboratory for Computer Science
MIT
Cambridge, MA 02139
klivans@theory.lcs.mit.edu

Abstract. Nisan [18] and Nisan and Wigderson [19] have constructed a pseudo-random generator which fools any family of polynomial-size constant depth circuits. At the core of their construction is the result due to Håstad [10] that no circuit of depth d and size $2^{n^{1/d}}$ can even weakly approximate (to within an inverse exponential factor) the parity function. We give a simpler proof of the inapproximability of parity by constant depth circuits which does not use the Håstad Switching Lemma. Our proof uses a well-known hardness amplification technique from derandomization: the XOR lemma. This appears to be the first use of the XOR lemma to prove an unconditional inapproximability result for an explicit function (in this case parity). In addition, we prove that BPAC_0 can be simulated by uniform quasipolynomial size constant depth circuits, improving on results due to Nisan [18] and Nisan and Wigderson [19].

1 Introduction

1.1 Background

The explicit derandomization of randomized complexity classes remains a central challenge in complexity theory. Considerable interest has of late been given to the seminal work of Nisan [18] and Nisan and Wigderson [19] in which the authors construct a pseudo-random generator powerful enough to derandomize well known complexity classes like BPP under the assumption that a boolean function exists with high average-case circuit complexity. Further work has shown how to amplify the worst-case circuit complexity of a boolean function to achieve the necessary hardness required by the Nisan-Wigderson generator [5,11,13,25]. Depending on the assumption taken, the Nisan-Wigderson generator (or its strengthened relative the Impagliazzo-Wigderson generator) can be used to derandomize a wide range of expressive complexity classes [14]. Still, it seems that the most expressive randomized complexity class we can unconditionally or explicitly derandomize is uniform RAC_0, randomized constant depth polynomial

* Supported in part by NSF grant CCR-97-01304.

M. Goemans et al. (Eds.): APPROX-RANDOM 2001, LNCS 2129, pp. 249–260, 2001.

size uniform circuit families [19]. In fact, the case of RAC_0 remains one of the very few examples where the Nisan-Wigderson generator can be used to obtain an unconditional derandomization (for another see [17]).

Roughly speaking, the reason we can derandomize RAC_0 is that we can prove strong lower bounds against constant depth circuits. The celebrated results of [1,9,10,21,23] show that parity cannot be computed by polynomial size constant depth circuits. This fact alone, however, is not sufficient to build a pseudorandom generator useful for derandomizing RAC_0. The Nisan-Wigderson generator requires the existence of a function highly *inapproximable* by constant depth circuits, and only the lower bound techniques due to Håstad [10] yield strong enough results for this type of derandomization. Nisan and Wigderson use the results of Håstad to prove that RAC_0 can be simulated in quasipolynomial time.

1.2 Our Results

In this paper, we give a new proof of the inapproximability of parity by constant depth circuits via a well known hardness amplification technique: the XOR lemma. The advantage of our proof is two-fold. First, it is a simpler proof than the one found in [10] and does not use the Håstad Switching Lemma, a sophisticated combinatorial tool for describing the effect of random restrictions on constant depth circuits. Second, it appears to be the first time the XOR lemma has been used to prove an unconditional inapproximability result for an explicit function (in this case parity). Although our results give a slightly weaker inapproximability result than the optimal ones found in [10], our lower bounds are strong enough to obtain the same qualitative derandomization, namely a quasipolynomial time simulation of RAC_0:

Theorem 1. *Let C be a circuit of depth $d-1$ and size $2^{n^{1/6d}}$. Then C cannot compute parity correctly on more than a $1/2 + 2^{-n^{1/6d}}$ fraction of inputs.*

Further, we will show that BPAC_0 (and RAC_0) has uniform quasipolynomial size constant depth circuit families, improving on results due to Nisan [18] and Nisan and Wigderson [19] who showed that BPAC_0 is contained in quasipolynomial *time*.

1.3 Our Approach

Our starting point is the voting polynomials paper of Aspnes et al. [4] which proves that any circuit of depth d computing parity on any constant fraction of inputs has size $2^{\Omega(n^{\frac{1}{4d}})}$. Their result uses only elementary linear algebra and the fact that constant depth circuits can be approximated by low degree polynomials. In addition, their result holds even if the circuit has a majority gate at the root. We wish to amplify this hardness from a constant to $1/2 - 2^{-n^{1/O(d)}}$. What we prove is that the XOR of several independent copies of parity increases in hardness exponentially with respect to the number of copies taken. To prove such a result, we appeal to the hard-core set construction of Impagliazzo [11] which gives a simple proof of an XOR lemma and works with respect to any model of computation closed under majority.

At first glance, this may seem useless for our case, as it is known that AC_0 circuits cannot compute the majority function [10]. Our main observation is that a lower bound against AC_0 with one majority gate (for example the one found in [4]) gives us enough leeway to carry out the XOR lemma due to Impagliazzo [11]. For clarity of exposition we view Impagliazzo's result from a *boosting* perspective [15] (although details about particular boosting algorithms are not important here).

Boosting is a technique from computational learning theory (see [15]) which combines small circuits that weakly approximate a boolean function f (in our case f is the parity function) to form a circuit that computes f on almost all inputs. The simple boosting algorithm implicit in Impagliazzo's construction [11,15] combines weakly approximating circuits by taking their majority to form a circuit computing f almost everywhere. From the voting polynomials result, however, we know that circuits with one majority gate at the root cannot compute the parity function. Hence no small circuit can even weakly approximate parity. This observation leads to the existence of a hard-core set, a set of inputs S such that no small circuit can even weakly approximate parity on a randomly chosen input from S. An XOR lemma follows almost immediately, and, since the XOR of several independent copies of parity is simply equal to the parity function on a larger input length, we obtain our desired inapproximability result.

1.4 Comparison with Previous Results

In [10], Håstad gives an optimal result for the inapproximability of parity:

Theorem 2. *[10] Let ϕ be the parity function on inputs of length n. Let C be a circuit of size 2^s and depth $d > 2$ such that $s \geqslant n^{1/d}$. Then $Pr_{x \in \{0,1\}^n}[C(x) = \phi(x)] \leqslant 1/2 + 2^{-\Omega(\frac{n}{s^{d-1}})}$. If $s < n^{1/d}$ then $Pr_{x \in \{0,1\}^n}[C(x) = \phi(x)] \leqslant 1/2 + 2^{-\Omega((\frac{n}{s})^{\frac{1}{d-1}})}$.*

Corollary 1. *[10] Let C be a circuit of size $2^{n^{1/d}}$ and depth d. Then C cannot compute parity correctly on more than a $1/2 + 2^{-n^{1/d+1}}$ fraction of inputs.*

1.5 Organization

We begin by reviewing the "voting polynomials" result of [4]. We then review the Impagliazzo construction from [11] and describe how the "voting polynomials" result implies it can be used in the constant depth circuit setting. In Section 5 we show how to improve known containments of BPAC_0. Section 5 is independent from the other sections in this paper.

2 Preliminaries

For basic definitions regarding circuits and standard complexity classes we refer the reader to [20]. For a definition of a pseudorandom generator see [25].

Definition 1. *A circuit family* $\{C_n\}$ *is* uniform *if there exists a Turing Machine* M *such that* M *on input* 1^n *outputs* C_n *in polynomial time in* n. *A language* L *is in* RAC_0 *if there exists a uniform family of polynomial-size, constant depth circuits* $\{C_n\}$ *where each circuit* C_n *is a circuit taking two inputs* x *and* r *(*$|r| = s = n^k$ *for some* k*) such that if* $x \in L$, $Pr_{r \in \{0,1\}^s}[C(x,r) = 1] \geqslant 1/2$ *and if* $x \notin L$, $Pr_{r \in \{0,1\}^s}[C(x,r) = 1] = 0$. *A language* L *is in* BPAC_0 *if the probabilities* $1/2$ *and* 0 *are replaced with* $2/3$ *and* $1/3$.

Definition 2. *Let* f *be a Boolean function on* $\{0,1\}^n$ *and* $\mathcal{D}$ *a distribution on* $\{0,1\}^n$. *Let* $0 < \epsilon < 1/2$ *and let* $n \leqslant g \leqslant 2^n/n$. *We say that* f *is* δ-hard for size g and depth d under $\mathcal{D}$ *if for any boolean circuit* C *of depth* d *with at most* g *gates, we have* $\Pr_{\mathcal{D}}[f(x) = C(x)] \leqslant 1 - \delta$. *We say that* f *is* ϵ-hard-core for size g and depth d under $\mathcal{D}$ *if for any boolean circuit* C *of depth* d *with at most* g *gates, we have* $\Pr_{\mathcal{D}}[f(x) = C(x)] \leqslant 1/2 + \epsilon$. *For a set of inputs* S *we say that* f *is* δ-hard on S for size g and depth d *if* f *is* δ-hard *for size* g *and depth* d *with respect to the uniform distribution over* S *(a similar definition applies to* f *being* ϵ-hard-core *on* S*). If* S *is equal to the set of all inputs then we omit* S, *and if there are no depth restrictions we omit* d.

Throughout the paper we use $\mathcal{U}$ to denote the uniform distribution on $\{0,1\}^n$.

Definition 3. *For a distribution* $\mathcal{D}$, *the* minimum entropy *of* $\mathcal{D}$ *is equal to* $\log(L_\infty(\mathcal{D})^{-1})$.

3 Mild Hardness via Voting Polynomials

In this section we sketch the elegant "voting polynomials" result due to [4] which shows that parity cannot be approximated to within any constant by a constant depth circuit of size $2^{o(n^{\frac{1}{4d}})}$ with one majority gate at the root. At the heart of their result is a simple proof that parity cannot be sign represented well by any low degree polynomial. This idea stems from the now well known *approximation method* originally pioneered by Razborov [21] to give a novel proof that parity is not in AC_0. Smolensky [23] extended Razborov's original idea and showed that since polynomials cannot compute parity over finite fields, the lower bound result holds for constant depth circuits with a mod p gate for any prime p. Our interpretation of the voting polynomials result is that if we work with polynomials over the rationals, instead of finite fields, then we obtain a lower bound for parity against circuits with a *majority* gate, rather than mod p gates, since one can easily embed majority within the rationals.[1] The addition

[1] Beigel [7] has shown how to improve this from one majority gate to $n^{o(1/d)}$ majority gates. His result is tight.

of the majority gate will be essential, as it enables us to carry out the necessary hardness amplification reductions. We begin with some definitions:

Definition 4. *Let f a boolean function on n variables be represented by a function mapping $\{-1,1\}^n$ to $\{-1,1\}$. Notice that in this representation, the parity function on n variables is simply equal to the monomial $x_1, \ldots, x_n$. A rational polynomial $p(x_1, \ldots, x_n)$* weakly represents *f if p is nonzero and for every x such that $p(x)$ is nonzero, $sgn(f(x)) = sgn(p(x))$. The* weak degree *of a function f is the degree of the lowest degree polynomial which weakly represents f. We sometimes refer to a polynomial which weakly represents f as a* voting polynomial *for f.*

Lemma 1. *The weak degree of the parity function on n variables is n.*

Proof. Let ϕ denote the parity function on n variables. Assume that there exists a polynomial p of degree less than n which weakly represents parity. Consider the vector space of rational polynomials in n variables of total degree at most n equipped with the inner product $(f, g) = \sum_{x \in \{-1,1\}^n} f(x)g(x)$. The set of all monomials of degree less than or equal to n forms a basis. Notice that for any monomial m of degree strictly less than n, m does not contain some variable x_i and so (m, ϕ) must equal 0. In other words, ϕ is orthogonal to all monomials of degree less than n. But, since p weakly represents parity, (p, ϕ) must be strictly greater than 0. Hence parity has weak degree n.

Having established this, we can further prove that no low degree polynomial can *approximate* parity well:

Theorem 3. *[4] Let p be a nonzero polynomial of degree d. Let S be the set of inputs for which $sgn(x) \neq \phi(x)$. Then*

$$|S| \geqslant \sum_{i=1}^{\frac{n-d-1}{2}} \binom{n}{i}$$

Proof. Assume $|S|$ is smaller than the above bound. Let q be the degree $\frac{n-d-1}{2}$ polynomial such that for any $x \in S$, $q(x) = 0$. Notice that any polynomial of degree $\frac{n-d-1}{2}$ has, potentially, $\sum_{i=1}^{\frac{n-d-1}{2}} \binom{n}{i}$ nonzero monomials (or degrees of freedom). Thus q is a well defined, non-zero polynomial whose coefficients can be computed by solving an appropriate system of linear equations. Now the polynomial q^2p weakly represents parity and is of degree at most $n-1$, contradicting Lemma 1.

Interestingly, the authors of [4] prove that this theorem is tight by giving a simple construction of a polynomial approximating parity which makes exactly the above number of errors.

Corollary 2. *Any polynomial which weakly represents parity and makes less than $\gamma 2^n$ errors for any $\gamma < 1/2$ must have degree $\Omega(\sqrt{n})$.*

The following theorem builds on work of Valiant and Vazirani [27] and states that constant depth circuits can be accurately approximated by low degree polynomials. The result we state here is succinctly proved in [4] and is a refinement of a theorem from [8].

Theorem 4. *[27] [8] [4] For any $\gamma > 0$ and any function f computed by a boolean circuit of size s and depth d there exists a polynomial of degree $O((\log(s/\gamma)\log(s))^d)$ which* computes *f for all but $\gamma 2^n$ inputs.*

Now we can obtain a similar theorem for a circuit of size s and depth d with one majority gate at the root. For each input to the majority gate (let us say without loss of generality there are s of them) there exists some low degree polynomial approximating that input. Since we are working over the rationals, taking the sum of all of these polynomials and subtracting $s/2$ results in a polynomial weakly representing the output of the majority gate (and thus the entire circuit) on almost every input. From this discussion and Theorem 4 we obtain the following:

Theorem 5. *[4] For any $\gamma > 0$ and any function f computed by a boolean circuit of size s and depth d with one majority gate at the root there exists a polynomial of degree $O((\log(s^2/\gamma)\log(s))^d)$ which* weakly represents *f for all but $\gamma 2^n$ of the inputs.*

Now we can combine the above theorem with Corollary 2:

Theorem 6. *Parity is 1/8-hard for depth d circuits of size $2^{o(n^{1/4d})}$ with one majority gate at the root.*

4 Extreme Hardness via Hard-Core Sets

In this section we amplify the hardness from a constant to $1/2 + 2^{-n^{(1/O(d))}}$ via Impagliazzo's hard-core set construction [11]. We will show that if there exists a circuit approximating parity to within $1/2 + \epsilon$ (think of ϵ here as $2^{-n^{(1/O(d))}}$) then there exists a subexponential sized circuit with a majority gate at the root computing parity on a constant fraction of the inputs, contradicting Theorem 6.

To achieve this we would like to combine circuits with correctness $1/2+\epsilon$ with respect to parity to form a slightly larger circuit computing parity on say a 7/8 fraction of the inputs, i.e. *boost* our ϵ advantage.[2] The existence of these weakly approximating circuits will not be enough, however, to directly carry out such a reduction (this is because it is possible for every weakly approximating circuit to be wrong on the same set of inputs, in which case we cannot hope to boost our advantage). To obtain such a reduction, we will need to assume something stronger, namely that for any distribution with high *min-entropy*, there exists a small circuit approximating parity to within ϵ *with respect to that distribution.*

[2] Here again we frame this informal discussion of Impagliazzo's work in terms of boosting for clarity of exposition. The methodology of boosting is not relevant here.

As a consequence we will have that there exists a distribution D with large *min-entropy* such that no small circuit can weakly approximate parity with respect to D. From this it follows that there is a set of inputs S (of size proportional to the min-entropy of D) such that no small circuit can weakly approximate parity with respect to the uniform distribution on S. This set S is known as a *hard-core* set for parity. With S in hand, it is not hard to prove that computing the XOR of several independent copies of parity is a significantly harder function to approximate– with high probability, one of the inputs to the independent copies will fall within the hard-core set.

An important definition is that of a *measure* which we think of as an unnormalized distribution:

Definition 5. *A* measure *on $\{0,1\}^n$ is a function $M : \{0,1\}^n \to [0,1]$. The* absolute size *of a measure M is denoted by $|M|$ and equals $\sum_x M(x)$; the* relative size *of M is denoted $\mu(M)$ and equals $|M|/2^n$. The distribution $\mathcal{D}_M$ induced by a measure M is defined by $\mathcal{D}_M(x) = M(x)/|M|$.*

The relationship between measure size and min-entropy is established in [15]:

Fact 7 *For any measure M, the distribution induced by M has min-entropy at least $n - \log(1/\mu(M))$.*

4.1 Hard-Core Sets for Constant Depth Circuits

Here we prove that the hard-core set construction of Impagliazzo works with respect to constant depth circuits. We will need to assume initial hardness with respect to circuits with one majority gate at the root to obtain hard-core distributions:

Theorem 8. *[11] Let f be δ-hard for circuits of size g and depth d with one majority gate at the root with respect to $\mathcal{U}$ and let $0 < \epsilon < 1$. Then there is a measure M on $\{0,1\}^n$ with $\mu(M) \geqslant \delta$ such that f is ϵ-hard-core with respect to M (i.e. with respect to the distribution induced by M) for any circuit of size $g' = (1/4)\delta^2\epsilon^2 g$ and depth $d-1$ with one majority gate at the root.*

Proof. Consider the algorithm given in Figure 4.1. We will need the following lemma:

Lemma 2. *Algorithm* **IHA** *terminates after at most $i_0 = 2/\delta^2\epsilon^2$ iterations.*

We defer the proof of Lemma 2 to the appendix.

Once the algorithm terminates it is easy to see that $h \equiv MAJ(C_0, \ldots, C_{i-1})$ agrees with f on all inputs except those which have $N_i(x) \leqslant 0$ and hence $M_i(x) = 1$. Since $\mu(M_i) < \delta$, this implies that $\Pr_{\mathcal{U}}[f(x) = h(x)] \geqslant 1 - \mu(M_i) > 1 - \delta$. But since h is a majority circuit over at most i_0 circuits each of size at most g', and depth at most $d-1$, it follows that h is a depth d circuits with at most $g'i_0 + i_0 \leqslant g$ gates, which contradicts the original assumption that f is δ-hard for circuits of size g and depth d with one majority gate under $\mathcal{U}$.

Input: $\delta > 0$, $\epsilon > 0$, boolean function f
Output: a circuit h such that $\Pr_{\mathcal{U}}[h(x) = f(x)] \geqslant 1 - \delta$

1. **set** $i \leftarrow 0$
2. $M_0(x) \equiv 1$
3. **until** $\mu(M_i) < \delta$ **do**
4. let C_i be a circuit of size at most g' and depth d with $\Pr_{\mathcal{D}_{M_i}}[C(x) = f(x)] \geqslant 1/2 + \epsilon$
5. $R_{C_i}(x) \equiv 1$ if $f(x) = C_i(x)$, $R_{C_i}(x) \equiv -1$ otherwise
6. $N_i(x) \equiv \sum_{0 \leqslant j \leqslant i} R_{C_j}(x)$
7. $M_{i+1}(x) \equiv 0$ if $N_i(x) \geqslant 1/\delta\epsilon$, $M_{i+1}(x) \equiv 1$ if $N_i(x) \leqslant 0$, $M_{i+1}(x) \equiv 1 - \delta\epsilon N_i(x)$ otherwise
8. **set** $i \leftarrow i + 1$
9. $h \equiv MAJ(C_0, C_1, \ldots, C_{i-1})$
10. **return** h

Fig. 1. The IHA algorithm.

Hard-core measures of relative size δ correspond to distributions with min-entropy at least $n - \log(1/\delta)$. Since distributions with min-entropy k are convex combinations of uniform distributions over sets of size 2^k ([26]), one would expect the existence of a hard-core measure of relative size δ to imply the existence of a set of inputs S of size $\delta 2^n$ such that f is hard-core with respect to the uniform distribution over S. The following fact confirms this intuition:

Fact 9 *[11] Let f be $\epsilon/2$-hard-core with respect to a measure M of relative size δ for circuits of size g (where $2n < g < (1/8)(2^n/n)(\epsilon\delta)^2$) and depth d. Then there exists a set of inputs S of size $\delta 2^n$ such that f is ϵ-hard-core on S for size g and depth d.*

We note that in [15] it is shown how to improve the circuit size parameter from Theorem 8 from $O(\delta^2\epsilon^2 g)$ to $O(\epsilon^2(\log(1/\delta))^{-1}g)$. In [22] it is shown how to generalize the above algorithm to work with real valued functions.

4.2 Amplifying Hardness with the XOR Lemma

Theorem 10. *(XOR-Lemma) [11] Let f be ϵ-hard-core for some set of $\delta 2^n$ inputs for circuits of size g and depth d, then the function $f'(x_1, \ldots, x_k) = f(x_1) \oplus \cdots \oplus f(x_k)$ is $\epsilon + (1 - \delta)^k$-hard-core for circuits of size g and depth d.*

Proof. Let C be a circuit of size h obtaining advantage more than $\epsilon + (1 - \delta)^k$. Let A_l be C's advantage given that l inputs lie in the hard-core set. Since the probability that all k inputs lie in the hard-core set is $(1 - \delta)^k$, A_l must be greater than ϵ for some $l > 0$. Consider the circuit C' which randomly chooses $l-1$ instances from the hard-core set, $k-(l-1)$ from the complement of the hard-core and, on input x, is equal to $C(x)$ with all the inputs inserted in a random order (and is complemented if the XOR of f of the randomly chosen inputs is 1). Then this circuit C' has average advantage ϵ over the choice of random inputs, so we can fix the random choices non-uniformly to obtain a circuit computing f with advantage ϵ when the input is chosen uniformly from the hard-core, a contradiction.

Using the hard-core set construction and XOR lemma yields our main inapproximability result:

Theorem 11. *The parity function is* $2^{-n^{1/6d}}$-hard-core *for circuits of depth* $d-1$ *and size* $2^{n^{1/6d}}$.

Proof. Corollary 2 tells us that parity is $1/8$-hard for circuits of size $2^{o(n^{1/4d})}$ and depth $d-1$ with one majority gate at the root. Now apply Theorem 8 with $\delta = 1/8$ and $\epsilon/2 = 2^{-n^{1/5d}}$ to prove that parity is ϵ-hard-core on a set of size $(1/8)2^n$ for circuits of depth $d-1$ and size $2^{n^{1/5d}}$. The above XOR lemma amplifies our hardness to what is claimed and increases the input length to $O(n^{1+1/5d})$.

Nisan [18] and Nisan and Wigderson [19] show how to use this inapproximability result to build a pseudorandom generator which fools AC_0 and has seed size polylogarithmic in n (more precisely $\log^{O(d)} n$ where d is the depth of the circuit). Our inapproximability result can be used to obtain a similar pseudorandom generator and the following corollary:

Corollary 3. *Uniform* BPAC_0 *is contained in quasipolynomial time.*

5 Approximate Majority and Uniform Circuit Constructions

From the results in the previous section, we would like to conclude that BPAC_0 has uniform, quasipolynomial-size, constant depth circuits (rather than containment in quasipolynomial time). This is immediately true for RAC_0, since it is a one-sided error class, and we accept an input x if and only if all there exists an output from a suitable pseudorandom generator G, call it r, such that $C(x, r) = 1$ where C is the circuit deciding the language. This can be implemented with an OR gate. For BPAC_0, we need to take a majority vote over all possible seeds given to G, and, as we have stated earlier, it is known that majority is not in AC_0. Fortunately, since we can construct strong pseudorandom generators, for every input the fraction of seeds making C accept (or reject) will always be very close to $2/3$ (or $1/3$). Thus, we only need to compute an *approximate* majority. Ajtai and Ben-Or [3] building on work of Stockmeyer [24] proved the existence of non-uniform constant depth circuit families capable of computing an approximate majority as long as the fraction of ones is always bounded away from $1/2$ by at least an inverse polylogarithmic factor. In 1993, Ajtai [2] gave a *explicit* construction of such a circuit family:

Theorem 12. *[2] For any constant* $0 < \delta < 1/2$ *there exists a uniformly constructible polynomial-size, constant depth circuit family such that for any* $s \in \{0,1\}^n$:

- $|s| > (1/2+\delta)n$ *implies* $C(s) = 1$
- $|s| < (1/2-\delta)n$ *implies* $C(s) = 0$

(Ajtai proves that δ can be replaced with $1/(\log n)^i$ where the depth and size of the circuit will depend on i. This construction is optimal due to the lower bounds found in [10]).

With this construction we can obtain our desired inclusion for BPAC_0:

Theorem 13. BPAC_0 *can be computed by uniform, constant depth, quasipolynomial-size circuits.*

Proof. Let $L \in \mathsf{BPAC}_0$. Then L is computed by a uniform circuit family $\{C_n\}$ of depth d and size n^k for some constants d and k. Let G be a pseudorandom generator which fools $\{C_n\}$, has seed size $(\log n)^{O(d)}$, and approximates the acceptance probability of $\{C_n\}$ to an additive $O(1/n)$ factor. The uniform circuit family for L is obtained by enumerating over all seeds to G and obtaining $2^{(\log n)^{O(d)}}$ circuits where each circuit equals C_n with one of the outputs from G as its random bit sequence. From the approximate majority construction, there exists a circuit M of depth d' and size $p(2^{(\log n)^{O(d)}})$ for some polynomial p computing the approximate majority of any sequence of length $2^{(\log n)^{O(d)}}$ where the number of ones is bounded away from $1/2$. For any input x, the output sequence of all of the above circuits on input x has at least $2/3 - 1/n$ ones or at most $1/3+1/n$ ones. Combining the enumerated circuits as inputs to the approximate majority circuit yields the desired constant depth, quasipolynomial size circuit.

6 Conclusion

In this paper we have given a new proof of the inapproximability of parity by AC_0 without using the Håstad switching lemma. Since the Håstad switching lemma has many important applications in such diverse areas as computational learning theory [16] and proof complexity (see [6]), we hope that the techniques presented in this paper can be used to provide simpler and/or improved results in these areas. In [11], Impagliazzo proves an XOR-lemma where the inputs are chosen pairwise independently. Is it possible to prove such a derandomized XOR lemma in the constant depth circuit setting? What is the most restricted circuit class where this can be carried out? We leave these as interesting open problems.

Acknowledgments

Thanks to Rocco Servedio for important conversations about the Impagliazzo construction and its relationship to boosting. Thanks to Ryan O'Donnell and Salil Vadhan for an interesting conversation about approximate majority and to V. Guruswami for his encouragement.

A Proof of Lemma 2

Algorithm **IHA** terminates after at most $i_0 = 2/\delta^2\epsilon^2$ iterations.

Proof. Let $\gamma = \epsilon\delta$. Consider the quantity $A_i(x) = \sum_{0 \leqslant j \leqslant i-1} R_{C_j}(x) M_{j-1}(x)$. This is the cumulative advantage that the first i circuits have on input x. It is not hard to see that $\sum_x A_i(x) \geqslant 2^n \gamma i$. For a fixed x, imagine plotting $N_j(x)$ with respect to i (think of j as increasing steps through the iterative algorithm). Match every time $N_j(x)$ increases from k to $k+1$ with every time $N_j(x)$ decreases from $k+1$ to k (if k is less than 0 match changes from k to $k-1$ with changes of $k-1$ to k). An easy calculation reveals that each of these pairs contributes at most γ to $A_i(x)$. For each decreasing unmatched time step we actually subtract from $A_i(x)$ and for each increasing unmatched time step (of which there can be at most $1/\gamma$) we add at most 1 to $A_i(x)$. There are at most $(1/2)i$ pairs. Thus $A_i(x) \leqslant 1/\gamma + (1/2)\gamma i$. Solving for i yields the claimed bound on the number of iterations.

References

1. M. Ajtai. Σ_1^1-Formulae on Finite Structures. *Annals of Pure and Applied Logic*, Vol. 24, 1983.
2. M. Ajtai. Approximate Counting with Uniform Constant-Depth Circuits. *DIMACS Series in Discrete Mathematics and Theoretical Computer Science*, Vol. 13, 1993.
3. M. Ajtai and M. Ben-Or. A Theorem on Probabilistic Constant Depth Computations. In *ACM Symposium on Theory of Computing (STOC)*, 1984.
4. J. Aspnes, R. Beigel, M. Furst, and S. Rudich. The Expressive Power of Voting Polynomials. *Combinatorica*, 14(2), 1994.
5. L. Babai, L. Fortnow, N. Nisan, and A. Wigderson. BPP has subexponential time simulations unless EXPTIME has publishable proofs. *Computational Complexity*, 3:307–318, 1993.
6. P. Beame. A Switching Lemma Primer. Technical Report UW-CSE-95-07-01, Department of Computer Science and Engineering, University of Washington, November 1994.
7. R. Beigel When do Extra Majority Gates Help? Polylog(n) Majority Gates are Equivalent to One. *Computational Complexity*, 4, 314-324, 1994.
8. R. Beigel, N. Reingold, and D. Spielman. The Perceptron Strikes Back. In *Proceedings of the 6th Annual Conference on Structure in Complexity Theory (SCTC '91)*, June 1991.
9. M. Furst, J. Saxe, and M. Sipser. Parity, Circuits, and the Polynomial-Time Hierarchy. *Mathematical Systems Theory*, 17(1), 1984.
10. J. Håstad. *Computational Limitations of Small Depth Circuits*. MIT-PRESS, 1986.
11. R. Impagliazzo. Hard-core distributions for somewhat hard problems. In *Proceedings of the 36th IEEE Symposium on Foundations of Computer Science*, pages 538–545. IEEE, 1995.
12. R. Impagliazzo, R. Shaltiel, and A. Wigderson. Near Optimal Conversion of Hardness into Pseudorandomness. In *Proceedings of the 40th IEEE Symposium on Foundations of Computer Science*, pages 538–545. IEEE, 1999.

13. R. Impagliazzo and A. Wigderson. P=BPP unless E has sub-exponential circuits: Derandomizing the XOR lemma. In *Proceedings of the 29th ACM Symposium on the Theory of Computing*, pages 220–229. ACM, 1997.
14. A. Klivans and D. van Melkebeek. Graph nonisomorphism has subexponential size proofs unless the polynomial hierarchy collapses. In *Proceedings of the 31st ACM Symposium on the Theory of Computing*, pages 659–667. ACM, 1999.
15. A. Klivans and R. Servedio. Boosting and Hard-Core Sets. In *Proceedings of the 40th IEEE Symposium on Foundations of Computer Science (FOCS)*, 1999.
16. N. Linial, Y. Mansour, and N. Nisan. Constant Depth Circuits, Fourier Transforms, and Learnability. *Journal of the ACM*, 40(3), July 1993.
17. M. Luby, B. Velickovic, and A. Wigderson. Deterministic Approximate Counting of Depth-2 Circuits. In *Proceedings of the Second Israeli Symposium on Theory of Computing and Systems*, 1993.
18. N. Nisan. Pseudorandom Bits for Constant Depth Circuits. *Combinatorica*, 11, 1991.
19. N. Nisan and A. Wigderson. Hardness vs. Randomness. *Journal of Computer and System Sciences*, 49:149–167, 1994.
20. C. Papadimitriou. *Computational Complexity*. Addison-Wesley, 1994.
21. A. Razborov. Lower Bounds on the Size of Bounded Depth Circuits over a Complete Basis with Logical Addition. *MATHNAUSSR: Mathematical Notes of the Academ of Sciences of the USSR*, 41, 1987.
22. R. Servedio. Smooth Boosting and Learning with Malicious Noise. To Appear, 2001.
23. R. Smolensky. Algebraic Methods in the Theory of Lower Bounds for Boolean Circuit Complexity. In *Proceedings of the Nineteenth Annual ACM Symposium on Theory of Computing*, May 1987.
24. L. Stockmeyer. The Complexity of Approximate Counting. In *ACM Symposium on Theory of Computing (STOC)*, 1983.
25. M. Sudan, L. Trevisan, and S. Vadhan. Pseudorandom generators without the XOR lemma. Technical Report TR-98-074, Electronic Colloquium on Computational Complexity, 2000. Revision 2.
26. L. Trevisan. Construction of extractors using pseudo-random generators. In *Proceedings of the 31st ACM Symposium on the Theory of Computing*, pages 141–148. ACM, 1999.
27. L. Valiant and V. Vazirani. NP is as easy as detecting unique solutions. *Theoretical Computer Science*, 47:85–93, 1986.

Testing Parenthesis Languages

Michal Parnas[1], Dana Ron[2,⋆], and Ronitt Rubinfeld[3]

[1] The Academic College of Tel-Aviv-Yaffo
michalp@mta.ac.il
[2] Department of EE – Systems, Tel-Aviv University
danar@eng.tau.ac.il
[3] NEC Research Institute, Princeton
ronitt@research.nj.nec.com

Abstract. We continue the investigation of properties defined by formal languages. This study was initiated by Alon et al. [1] who described an algorithm for testing properties defined by regular languages. Alon et al. also considered several context free languages, and in particular Dyck languages, which contain strings of properly balanced parentheses. They showed that the first Dyck language, which contains strings over a single type of pairs of parentheses, is testable in time independent of n, where n is the length of the input string. However, the second Dyck language, defined over two types of parentheses, requires $\Omega(\log n)$ queries.
Here we describe a sublinear-time algorithm for testing all Dyck languages. Specifically, the running time of our algorithm is $\tilde{O}(n^{2/3}/\epsilon^3)$, where ϵ is the given distance parameter. Furthermore, we improve the lower bound for testing Dyck languages to $\tilde{\Omega}(n^{1/11})$ for constant ϵ. We also have a testing algorithm for the context free language $L_{\mathrm{REV}} = \{w = uu^r vv^r : w \in \Sigma^n\}$, where Σ is a fixed alphabet. The running time of our algorithm is $\tilde{O}(\sqrt{n}/\epsilon)$, which almost matches the lower bound given by Alon et al. [1].

1 Introduction

Property testing [9,2] is a relaxation of the standard notion of a decision problem: property testing algorithms distinguish between inputs that have a certain property and those that are *far* from having the property. More precisely, for any fixed property $\mathcal{P}$, a testing algorithm for $\mathcal{P}$ is given query access to the input and a distance parameter ϵ. The algorithm should output accept with high probability if the input has the property $\mathcal{P}$, and output reject if the input is ϵ-far from having $\mathcal{P}$. By ϵ-far we mean that more than an ϵ–fraction of the input should be modified so that the input obtains the desired property $\mathcal{P}$.

Testing algorithms whose query complexity is sublinear and even independent of the input size, have been designed for testing various algebraic and combinatorial properties (see [8] for a survey).

Motivated by the desire to understand in what sense the complexity of testing properties of strings is related to the complexity of formal languages, Alon et al

⋆ Supported by the Israel Science Foundation (grant number 32/00-1).

M. Goemans et al. (Eds.): APPROX-RANDOM 2001, LNCS 2129, pp. 261–272, 2001.

al. [1], have shown that all properties defined by regular languages are testable in time that is independent of the input size. Specifically, given a regular language L, they describe an algorithm that tests, using $\tilde{O}(1/\epsilon)$ queries, whether a given string s belongs to L or is ϵ-far from any string in L. This result was later extended by Newman [6] to properties defined by bounded-width branching programs. However, Alon et al. [1] showed that the situation changes quite dramatically for context-free languages. In particular, they prove that there are context-free languages that are not testable even in time square root in the input size. The question remains whether context-free languages can be tested in sublinear time. In this paper, we give evidence for an affirmative answer by presenting sublinear time testers for certain important subclasses of the context-free languages.

Dyck Languages. One important subclass of the context-free languages is the Dyck language, which includes strings of properly balanced parentheses. Strings such as "(( )( ))" belong to this class, whereas strings such as "(( )" or ") (" do not. If we allow more than one type of parentheses then "([ ])" is a balanced string but "([ )]" is not. Formally, the Dyck language D_m contains all balanced strings that contain at most m types of parentheses. Thus for example "(( )( ))" belongs to D_1 and "([ ])" belongs to D_2.

Dyck languages appear in many contexts. For example, these languages describe a property that should be held by commands in most commonly used programming languages, as well as various subsets of the symbols/commands used in latex. Furthermore, Dyck languages play an important role in the theory of context-free languages. As stated by the Chomsky-Schötzenberger Theorem, every context-free language can be mapped to a restricted subset of D_m [10]. A comprehensive discussion of context free languages and Dyck languages can be found in [3,5].

Thus testing membership in D_m is a basic and important problem. Alon et al. [1], have shown that membership in D_1 can be tested in time $\tilde{O}(1/\epsilon)$, whereas membership in D_2 cannot be tested in less than a logarithmic time in the length n of the string.

Our Results.

- We present an algorithm that tests whether a string s belongs to D_m. The query complexity and running time of the algorithm are $\tilde{O}\left(n^{2/3}/\epsilon^3\right)$, where n is the length of s. The complexity does not depend on m, the number of different types of parentheses.
- We prove a lower bound of $\Omega(n^{1/11}/\log n)$ on the query complexity of any algorithm for testing D_m for $m > 1$.
- We consider the context free language $L_{\mathrm{REV}} = \{w = uu^r vv^r : w \in \Sigma^n\}$ where Σ is any fixed alphabet and u^r denotes the string u in reverse order. We show that L_{REV} can be tested in $\tilde{O}(\frac{1}{\epsilon}\sqrt{n})$ time. Our algorithm almost matches the $\Omega(\sqrt{n})$ lower bound of Alon et al. [1] on the number of queries required for testing L_{REV}.

The structure of our testing algorithm for D_m. Our testing algorithm for D_m combines *local* checks with *global* checks. Specifically, the first part of the test randomly selects consecutive substrings of the given input string, and checks that they do not constitute a witness to the string not belonging to D_m. The second, more elaborate part of the test, verifies that non-consecutive pairs of substrings that are supposed to contain matching parentheses, in fact do. In particular, the string is partitioned into fixed *blocks* (consecutive substrings), and the algorithm computes various statistics concerning the numbers of opening and closing parentheses in the different blocks. Using these statistics it is possible to determine which pairs of blocks should contain many matching parentheses in case that the string in fact belongs to D_m. The testing algorithm then randomly selects such pairs of blocks and verifies the existence of such a matching between opening parentheses in one block and closing parentheses in the other block.

Organization. In Section 2 we describe the necessary preliminaries. Our testing algorithm for Dyck Languages is presented in Section 3. Most proofs, as well as the lower bound for Dyck languages and the algorithm for testing L_{REV} can be found in the full version of this paper [7].

2 Preliminaries

Let $s = s_1 \dots s_n$ be a string over an alphabet $\Sigma_m = \{0, \dots, 2m-1\}$ where $2i, 2i+1$ correspond to the ith type of opening and closing parentheses. We will use the following notation for strings and substrings.

Definition 1 (Substrings) *For a string $s = s_1 \dots s_n$ and $i \le j$, we let $s_{i,j}$ denote the substring $s_i, s_{i+1}, \dots, s_j$. If s', s'' are two strings, then $s's''$ denotes the concatenation of the two strings.*

Definition 2 (Dyck Language) *The Dyck language D_m can be defined recursively as follows:*

1. *The empty string belongs to D_m.*
2. *If $s' \in D_m$, $\sigma = 2i$ is an opening parenthesis and $\tau = 2i+1$ is a matching closing parenthesis (for some $0 \le i \le m-1$), then $\sigma s' \tau \in D_m$.*
3. *If $s', s'' \in D_m$, then $s's'' \in D_m$.*

It is clear from the recursive definition of D_m that the parentheses in a string s have a nested structure and are balanced. The first step of our algorithm will test if the string s is a legal string when we view it as a string in D_1, using the test given by [1]. Furthermore, the algorithm will test if consecutive substrings of s can be extended to a legal string in D_m. The following definitions address these aspects formally.

Definition 3 (Single-Parentheses Mapping) *Given a string s over Σ_m, we can map it to a string $\mu(s)$ over $\Sigma_1 = \{0, 1\}$ in the obvious manner: Every opening parenthesis is mapped to 0, and every closing parentheses is mapped to 1. We denote this mapping by μ.*

Definition 4 (Parentheses Matching) *For every string s such that $\mu(s) \in D_1$, there exists a* unique perfect matching *$M(s)$ between opening and closing parentheses in s, such that each opening parenthesis s_j is matched to a closing parenthesis s_k, and no two matched pairs "cross". That is, if s_{j_1} is matched to s_{k_1}, and s_{j_2} to s_{k_2} where $j_1 < k_1$, and $j_1 < j_2 < k_2$, then either $k_1 < j_2$ or $k_1 > k_2$.*

Definition 5 (Consistency of Substrings) *We say that a substring s' over Σ_m is* Dyck Consistent, *if there exists a string $s \in D_m$ such that s' is a (consecutive) substring of s.*

The second part of our algorithm finds disjoint pairs of substrings such that there exist opening parentheses in the first substring that should be matched to closing parentheses in the second substring. The algorithm verifies that these pairs of parentheses match in type as required. The following concepts will be needed for this part of the algorithm.

Definition 6 (Parentheses Numbers) *For any substring s' of s, define $n_0(s')$ to be the number of opening parentheses in s', and define $n_1(s')$ to be the number of closing parentheses in s'.*

Fact 1 *A string s belongs to D_1 if and only if: (1) For every prefix s' of s, $n_0(s') \geq n_1(s')$; (2) $n_0(s) = n_1(s)$.*

The above fact implies that any string s' over $\Sigma_1 = \{0, 1\}$ is Dyck Consistent, since for such a string there exist integers k and ℓ such that $0^k s' 1^\ell \in D_1$. In this case we can view s' as having an *excess* of k closing parentheses and ℓ opening parentheses (assuming k and ℓ are the smallest integers such that $0^k s' 1^\ell \in D_1$). The following definition extends this notion of excess parentheses in a substring to any alphabet Σ_m.

Definition 7 (Excess Numbers) *Let s' be a substring over Σ_m, and let k and ℓ be the smallest integers such that $0^k \mu(s') 1^\ell \in D_1$. Then k is called the* excess number *of closing parentheses in s', and ℓ is the excess number of opening parentheses in s'. Denote k by $e_1(s')$ and ℓ by $e_0(s')$.*

For example if s' = "] [( )]) (", then $e_1(s') = 2$ and $e_0(s') = 1$, where for the sake of the presentation we denote the pairs of parentheses by ( ) and [ ]. It is possible to compute the excess numbers from the parentheses numbers as follows.

Claim 2 *The following two equalities hold for every substring s',*

$$e_1(s') = \max_{s'' \text{ prefix of } s'} (n_1(s'') - n_0(s''))$$

$$e_0(s') = \max_{s'' \text{ suffix of } s'} (n_0(s'') - n_1(s'')) \tag{1}$$

In both cases the maximum is also over the empty prefix (suffix) s'', for which $n_1(s'') - n_0(s'') = 0$.

3 The Algorithm for Testing D_m

In the following subsections we describe several building blocks of our algorithm. Recall that the algorithm has two main parts. First we test that $\mu(s) \in D_1$, and that consecutive substrings of s are Dyck consistent. Then, by estimating the excess numbers for substrings of s, we find pairs of substrings that contain a significant number of matched pairs of parentheses according to $M(s)$, and check that these pairs match in type. To do the latter, we break the string into $n^{1/3}$ substrings each of length $n^{2/3}$, which we refer to as *blocks*. Assume for simplicity that $\mu(s) \in D_1$ (where we of course deal with the general case). Then there exists a weighted graph, whose vertices correspond to these blocks, and in which there is an edge between block i and block $j > i$ if and only if the matching $M(s)$ (as in Definition 4) matches between excess opening parentheses in block i to excess closing parentheses in block j. The weight of each edge is simply the number of corresponding matched pairs of excess parentheses. As we show subsequently, this weight can be determined by the values of the excess numbers for every consecutive sequence of blocks. Hence, if we were provided with these *exact* values, we could verify, for randomly selected pairs of blocks that are connected by an edge in the graph, whether their excess parentheses match as required. Since we do not have these exact values, but rather approximate values, we use our estimates of the excess values to construct an approximation of the above graph, and to perform the above verification of matching excess parentheses.

3.1 Checking Consistency

It is well known that it is possible to check in time $O(n)$ using a stack whether a string s of length n belongs to D_m. This is done as follows: The symbols of s are read one by one. If the current symbol read is an opening parenthesis then it is pushed onto the stack. If it is a closing parenthesis, then the top symbol on the stack is popped and compared to the current symbol. The algorithm rejects if the symbol popped (which must be an opening parenthesis) does not match the current symbol. The algorithm also rejects if the stack is empty when trying to pop a symbol. The algorithm accepts if after reading all symbols the stack is empty.

The above algorithm can be easily modified to check whether a substring s' is Dyck consistent. The only two differences are: (1) When reading a closing parenthesis and finding that the stack is empty, the algorithm does not reject but rather continues with the next symbol. (2) If the algorithm has completed reading the string without finding a mismatched pair of parentheses, then it accepts even if the stack is not empty. Thus the algorithm rejects only if it finds a mismatch in the type of parentheses.

3.2 A Preprocessing Stage

An important component of our algorithm is acquiring good estimates of the excess numbers of different substrings of the given input string s. We start by describing a preprocessing step based on which we can obtain such estimates for a fixed set of *basic* substrings of s (having varying sizes). By sampling from such a substring s', we obtain estimates of $n_0(s')$ and $n_1(s')$. Using these estimates we can derive estimates for the excess numbers of any given substring of s.

Let $r = \log(n^{1/3}/\delta)$, where $0 < \delta < 1$ is a parameter that is set subsequently. For each $j \in \{0, 1, \ldots, r\}$, we consider the partition of s into 2^j consecutive substrings each of length $n/2^j$. We assume for simplicity that n is divisible by $2^r = n^{1/3}/\delta$. Thus the total number of substrings is $O(n^{1/3}/\delta)$, where the longest is the whole string s, and the shortest ones are of length $\delta \cdot n^{2/3}$. We refer to these substrings as the *basic substrings* of s.

For each basic substring s' of length $n/2^j$, we uniformly and independently select a sample of m_j symbols from s', where $m_j = \Theta\left(\frac{n^{2/3}}{2^{2j}} \cdot \frac{\log^3(n/\delta)}{\delta^2}\right)$. Let m_j^0 be the number of opening parentheses in the sample, and m_j^1 be the number of closing parentheses in the sample. Our estimates of the number of opening and closing parentheses in s' are respectively: $\hat{n}_0(s') = \frac{m_j^0}{m_j} \cdot |s'| = \frac{m_j^0}{m_j} \cdot \frac{n}{2^j}$ and $\hat{n}_1(s') = \frac{m_j^1}{m_j} \cdot \frac{n}{2^j}$.

Lemma 3 *With probability at least* $1 - o(1)$, *for each of the basic substrings* $s' \subseteq s$, $|\hat{n}_0(s') - n_0(s')| \leq \frac{\delta}{24\log(n/\delta)} \cdot n^{2/3}$ *and* $|\hat{n}_1(s') - n_1(s')| \leq \frac{\delta}{24\log(n/\delta)} \cdot n^{2/3}$. *The total size of the sample is* $O\left(\frac{n^{2/3} \cdot \log^3(n/\delta)}{\delta^2}\right)$.

We assume from now on that the quality of our estimates $\hat{n}_0(s')$ and $\hat{n}_1(s')$ are in fact as stated in the lemma for every basic substring s'. We refer to this as the *successful preprocessing assumption*.

Assumption 4 (Successful Preprocessing Assumption) *For each of the basic substrings* s', $|\hat{n}_0(s') - n_0(s')| \leq \frac{\delta}{24\log(n/\delta)} \cdot n^{2/3}$, *and the same bound holds for* $\hat{n}_1(s')$.

3.3 Obtaining Estimates of Excess numbers

We first consider obtaining estimates for $n_0(s')$ and $n_1(s')$ for substrings s' of s of the form defined in the next claim.

Claim 5 *Let* $s' = s_{k,\ell}$ *be any substring of* s *such that* $k = t_1 \cdot (\delta \cdot n^{2/3}) + 1$ *and* $\ell = t_2 \cdot (\delta \cdot n^{2/3})$, *for* $0 \leq t_1 < t_2 \leq n^{1/3}/\delta$. *Then* s' *is the concatenation of at most* $2\log|s'| + 1$ *of the basic substrings.*

Assume the substring s' is the concatenation of the basic substrings $s^1, \ldots, s^t$. Then we can estimate $n_0(s')$ by $\hat{n}_0(s') = \sum_{i=1}^{t} \hat{n}_0(s^i)$, where $\hat{n}_0(s^i)$ is the estimate we got above for the basic substring s^i. Similarly, we can estimate $n_1(s')$.

Corollary 6 *Under Assumption 4, for every substring* s' *as in Claim 5,* $|\hat{n}_0(s') - n_0(s')| < \frac{\delta}{4} \cdot n^{2/3}$ *and* $|\hat{n}_1(s') - n_1(s')| < \frac{\delta}{4} \cdot n^{2/3}$.

We next consider how to obtain estimates for the excess number of opening parentheses of a given substring $s' = s_{k,\ell}$ (where k, ℓ, t_1, t_2 are assumed to be as in Claim 5), and similarly for the excess number of closing parentheses. To this end we appeal to Claim 2, and use our estimates for the total number of opening and closing parentheses in certain prefixes and suffixes of s'. As we show below, for the purpose of getting an additive estimate of the excess within $\delta \cdot n^{2/3}$ for any substring, it is enough to use estimates of n_0 and n_1 for prefixes and suffixes of the substring that are multiples of $\delta \cdot n^{2/3}$. Specifically,

Claim 7 *Let $s' = s_{k,\ell}$ be as in Claim 5, and define two sets*

$$\textit{Prefix} = \{s'' | s'' = s_{k,\ell'},\ \ell' = t'_2 \cdot (\delta \cdot n^{2/3}) + 1,\ t_1 \le t'_2 < t_2\}$$

$$\textit{Suffix} = \{s'' | s'' = s_{k',\ell},\ k' = t'_1 \cdot (\delta \cdot n^{2/3}) + 1,\ t_1 < t'_1 \le t_2\}.$$

Let

$$\hat{e}_0(s') = \max_{s'' \in \textit{Suffix}} (\hat{n}_0(s'') - \hat{n}_1(s'')), \qquad \hat{e}_1(s') = \max_{s'' \in \textit{Prefix}} (\hat{n}_1(s'') - \hat{n}_0(s'')).$$

Then, under Assumption 4, $|\hat{e}_0(s') - e_0(s')| \le \delta \cdot n^{2/3}$ and $|\hat{e}_1(s') - e_1(s')| \le \delta \cdot n^{2/3}$.

3.4 The Matching Graph

Before defining the matching graph, we extend the notion of the matching $M(s)$ (see Definition 4) to strings s such that $\mu(s) \notin D_1$. In this case we do not obtain a *perfect* matching, but rather a matching of all the parentheses in the string that are not excess parentheses with respect to the whole string. Specifically, by definition of the excess numbers, the string $\tilde{s} = 0^{e_1(s)} \mu(s) 1^{e_0(s)}$ belongs to D_1. Thus we let $M(s)$ be the restriction of $M(\tilde{s})$ to pairs of parentheses that are *both in* s. For example, if $s =$ "( ]] ([ )", then $M(s)$ matches between s_1 and s_2 and between s_5 and s_6.

In all that follows we assume that $n^{2/3}$ is an even integer. It is not hard to verify that this can be done without loss of generality. We partition the given string s into $n^{1/3}$ consecutive and disjoint substrings, each of length $n^{2/3}$, which we refer to as *blocks*.

Definition 8 (Neighbor Blocks) *We say that two blocks i and j are* neighbors *in a string s, if the matching $M(s)$ matches between excess opening parentheses in block i and excess closing parentheses in block j.*

Definition 9 (The Matching Graph of a String) *Given a string s, we define a weighted graph as follows. The vertices of the graph are the $n^{1/3}$ blocks of s. Two blocks $i < j$ are connected by an edge (i, j) if and only if they are neighbor blocks (as defined above). The weight $w(i, j)$ of the edge (i, j) is the number of excess opening parentheses in block i that are matched by $M(s)$ to excess closing parentheses in block j. The resulting graph is called the* matching graph *of s, and is denoted by $G(s)$. The set of edges of the graph is denoted by $E(G(s))$.*

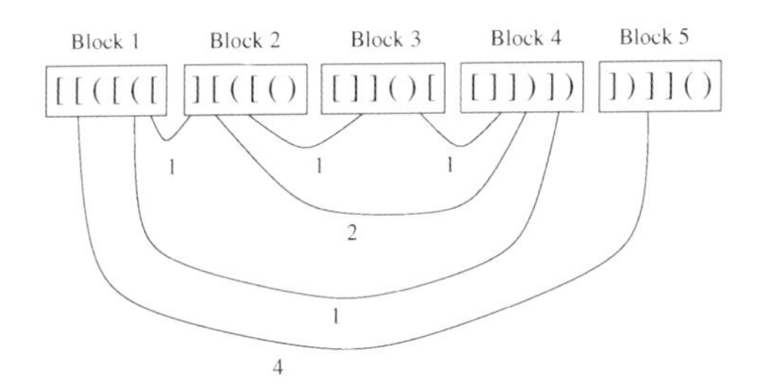

Fig. 1. An example of the matching graph of a string (in D_2). The string consists of 5 blocks (outlined by rectangles), with 6 symbols in each block. The numbers below the edges are the weights of the edges.

By definition of the matching $M(s)$ between closing and opening parentheses we get:

Claim 8 *For every string s, the matching graph $G(s)$ is planar, and therefore* $|E(G(s))| \leq 3n^{1/3}$.

It is possible to determine which blocks are neighbors in $G(s)$, and what is the weight of the edge between them, using the excess numbers e_1 and e_0 as follows. We first introduce one more definition.

Definition 10 (Intervals) *For a given string s, let $I_{i,j}$ denote the substring, which we refer to as* interval, *that starts at block i and ends at block j (including both of them).*

Claim 9 *Let s be a given string and let $i < j$ be two blocks in $G(s)$. Define:*

$$x(i,j) \stackrel{\text{def}}{=} \min\{e_1(I_{i+1,j}), e_0(I_{i,i})\} - e_1(I_{i+1,j-1}) \ . \tag{2}$$

If i and j are neighbors in $G(s)$ then $w(i,j) = x(i,j)$, and if $x(i,j) > 0$ then i and j are neighbors in $G(s)$.

Proof: We first observe that for both parts of the claim the premise implies that $e_0(I_{i,i}) > 0$. That is, the i'th block has excess opening parentheses. We next observe that the sequence $e_1(I_{i+1,j})$ is monotonically non-decreasing with j. Let $t > i$ be the maximum index such that $e_1(I_{i+1,t-1}) < e_0(I_{i,i})$, (so that in particular $e_1(I_{i+1,t}) \geq e_0(I_{i,i})$), and for every $j > t$, $e_1(I_{i+1,j}) > e_0(I_{i,i})$. Then all the neighbors $j > i$ of block i are found in the interval $I_{i+1,t}$. The number of excess closing parentheses of block j where $i+1 \leq j \leq t-1$, that are matched to block i, is $e_1(I_{i+1,j}) - e_1(I_{i+1,j-1})$. Therefore, if $e_1(I_{i+1,j}) - e_1(I_{i+1,j-1}) > 0$ then blocks i and j are neighbors. The number of parentheses matched between block i and block t is $e_0(I_{i,i}) - e_1(I_{i+1,t-1})$. The claim follows. □

It is not hard to verify that based on the symmetry of the matching, if i and j are neighbors in $G(s)$ then also $w(i,j) = \min\{e_0(I_{i,j-1}), e_1(I_{j,j})\} - e_0(I_{i+1,j-1})$.

We next turn to the case in which we only have estimates of the excess numbers. Here we define a graph based on the estimates we have for the excess numbers. This graph contains only relatively "heavy" edges in order to overcome approximation errors.

Definition 11 (The Approximate Matching Graph) *Given a string s, we partition it into blocks of size $n^{2/3}$, and define a graph $\hat{G}(s)$ whose vertices are the $n^{1/3}$ blocks of s. A pair of blocks $i < j$ will be connected by an edge (i, j) if and only if $\min\{\hat{e}_1(I_{i+1,j}), e_0(I_{i,i})\} - \hat{e}_1(I_{i+1,j-1}) \geq 4\delta n^{2/3}$, where $\hat{e}_0$ and $\hat{e}_1$ are as defined in Claim 7.*

The reason that in the definition we use the approximate values $\hat{e}_1(I_{i+1,j})$ but the exact value $e_0(I_{i,i})$, is that in the algorithm the value of $e_0(I_{i,i})$ is known exactly. The following lemma is central to our algorithm and its analysis.

Lemma 10 *Suppose Assumption 4 holds. Then for any given string s, the graph $\hat{G}(s)$ is a subgraph of $G(s)$, and every vertex in $\hat{G}(s)$ has degree at most $1/(2\delta)$. Furthermore, if $\mu(s)$ is δ-close to a string in D_1, then $\hat{G}(s)$ "accounts for most of the excess" in s. Namely,*

$$\sum_{(k,\ell) \in E(\hat{G}(s)),\ k<\ell} x(k, \ell) \geq \frac{1}{2} \sum_{i=1}^{n^{1/3}} (e_0(I_{i,i}) + e_1(I_{i,i})) \quad - 19\delta n.$$

3.5 Matching between Neighbors

Definition 12 (Matching Substrings) *Let s' be a substring of opening parentheses and let s'' be a substring of closing parentheses. We say that s' and s''* match *if $|s'| = |s'|$ and $s's'' \in D_m$.*

Given a string s, it is possible to determine for any two neighbor blocks $i < j$ in $G(s)$, which pairs of excess parentheses within these blocks should match. Let $\mathcal{E}_0(i)$ denote the (non-consecutive) substring of excess opening parentheses in block i, and let $\mathcal{E}_1(j)$ denote the substring of excess closing parentheses in block j. By definition, $|\mathcal{E}_0(i)| = e_0(I_{i,i})$ and $|\mathcal{E}_1(j)| = e_1(I_{j,j})$.

We first find $\mathcal{E}_0(i)$ and $\mathcal{E}_1(j)$. This is done by slight modifications of the Dyck-consistency procedure. Namely, when reading block i, the substring $\mathcal{E}_0(i)$ consists of those opening parentheses that are left on the stack when the procedure terminates. On the other hand, the substring $\mathcal{E}_1(j)$ consists of those closing parentheses, that when read, the stack is found to be empty.

Recall that by Claim 9, for every two blocks $i < j$ that are neighbors in $G(s)$, there are $w(i, j) = x(i, j)$ excess opening parentheses in block i that are matched to excess closing parentheses in block j (where $x(i, j)$ is as defined in Claim 9, Equation (2)). Note that there are $e_1(I_{i+1,j-1})$ excess opening parentheses in block i that are matched to excess closing parentheses in the interval $I_{i+1,j-1}$. Similarly, there are $e_0(I_{i+1,j-1})$ closing parentheses in block j that are matched to opening parentheses in $I_{i+1,j-1}$. Observe that either all $e_0(I_{i,i})$ excess opening parentheses in block i get matched to excess closing parentheses in blocks

$i+1, \cdots, j$, or all $e_1(I_{j,j})$ excess closing parentheses in block j get matched to opening parentheses in blocks $i, \cdots, j-1$. This leads to the following exact matching procedure, with two cases: The first corresponds to the situation when all of the excess closing parenthesis in block j are matched to parentheses in the interval $I_{i,j-1}$. In particular this implies that those parentheses in block j that are matched to parentheses in block i constitute a *suffix* of the excess $\mathcal{E}_1(j)$. The second case corresponds to the situation when all of the excess opening parentheses in block i are matched to the interval $I_{i+1,j}$ and so a *prefix* of $\mathcal{E}_0(i)$ is matched to a substring of $\mathcal{E}_1(j)$.

In what follows, for a (consecutive) substring s' of $\mathcal{E}_0(i)$, we denote by $F_i(s')$ and $L_i(s')$ the positions in $\mathcal{E}_0(i)$ of the first and last symbols of s', respectively. Similarly, for a substring s'' of $\mathcal{E}_1(j)$, we denote by $F_j(s'')$ and $L_j(s'')$ the positions of the first and last symbols of s'' in $\mathcal{E}_1(j)$ respectively.

Exact Parentheses Matching Procedure(i, j)

1. If $e_1(I_{i+1,j}) < e_0(I_{i,i})$: Let s'' be the suffix of $\mathcal{E}_1(j)$ of length $x(i,j)$, and let s' be the substring of $\mathcal{E}_0(i)$ such that $L_i(s') = e_0(I_{i,i}) - e_1(I_{i+1,j-1})$, where $|s'| = |s''|$.
2. If $e_1(I_{i+1,j}) \geq e_0(I_{i,i})$: Let s' be the prefix of $\mathcal{E}_0(i)$ of length $x(i,j)$, and let s'' be the substring of $\mathcal{E}_1(j)$ such that $F_j(s'') = e_0(I_{i+1,j-1}) + 1$, where $|s''| = |s'|$.
3. If s' and s'' match, then return success, otherwise return fail.

It may be verified that $L_i(s') = e_0(I_{i,i}) - e_1(I_{i+1,j-1})$ and $F_j(s'') = e_0(I_{i+1,j-1}) + 1$, no matter which step of the procedure is applied. Hence, the two cases can actually be merged into one, but the above formulation will be helpful in understanding a variant of this procedure that is presented subsequently.

An Example. Consider for example the string from Figure 1, and the neighboring blocks $i = 1$ and $j = 4$. Then $\mathcal{E}_0(1) =$ "[[([([", that is, the block consists only of excess parentheses, and $\mathcal{E}_1(4) =$ "])])". Thus, $e_0(I_{1,1}) = 6$. The other relevant values are: $x(1,4) = 2$, $e_1(I_{2,4}) = 3$, and $e_1(I_{2,3}) = 1$. Hence s'' is the suffix of length 2 of $\mathcal{E}_1(4)$, that is, $s'' =$ "])". We also get that $L_i(s') = 6 - 1 = 5$, and so s' is the substring of $\mathcal{E}_0(1)$ of length 2 that ends at position 5, that is $s' =$ "([". The substrings s' and s'' match of course since $s's'' \in D_2$.

Since we only have estimates $\hat{e}_1(I_{i+1,j-1})$ and $\hat{e}_0(I_{i+1,j-1})$ of the excess numbers in the interval $I_{i+1,j-1}$, then we apply the following *partial* matching procedure to any pair of neighbor blocks $i < j$ in $\hat{G}(s)$. The procedure is basically the same as the exact matching procedure, but it searches for a possibly smaller match in a larger range (where the size of the match and the range are determined by the quality of the approximation we have). Thus we define

$$\hat{x}(i,j) \stackrel{\text{def}}{=} \min\{\hat{e}_1(I_{i+1,j}), e_0(I_{i,i})\} - \hat{e}_1(I_{i+1,j-1}) - 2\delta n^{2/3},$$

and look for matching substrings of length $\hat{x}(i,j)$. Furthermore, we only allow matches of locations that have an even number of symbols between them. If

$s \in D_m$ and blocks i and j are neighbors in $G(s)$, then the existing matching between excess opening parentheses in block i and excess closing parentheses in block j, should in fact obey this constraint.

Partial Parentheses Matching Procedure(i, j)

1. If $\hat{e}_1(I_{i+1,j}) < e_0(I_{i,i}) - \delta n^{2/3}$: Let $\hat{s}''$ be the suffix of $\mathcal{E}_1(j)$ of length $\hat{x}(i,j)$. Search for a matching substring $\hat{s}'$ of $\mathcal{E}_0(i)$ such that $L_i(\hat{s}')$ is in the range
$$\left(e_0(I_{i,i}) - \hat{e}_1(I_{i+1,j-1}) - 3\delta n^{2/3}\ ,\ e_0(I_{i,i}) - \hat{e}_1(I_{i+1,j-1}) + \delta n^{2/3}\right) \quad (3)$$
and such that $L_i(\hat{s}')$ has opposite parity from the parity of $F_j(\hat{s}'')$. (If $|\hat{x}(i,j)| \leq 0$ then $\hat{s}''$ is the empty string, and a matching exists trivially.)
2. If $\hat{e}_1(I_{i+1,j}) \geq e_0(I_{i,i}) + \delta n^{2/3}$: Let $\hat{s}'$ be the prefix of $\mathcal{E}_0(i)$ of length $\hat{x}(i,j)$. Search for a matching substring $\hat{s}''$ of $\mathcal{E}_1(j)$ such that $F_j(\hat{s}'')$ is in the range
$$(\hat{e}_0(I_{i+1,j-1}) + 1 - \delta n^{2/3}\ ,\ \hat{e}_0(I_{i+1,j-1}) + 1 + 3\delta n^{2/3}) \quad (4)$$
where again, $F_j(\hat{s}'')$ should have opposite parity from that of $L_i(\hat{s}')$.
3. If $|\hat{e}_1(I_{i+1,j}) - e_0(I_{i,i})| \leq \delta n^{2/3}$: Search for a matching as described in Step 1 above. If a matching is not found, search for a matching as described in Step 2 above.
4. If a matching was found, then return success, otherwise return fail.

To implement either step in the above procedure, we run a linear time string matching algorithm [4].

Lemma 11 *Assume that Assumption 4 holds. Then we have the following.*

1. *If $s \in D_m$, then for every two neighbor blocks $i < j$ in $\hat{G}(s)$, the partial matching procedure described above succeeds in finding a matching.*
2. *Let s be any given string and consider any three blocks $i < j_1 < j_2$ such that j_1 and j_2 are both neighbors of i in $\hat{G}(s)$. Suppose that the partial matching procedure succeeds in finding a matching between substrings $\hat{s}'$ and $\hat{t}'$ of $\mathcal{E}_0(i)$ and substrings $\hat{s}''$ of $\mathcal{E}_1(j_1)$ and $\hat{t}''$ of $\mathcal{E}_1(j_2)$, respectively. Then, under Assumption 4, $\hat{s}'$ and $\hat{t}'$ overlap by at most $6\delta n^{2/3}$. An analogous statement holds for triples $i_1 < i_2 < j$ such that i_1 and i_2 are both neighbors of j in $\hat{G}(s)$.*

3.6 Putting It All Together

Algorithm 1 Test for D_m

1. *Let $\delta = \frac{\epsilon}{200}$. Test that $\mu(s) \in D_1$ with distance parameter δ and confidence 9/10. If the D_1 test rejects, then reject.*
2. *Partition the string into $n^{1/3}$ substrings of length $n^{2/3}$ each, which we refer to as "blocks".*
3. *Select $100/\epsilon$ blocks uniformly, and check that they are D_m consistent.*

4. *Perform the preprocessing step on the basic substrings of s (defined based on the above setting of δ).*
5. *Uniformly select $100/\epsilon$ blocks and for each find its neighboring blocks in $\hat{G}(s)$. For each selected block, and for each of its neighbors, check that their excess parentheses match correctly by invoking the partial matching procedure.*

Theorem 1 *If $s \in D_m$ then the above testing algorithm accepts with probability at least 2/3, and if s is ϵ-far from D_m then the above test rejects with probability at least 2/3.*
The query complexity and running time of the algorithm are $O\left(\frac{n^{2/3} \cdot \log^3(n/\epsilon)}{\epsilon^3}\right)$.

Proof Sketch (the complete proof appears in the full version of this paper [7]): The main part of the proof is showing that if s is accepted with probability greater than 1/3 then it is ϵ-close to D_m. If s is accepted with probability greater than 1/3 then necessarily it must pass each part of the test with probability greater than 1/3. This implies that: (1) $\mu(s)$ is δ-close to D_1; (2) All but at most an $\frac{\epsilon}{4}$-fraction of the blocks of s are D_m-consistent; (3) The fraction of blocks i that have a neighbor j in $\hat{G}(s)$ for which the matching procedure would fail if executed on i and j is at most $\frac{\epsilon}{4}$; Using these assertions, together with Assumption 4 (which holds with sufficiently high probability), we are able to show how to modify s in at most ϵn positions so that it becomes a string in D_m. □

References

1. N. Alon, M. Krivelevich, I. Newman, and M Szegedy. Regular languages are testable with a constant number of queries. In *Proceedings of the Fortieth Annual Symposium on Foundations of Computer Science*, pages 656–666, 1999.
2. O. Goldreich, S. Goldwasser, and D. Ron. Property testing and its connection to learning and approximation. *JACM*, 45(4):653–750, 1998.
3. M. Harrison. *Introduction to formal language theory*. Addison-Wesley, 1978.
4. D. E. Knuth, J. H. Morris, and V. R. Pratt. Fast pattern matching in strings. *SIAM Journal on Computing*, 6(2):323–350, 1977.
5. D. Kozen. *Automata and Computability*. Springer Verlag, 1997.
6. I. Newman. Testing of functions that have small width branching programs. In *Proceedings of the Forty-First Annual Symposium on Foundations of Computer Science*, pages 251–258, 2000.
7. M. Parnas, D. Ron, and R. Rubinfeld. Testing parenthesis languages. Available from: `http://www.eng.tau.ac.il/~danar`, 2001.
8. D. Ron. Property testing. To appear in the Handbook on Randomization. Currently available from: `http://www.eng.tau.ac.il/~danar`, 2000.
9. R. Rubinfeld and M. Sudan. Robust characterization of polynomials with applications to program testing. *SIAM Journal on Computing*, 25(2):252–271, 1996.
10. N. Chomsky M. P. Schotzenberger. The algebraic theory of context-free languages. In *Computer Programming and Formal Languages, P. Braffort and D. Hirschberg, Eds, North Holland*, pages 118–161, 1963.

Proclaiming Dictators and Juntas or Testing Boolean Formulae

Michal Parnas[1], Dana Ron[2,*], and Alex Samorodnitsky[3,**]

[1] The Academic College of Tel-Aviv-Yaffo
michalp@mta.ac.il
[2] Department of EE – Systems, Tel-Aviv University
danar@eng.tau.ac.il
[3] Institute for Advanced Study, Princeton
asamor@ias.edu

Abstract. We consider the problem of determining whether a given function $f : \{0,1\}^n \rightarrow \{0,1\}$ belongs to a certain class of Boolean functions $\mathcal{F}$ or whether it is *far* from the class. More precisely, given query access to the function f and given a distance parameter ϵ, we would like to decide whether $f \in \mathcal{F}$ or whether it differs from every $g \in \mathcal{F}$ on more than an ϵ-fraction of the domain elements. The classes of functions we consider are singleton ("dictatorship") functions, monomials, and monotone DNF functions with a bounded number of terms. In all cases we provide algorithms whose query complexity is independent of n (the number of function variables), and polynomial in the other relevant parameters.

1 Introduction

The newly founded country of Eff is interested in joining the international organization Pea. This organization has one rule: It does not admit dictatorships. Eff claims it is not a dictatorship but is unwilling to reveal the procedure by which it combines the votes of its government members into a final decision. However, it agrees to allow Pea's special envoy, Tee, to perform a small number of experiments with its voting method. Namely, Tee may set the votes of the government members (using Eff's advanced electronic system) in any possible way, and obtain the final decision given these votes. Tee's mission is not to actually identify the dictator among the government members (if such exists), but only to discover *whether* such a dictator exists. Most importantly, she must do so by performing as few experiments as possible. Given this constraint, Tee may decline Eff's request to join Pea even if Eff is not exactly a dictatorship but only behaves like one most of the time.

The above can be formalized as a *Property Testing Problem*: Let $f : \{0,1\}^n \rightarrow \{0,1\}$ be a fixed (but unknown) function, and let $\mathcal{P}$ be a fixed property of

* Supported by the Israel Science Foundation (grant number 32/00-1).
** Supported by the NSF (grant number CCR-9987845)

M. Goemans et al. (Eds.): APPROX-RANDOM 2001, LNCS 2129, pp. 273–285, 2001.

functions. We would like to determine, by querying f, whether f has the property $\mathcal{P}$, or whether it is ϵ-*far* from having the property for a given distance parameter ϵ. By ϵ-*far* we mean that more than an ϵ–fraction of its values should be modified so that it obtains the property $\mathcal{P}$. For example, in the above setting we would like to test whether a given function f is a "dictatorship function". That is, whether there exists an index $1 \leq i \leq n$, such that $f(x) = x_i$ for every $x \in \{0,1\}^n$.

Previous work on testing properties of functions mainly focused on algebraic properties (e.g., [5,17,16]), or on properties defined by relatively rich families of functions such as the family of all monotone functions [9,8]. Here we are interested in studying the most basic families of Boolean functions: singletons, monomials, and DNF functions.

One possible approach to testing whether a function f has a certain property $\mathcal{P}$, is to try and actually *find* a good approximation for f from within the family of functions $\mathcal{F}_{\mathcal{P}}$ having the tested property $\mathcal{P}$. For this task we would use a *learning algorithm* that performs queries and works under the uniform distribution. Such an algorithm ensures that if f has the property (that is, $f \in \mathcal{F}_{\mathcal{P}}$), then with high probability the learning algorithm outputs a *hypothesis* $h \in \mathcal{F}_{\mathcal{P}}$ such that $\Pr[f(x) \neq h(x)] \leq \epsilon$, where ϵ is a given distance (or error) parameter. The testing algorithm would run the learning algorithm, obtain the hypothesis $h \in \mathcal{F}_{\mathcal{P}}$, and check that h and f in fact differ only on a small fraction of the domain. This last step is performed by taking a sample of size $\Theta(1/\epsilon)$ from $\{0,1\}^n$ and comparing f and h on the sample. Thus, if f has property $\mathcal{P}$ then it will be accepted with high probability, and if f is ϵ-far from having $\mathcal{P}$ (so that $\Pr[f(x) \neq h(x)] > \epsilon$ for every $h \in \mathcal{F}_{\mathcal{P}}$, then it will be rejected with high probability.

Hence, provided there exists a learning algorithm for the tested family $\mathcal{F}_{\mathcal{P}}$, we obtain a testing algorithm whose complexity is of the same order of the learning algorithm. To be more precise, the learning algorithm should be a *proper* learning algorithm. That is, the hypothesis h it outputs must belong to $\mathcal{F}_{\mathcal{P}}$.[1]

A natural question that arises is whether we can do better by using a different approach. Recall that we are not interested in actually finding a good approximation for f in $\mathcal{F}_{\mathcal{P}}$ but we only want to know whether such an approximation *exists*. Therefore, perhaps we can design a different and more efficient testing algorithm than the one based on learning. In particular, the complexity measure we would like to improve is the *query complexity* of the algorithm.

As we show below, for all the properties we study, we describe algorithms whose query complexity is polynomial in $1/\epsilon$, where ϵ is the given distance parameter, and *independent* of the input size n.[2] As we discuss shortly, the cor-

---

[1] This is as opposed to *non-proper* learning algorithms that given query access to $f \in \mathcal{F}_{\mathcal{P}}$ are allowed to output a hypothesis h that belongs to a more general hypothesis class $\mathcal{F}' \supset \mathcal{F}_{\mathcal{P}}$. Non-proper learning algorithms are not directly applicable for our purposes.

[2] The running times of the algorithms are all linear in the number of queries performed and in n. This dependence on n in the running time is clearly unavoidable, since even writing down a query takes time n.

responding proper learning algorithms have query complexities that depend on n, though only polylogarithmically. Thus our improvement is not so much of a quantitative nature (and we also note that our dependence on $1/\epsilon$ in some cases is worse). However, we believe that our results are of interest both because they completely remove the dependence on n in the query complexity, and also because in certain aspects they are inherently different from the corresponding learning algorithms. Hence they may shed new light on the structure of the properties studied.

1.1 Our Results

We present the following testing algorithms:

- An algorithm that tests whether f is a singleton function. That is, whether there exists an index $1 \leq i \leq n$, such that $f(x) = x_i$ for every $x \in \{0,1\}^n$, or $f(x) = \bar{x}_i$ for every $x \in \{0,1\}^n$. This algorithm has query complexity $O(1/\epsilon)$.
- An algorithm that tests whether f is a monomial with query complexity $\tilde{O}(1/\epsilon^3)$.
- An algorithm that tests whether f is a monotone DNF having at most ℓ terms, with query complexity $\tilde{O}(\ell^4/\epsilon^3)$.

Due to space constraints, most proofs and some details are deferred to the full version of this paper [14].

Techniques. Our algorithms for testing singletons and for testing monomials have a similar structure. In particular, they combine two tests. One test is a "natural" test that arises from an exact logical characterization of these families of functions. In the case of singletons this test uniformly selects pairs $x, y \in \{0,1\}^n$ and verifies that $f(x \wedge y) = f(x) \wedge f(y)$, where $x \wedge y$ denotes the bitwise 'and' of the two strings. The corresponding test for monomials performs a slight variant of this test. The other test in both cases is a seemingly less evident test with an algebraic flavor. In the case of singletons it is a linearity test [5] and in the case of monomials it is an affinity test. This test ensures that if f passes it then it has (or is close to having) a certain structure. This structure aids us in analyzing the logical test.

The testing algorithm for monotone DNF functions uses the test for monomials as a sub-routine. Recall that a DNF function is a disjunction of monomials (the terms of the function). If f is a DNF function with a bounded number of terms, then the test will isolate the different terms of the function and test that each is in fact a monomial. If f is far from being such a DNF function, then at least one of these tests will fail with high probability.

It is worthwhile noting that, given the structure of the monotone DNF tester, any improvement in the complexity of the monomial testing algorithm will imply an improvement in the DNF tester.

1.2 Related Work

Property Testing. Property testing was first defined and applied in the context of algebraic properties of functions [17], and has since been extended to various domains, perhaps most notably those of graph properties (e.g. [10,11,1]). (For a survey see [15]). The relation between testing and learning is discussed at length in [10]. In particular, that paper suggests that testing may be applied as a preliminary stage to learning. Namely, efficient testing algorithms can be used in order to help in determining what hypothesis class should be used by the learning algorithm.

As noted above, we use linearity testing [5] in our test for singletons, and affinity testing, which can be viewed as an extension of linearity testing, for testing monomials. Other works in which improvements and variants of linearity testing are analyzed include [4,3]. In particular, the paper by Bellare et. al. [4] is the first to establish the connection between linearity testing and Fourier analysis.

Learning Boolean Formulae. Singletons, and more generally monomials, can be easily learned under the uniform distribution. The learning algorithm uniformly selects a sample of size $\Theta(\log n/\epsilon)$ and queries the function f on all sample strings. It then searches for a monomial that is consistent with f on the sample. Finding a consistent monomial (if such exists) can be done in time linear in the sample size and in n. A simple probabilistic argument (that is a slight variant of Occam's Razor [6][3]) can be used to show that a sample of size $\Theta(\log n/\epsilon)$ is sufficient to ensure that with high probability any monomial that is consistent with the sample is an ϵ-good approximation of f.

There is a large variety of results on learning DNF functions (and in particular monotone DNF), in several different models. We restrict our attention to the model most relevant to our work, namely when membership queries are allowed and the underlying distribution is uniform. The best known algorithm results from combining the works of [7] and [13], and builds on Jackson's celebrated Harmonic Sieve algorithm [12]. This algorithm has query complexity $\tilde{O}\left(r \cdot \left(\frac{\log^2 n}{\epsilon} + \frac{\ell^2}{\epsilon^2}\right)\right)$, where r is the number of variables appearing in the DNF formula, and ℓ is the number of terms. However, this algorithm does not output a DNF formula as its hypothesis. On the other hand, Angluin [2] describes a proper learning algorithm for monotone DNF formula that uses membership queries and works under arbitrary distributions. The query complexity of her algorithm is $\tilde{O}(\ell \cdot n + \ell/\epsilon)$. Using the same preprocessing technique as suggested in [7], if the underlying distribution is uniform then the query complexity can be reduced to $\tilde{O}\left(\frac{r \cdot \log^2 n}{\epsilon} + \ell \cdot \left(r + \frac{1}{\epsilon}\right)\right)$. Recall that the query complexity of our testing algorithm is a faster growing function of ℓ and $1/\epsilon$, but does not depend on n. Hence we get better results when ℓ and $1/\epsilon$ are sub-logarithmic in n, and in particular when they are constant.

Finally, we note that similarly to the Harmonic-Sieve based results for learning DNF, we appeal to the Fourier coefficients of the tested function f. However,

[3] Applying the theorem known as Occam's Razor would give a stronger result in the sense that the underlying distribution may be arbitrary (that is, not necessarily uniform). This however comes at a price of a linear, as opposed to logarithmic, dependence of the sample/query complexity on n.

somewhat differently, these do not appear explicitly in our algorithms but are only used in part of our analysis.

2 Preliminaries

Definition 1 *Let $x, y \in \{0,1\}^n$, and let $[n] \stackrel{\text{def}}{=} \{1,\ldots,n\}$. We let $x \wedge y$ denote the string $z \in \{0,1\}^n$ such that for every $i \in [n]$, $z_i = x_i \wedge y_i$, and we let $x \oplus y$ denote the string $z \in \{0,1\}^n$ such that for every $i \in [n]$, $z_i = x_i \oplus y_i$.*

Definition 2 (Singletons, Monomials, and DNF Functions) *A function $f : \{0,1\}^n \to \{0,1\}$ is a* singleton *function, if there exists an $i \in [n]$ such that $f(x) = x_i$ for every $x \in \{0,1\}^n$, or $f(x) = \bar{x}_i$ for every $x \in \{0,1\}^n$.*

We say that f is a monotone k-monomial *for $1 \le k \le n$ if there exist k indices $i_1, \ldots, i_k \in [n]$, such that $f(x) = x_{i_1} \wedge \cdots \wedge x_{i_k}$ for every $x \in \{0,1\}^n$. If we allow to replace some of the x_{i_j}'s above with $\bar{x}_{i_j}$, then f is a* k-monomial. *The function f is a* monomial *if it is a k-monomial for some $1 \le k \le n$.*

A function f is an ℓ-term DNF *function if it is a disjunction of ℓ monomials. If all monomials are monotone, then it is a* monotone DNF *function.*

Whenever the identity of the function f is clear from the context, we may use the following notation.

Definition 3 *Define $F_0 \stackrel{\text{def}}{=} \{x | f(x) = 0\}$ and $F_1 \stackrel{\text{def}}{=} \{x | f(x) = 1\}$.*

Definition 4 (Distance between Functions) *The distance (according to the uniform distribution), between two functions $f, g : \{0,1\}^n \to \{0,1\}$ is denoted by $\text{dist}(f,g)$, and is defined as follows: $\text{dist}(f,g) \stackrel{\text{def}}{=} \Pr_{x \in \{0,1\}^n}[f(x) \neq g(x)]$.*

The distance between a function f and a family of functions $\mathcal{F}$ is $\text{dist}(f,\mathcal{F}) \stackrel{\text{def}}{=} \min_{g \in \mathcal{F}} \text{dist}(f,g)$. If $\text{dist}(f,\mathcal{F}) > \epsilon$ for some $0 < \epsilon < 1$, then we say that f is ϵ-far *from $\mathcal{F}$. Otherwise, it is* ϵ-close.

Definition 5 (Testing Algorithms) *A testing algorithm for a family of boolean functions $\mathcal{F}$ over $\{0,1\}^n$, is given a distance parameter ϵ, $0 < \epsilon < 1$, and is provided with query access to an arbitrary function $f : \{0,1\}^n \to \{0,1\}$.*

If $f \in \mathcal{F}$ then the algorithm must output accept *with probability at least 2/3, and if f is ϵ-far from $\mathcal{F}$ then it must output reject with probability at least 2/3.*

3 Testing Singletons

We start by presenting an algorithm for testing singletons. The testing algorithm for k-monomials will generalize this algorithm. More precisely, we present an algorithm for testing whether a function f is a *monotone* singleton. In order to

test whether f is a singleton we may check whether either f or $\bar{f}$ passes the monotone singleton test. For sake of succinctness, in what follows we refer to monotone singletons simply as singletons.
The following characterization of monotone k-monomials motivates our tests. We later show that the requirement of monotonicity can be removed.

Claim 1 *Let $f : \{0,1\}^n \to \{0,1\}$. Then f is a monotone k-monomial if and only if the following three conditions hold:*
(1) $\Pr[f = 1] = \frac{1}{2^k}$; (2) $\forall x, y$, $f(x \wedge y) = f(x) \wedge f(y)$; (3) $f(x) = 0$ for all $|x| < k$, where $|x|$ denotes the number of ones in x.

Definition 6 *We say that $x, y \in \{0,1\}^n$ are a* violating pair *with respect to a function $f : \{0,1\}^n \to \{0,1\}$, if $f(x) \wedge f(y) \neq f(x \wedge y)$.*

Given the above definition, Claim 1 states that a basic property of (monotone) singletons (and more generally of monotone k-monomials), is that there are no violating pairs with respect to f. A natural candidate for a testing algorithm for singletons would take a sample of uniformly selected pairs x, y, and for each pair verify that it is not violating with respect to f. In addition, the test would check that $\Pr[f = 0]$ is roughly $1/2$ (or else any monotone k-monomial would pass the test).

We note that we were unable to give a complete proof for the correctness of this test. Somewhat unintuitively, the difficulty with the analysis lies in the case when the function f is *very far* from being a singleton. More precisely, the analysis is quite simple when the distance δ between f and the closest singleton is bounded away from $1/2$. However, the argument does not directly apply to δ arbitrarily close to $1/2$. We believe it would be interesting to prove that this simple test is in fact correct (or to come up with a counterexample of a function f that is almost $1/2$-far from any singleton, but passes the test).

In the algorithm described below we circumvent the above difficulty by "forcing more structure" on f. Specifically, we first perform another test that only accepts functions that have, or more precisely, that are close to having a certain structure. In particular, every singleton will pass the test. We then perform a slight variant of our original test. Provided that f passes the first test, it will be easy to show that f passes the second test with high probability only if it is close to a singleton function. Details follow.

The algorithm begins by testing whether the function f belongs to a larger family of functions that contains singletons as a sub-family. This is the family of *parity functions*.

Definition 7 *A function $f : \{0,1\}^n \to \{0,1\}$ is a* parity function *(a linear function over* GF(2)*) if there exists a subset $S \subseteq [n]$ such that $f(x) = \oplus_{i \in S} x_i$ for every $x \in \{0,1\}^n$.*

The test for parity functions is a special case of the linearity test over general fields due to Blum, Luby and Rubinfeld [5]. If the tested function f is a parity function, then the test always accepts, and if f is ϵ-far from any parity function then the test rejects with probability at least $9/10$. The query complexity of this test is $O(1/\epsilon)$. Assuming this test passes, we still need to verify that f is actually close to a singleton function and not close to some other parity function. Suppose that the parity test only accepted parity functions and not functions that are only close to parity functions. Then, by the following claim, if f passes the parity test but *is not* a singleton, then a constant size sample of pairs x, y would, with high probability, contain a violating pair with respect to f.

Claim 2 *Let* $g = \oplus_{i \in S} x_i$ *for* $S \subseteq [n]$. *If* $|S|$ *is even then* $\Pr[g(x \wedge y) = g(x) \wedge g(y)] = \frac{1}{2} + \frac{1}{2^{|S|+1}}$, *and if* $|S|$ *is odd then* $\Pr[g(x \wedge y) = g(x) \wedge g(y)] = \frac{1}{2} + \frac{1}{2^{|S|}}$.

Hence, if f is a parity function that is not a singleton (that is $|S| \geq 2$), then the probability that a uniformly selected pair x, y is violating with respect to f is at least $1/8$. In this case, a sample of at least 16 such pairs will contain a violating pair with probability at least $1 - (1 - 1/8)^{16} \geq 1 - e^{-2} > 2/3$.

However, what if f passes the parity test but is only close to being a parity function? Let g denote the parity function that is closest to f and let δ be the distance between them. (Where g is unique, given that f is sufficiently close to a parity function). What we would like to do is check whether g is a singleton, by selecting a sample of pairs x, y and checking whether it contains a violating pair with respect to g. Observe, that since the distance between functions is measured with respect to the uniform distribution, then for a uniformly selected pair x, y, with probability at least $(1 - \delta)^2$, both $f(x) = g(x)$ and $f(y) = g(y)$. However, we cannot make a similar claim about $f(x \wedge y)$ and $g(x \wedge y)$, since $x \wedge y$ is *not* uniformly distributed. Thus it is not clear that we can replace the violation test for g with a violation test for f.

However, we can use what is known as a *self-corrector* for linear (parity) functions [5]. Given query access to a function $f : \{0,1\}^n \to \{0,1\}$ and any input $x \in \{0,1\}^n$, if f is strictly closer than $1/4$ to some parity function g, then the procedure Self-Correct(f, x) returns the value of $g(x)$, with probability at least $9/10$. The query complexity of the procedure is constant.
The above discussion suggests the following testing algorithm.

Algorithm 1 Test for Singleton Functions

1. *Apply the parity test to* f *with distance parameter* $\min(1/5, \epsilon)$.
2. *Uniformly and independently select* $m = 32$ *pairs of points* x, y.
   - *For each such pair, let* $b_x =$ *Self-Correct*(f, x), $b_y =$ *Self-Correct*(f, y) *and* $b_{x \wedge y} =$ *Self-Correct*$(f, x \wedge y)$.
   - *Check that* $b_{x \wedge y} = b_x \wedge b_y$.
3. *If one of the above fails* - reject. *Otherwise,* accept.

Theorem 1 *Algorithm 1 is a testing algorithm for monotone singletons. Furthermore, it has one sided error. That is, if* f *is a monotone singleton then the algorithm always accepts. The query complexity of the algorithm is* $O(1/\epsilon)$.

4 Testing Monomials

In what follows we describe an algorithm for testing *monotone* k-monomials, where k is provided to the algorithm. As we show in the full version of this paper [14], it is no hard to extend this to testing monomials when k is not specified. As for the monotonicity requirement, the following observation and corollary show that this requirement can be easily removed if desired.

Observation 3 *Let $f : \{0,1\}^n \rightarrow \{0,1\}$, and let $z \in \{0,1\}^n$. Consider the function $f_z : \{0,1\}^n \rightarrow \{0,1\}$ which is defined as follows: $f_z(x) = f(x \oplus z)$. Then the following are immediate:*

1. *The function f is a k-monomial if and only if f_z is a k-monomial.*
2. *Let $y \in F_1$. If f is a (not necessarily monotone) k-monomial, then $f_{\bar{y}}$ is a monotone k-monomial.*

Corollary 4 *If f is ϵ-far from every (not necessarily monotone) k-monomial, then for every $y \in F_1$, $f_{\bar{y}}$ is ϵ-far from every monotone k-monomial.*

We now present the algorithm for testing monotone k-monomials. The first two steps of the algorithm are an attempt to generalize the use of parity testing in Algorithm 1. In particular, we test whether F_1 is an *affine subspace*.

Definition 8 (Affine Subspaces) *A subset $H \subseteq \{0,1\}^n$ is an* affine subspace *of $\{0,1\}^n$ if and only if there exist an $x \in \{0,1\}^n$ and a linear subspace V of $\{0,1\}^n$, such that $H = V \oplus x$. That is, $H = \{y \mid y = v \oplus x, for\ some\ v \in V\}$.*

The following is an alternative characterization of affine subspaces, which is a basis for our test.

Fact 5 *H is an affine subspace if and only if for every $y_1, y_2, y_3 \in H$ we have $y_1 \oplus y_2 \oplus y_3 \in H$.*

In the description of the algorithm, there is an explicit (exponential) dependence on the size k of the monomial. However, as we observe momentarily, we can assume, without loss of generality, that $2^k = O(1/\epsilon)$, and so the query complexity of the algorithm is actually polynomial in $1/\epsilon$.

Algorithm 2 Test for monotone k-monomials

1. Size Test: *Uniformly select a sample of $m = \Theta(2^k + 1/\epsilon^2)$ strings in $\{0,1\}^n$. For each x in the sample, obtain $f(x)$. Let α be the fraction of sample strings x such that $f(x) = 1$. If $|\alpha - 2^{-k}| > \min(2^{-k-5}, \epsilon/4)$ then reject, otherwise continue.*
2. Affinity Test: *Repeat the following $3k^2 \cdot \max(2/\epsilon^2, 2^{2(k+4)})$ times: Uniformly select $x, y \in F_1$ and $z \in \{0,1\}^n$, and check whether $f(x \oplus y \oplus z) = f(x) \oplus f(y) \oplus f(z)$. If some triple does not satisfy this constraint then reject. (Since $f(x) = f(y) = 1$, we are actually checking whether $f(x \oplus y \oplus z) = f(z)$. As we show in our analysis, this step will ensure that f is close to some function g for which $g(x) \oplus g(y) \oplus g(z) = g(x \oplus y \oplus z)$ for all $x, y, z \in G_1 = \{x | g(x) = 1\}$.)*

3. Closure-Under-Intersection Test: *Repeat the following* 2^{k+5} *times:*
   (a) *Uniformly select* $x \in F_1$ *and* $y \in \{0,1\}^n$. *If* x *and* y *are a violating pair, then reject. (Note that since* $x \in F_1$, *this test actually checks that* $f(y) = f(x \wedge y)$.*)*
4. *If no step caused rejection, then accept.*

In both the affinity test and the closure-under-intersection test, we need to select strings in F_1 uniformly. This is simply done by sampling from $\{0,1\}^n$ and using only x's for which $f(x) = 1$. This comes at an additional multiplicative cost of $O(2^k)$ in the query complexity. The query complexity of the algorithm is hence $O(2^k \cdot k^2 \cdot \max(1/\epsilon^2, 2^{2k}))$. As the following observation implies, we may assume, without loss of generality, that $\epsilon < 2^{-k+2}$, and so the complexity is actually $\tilde{O}(1/\epsilon^3)$.

Observation 6 *Suppose that* $\epsilon \geq 2^{-k+2}$. *Then:*

1. *If* $\Pr[f = 1] \leq \frac{\epsilon}{2}$, *then* f *is* ϵ*-close to every* k*-monomial (and in particular to every monotone* k*-monomial).*
2. *If* $\Pr[f = 1] > \frac{\epsilon}{4}$, *then* f *is not a* k*-monomial.*

Theorem 2 *Algorithm 2 is a testing algorithm for monotone* k*-monomials. The query complexity of the algorithm is* $\tilde{O}(1/\epsilon^3)$.

The proof of the theorem is based on the following two lemmas.

Lemma 7 *Let* $\eta \stackrel{\text{def}}{=} \Pr_{x,y \in F_1, z \in \{0,1\}^n}[f(x \oplus y \oplus z) \neq f(z)]$. *If* $\eta \leq 2^{-2k-1}$ *and* $\left|\frac{|F_1|}{2^n} - 2^{-k}\right| \leq 2^{-(k+3)}$, *then there exists a function* g *such that* $G_1 \stackrel{\text{def}}{=} \{x : g(x) = 1\}$ *is an affine subspace of dimension* $n - k$ *and which satisfies:*

$$\text{dist}(f,g) \leq \left|\frac{|F_1|}{2^n} - 2^{-k}\right| + k\eta^{\frac{1}{2}}.$$

Lemma 8 *Let* $f : \{0,1\}^n \to \{0,1\}$ *be a function for which* $|\Pr[f = 1] - 2^{-k}| < 2^{-k-3}$. *Suppose that there exists a function* $g : \{0,1\}^n \to \{0,1\}$ *such that:*

1. $\text{dist}(f,g) \leq 2^{-k-3}$.
2. $G_1 \stackrel{\text{def}}{=} \{x : g(x) = 1\}$ *is an affine subspace of dimension* $n - k$.

If g *is not a monotone* k*-monomial, then* $\Pr_{x \in F_1, y}[f(y) \neq f(x \wedge y)] \geq 2^{-k-3}$.

5 Testing Monotone DNF Formulae

In this section we describe an algorithm for testing whether a function f is a monotone DNF formula with at most ℓ terms, for a given integer ℓ. Namely, that $f = T_1 \vee T_2 \vee \cdots \vee T_{\ell'}$ for $\ell' \leq \ell$, and each term T_i is of the form $T_i =$

$x_{j_1} \wedge x_{j_2} \wedge \cdots \wedge x_{j_{k(i)}}$, where the size of the terms may vary. We assume, without loss of generality, that no term contains the set of variables of any other term (or else we can ignore the more specific term), though the same variable can of course appear in several terms. The basic idea underlying the algorithm is to test whether the set $F_1 \stackrel{\text{def}}{=} \{x : f(x) = 1\}$ can be "approximately covered" by at most ℓ terms (monomials). To this end, the algorithm finds strings $x^i \in \{0,1\}^n$ and uses them to define functions f^i that are tested for being monomials. If the original function f is in fact an ℓ-term DNF, then, with high probability, each such function f^i corresponds to one of the terms of f.

Let f be a monotone ℓ-term DNF, and let its terms be $T_1, \ldots, T_\ell$. Then, for any $x \in \{0,1\}^n$, we let $S(x) \subseteq \{1, \ldots, \ell\}$ denote the subset of indices of the terms satisfied by x. That is: $S(x) \stackrel{\text{def}}{=} \{i : T_i(x) = 1\}$. In particular, if $f(x) = 0$ then $S(x) = \emptyset$. This notion extends to a set $R \subseteq F_1$, were $S(R) \stackrel{\text{def}}{=} \bigcup_{x \in R} S(x)$. We observe that if f is a monotone ℓ-term DNF, then for every $x, y \in \{0,1\}^n$ we have $S(x \wedge y) = S(x) \cap S(y)$.

Definition 9 (Single-Term Representatives) *Let f be a monotone ℓ-term DNF. We say that $x \in F_1$ is a* single-term representative *for f if $|S(x)| = 1$. That is, x satisfies only a single term in f.*

Definition 10 (Neighbors) *Let $x \in F_1$. The* set of neighbors of x, *denoted $N(x)$, is defined as follows:*

$$N(x) \stackrel{\text{def}}{=} \{y \mid f(y) = 1 \text{ and } f(x \wedge y) = 1\} \ .$$

The notion of neighbors extends to a set $R \subseteq F_1$, where $N(R) \stackrel{\text{def}}{=} \bigcup_{x \in R} N(x)$.

Consider the case in which x is a single-term representative of f, and $S(x) = \{i\}$. Then, for every neighbor $y \in N(x)$, we must have $i \in S(y)$ (or else $S(x \wedge y)$ would be empty, implying that $f(x \wedge y) = 0$). Notice that the converse statement holds as well, that is, $i \in S(y)$ implies that x and y are neighbors. Therefore, the set of neighbors of x is exactly the set of all strings satisfying the term T_i. The goal of the algorithm will be to find at most ℓ such single-term representatives $x \in \{0,1\}^n$, and for each such x to test that its set of neighbors $N(x)$ satisfies some common term. We shall show that if f is in fact a monotone ℓ-term DNF, then all these tests pass with high probability. On the other hand, if all the tests pass with high probability, then f is close to some monotone ℓ-term DNF.

Algorithm 3 Test for Monotone ℓ-term DNF

1. *$R \leftarrow \emptyset$. R is designed to be a set of single-term representatives for f.*
2. *For $i = 1$ to $\ell + 1$ (Try to add ℓ single-term representatives to R):*
   (a) *Take a uniform sample U^i of size $m_1 = \Theta\left(\frac{\ell \log \ell}{\epsilon}\right)$ strings. Let $W^i = (U^i \cap F_1) \setminus N(R)$. That is, W^i consists of strings x in the sample such that $f(x) = 1$, and x is not a neighbor of any string already in R.*
   Observe that if the strings in R are in fact single-term representatives, then every $x \in W^i$ satisfies only terms not satisfied by the representatives in R.

(b) *If* $i = \ell + 1$ *and* $W^i \neq \emptyset$, *then* reject.
If there are more than ℓ *single term representatives for* f *then necessarily* f *is not an* ℓ*-term DNF.*

(c) *Else, if* $\frac{|W^i|}{m_1} < \frac{\epsilon}{4}$ *then* go to Step 3.
The current set of representatives already "covers" almost all of F_1.

(d) *Else* ($\frac{|W^i|}{m_1} \geq \frac{\epsilon}{4}$ *and* $i \leq \ell$), *use* W^i *in order to find a string* x^i *that is designed to be a single-term representative of a term not yet represented in* R. *This step will be described subsequently.*

3. *For each string* $x^i \in R$, *let the function* $f^i : \{0,1\}^n \mapsto \{0,1\}$ *be defined as follows:* $f^i(y) = 1$ *if and only if* $y \in N(x^i)$.
As observed previously, if x^i *is in fact a single-term representative, then* f^i *is a monomial.*
4. *For each* f^i, *test that it is monomial, using distance parameter* $\epsilon' = \frac{\epsilon}{2\ell}$ *and confidence* $1 - \frac{1}{6\ell}$ *(instead of* $\frac{2}{3}$ *— this can simply be done by* $O(\log \ell)$ *repeated applications of each test).*
Note that we do not specify the size of the monomial, and so we need to apply the appropriate variant of our test that does not assume the knowledge of k.
5. *If any of the tests fails then* reject, *otherwise* accept.

The heart of the algorithm and its correctness lie in finding a new representative in each iteration of Step 2. We now sketch how this is done.

Suppose that f is a monotone ℓ-term DNF, and consider any fixed iteration i in Step 2 of the algorithm. Assume that $R \subset \{0,1\}^n$ is subset of single-term representatives for f, such that $\Pr[x \in F_1 \setminus N(R)] \geq \epsilon/8$. The idea is that, given a string $x_0 \in W^i$, we shall try and "remove" terms from $S(x_0)$, until we are left with a single term. More precisely, we try and produce a sequence of strings $x_0, \ldots, x_r$, where $x_0 \in W^i$, such that $\emptyset \neq S(x_{j+1}) \subseteq S(x_j)$, and in particular $|S(x_r)| = 1$. The aim is to decrease the size of $S(x_j)$ by a constant factor for most j's. This will ensure that for $r = \Theta(\log \ell)$, the final string x_r is a single-term representative, as desired. How is such a sequence obtained? Given a string $y_j \in N(x_j)$, define $x_{j+1} = x_j \wedge y_j$. Then $f(x_{j+1}) = 1$ (i.e., $S(x_{j+1}) \neq \emptyset$) and $S(x_{j+1}) = S(x_j) \cap S(y_j) \subseteq S(x_j)$, as desired. The string y_j is acquired simply by uniformly selecting a sufficiently large sample from $\{0,1\}^n$, and picking the first string in the sample that belongs to $N(x_j)$ (if such exists).

Theorem 3 *Algorithm 3 is a testing algorithm for monotone* ℓ*-term DNF. The query complexity of the algorithm is* $\tilde{O}(\ell^4/\epsilon^3)$.

Further Research

Our results raise several questions that we believe may be interesting to study.

- Our algorithms for testing singletons and, more generally, monomials, apply two tests. The role of the first test is essentially to facilitate the analysis of the second, natural test (the closure under intersection test). The question is whether the first test is necessary.

- Our algorithm for testing monomials has a cubic dependence on $1/\epsilon$, as opposed to the linear dependence of the singleton testing algorithm. Can this dependence be improved?
- The query complexity of our algorithm for testing ℓ-term DNF grows like ℓ^4. While some dependence on ℓ seems necessary, we conjecture that a lower dependence is achievable. In particular, suppose we slightly relax the requirements of the testing algorithm and only ask that it reject functions that are ϵ-far from any monotone DNF with at most $c \cdot \ell$ (or possibly ℓ^c) terms, for some constant c. Is it possible, under this relaxation, to devise an algorithm that has only polylogarithmic dependence on ℓ?
- Finally, can our algorithm for testing monotone DNF functions be extended to testing general DNF functions?

References

1. N. Alon, E. Fischer, M. Krivelevich, and M Szegedy. Efficient testing of large graphs. In *Proceedings of FOCS*, pages 645–655, 1999.
2. D. Angluin. Queries and concept learning. *Machine Learning*, 2:319–342, 1988.
3. Y. Aumann, J. Håstad, M. Rabin, and M. Sudan. Linear consistency testing. In *Proceedings of RANDOM*, pages 109–120, 1999.
4. M. Bellare, D. Coppersmith, J. Håstad, M. Kiwi, and M. Sudan. Linearity testing in characteristic two. In *Proceedings of FOCS*, pages 432–441, 1995.
5. M. Blum, M. Luby, and R. Rubinfeld. Self-testing/correcting with applications to numerical problems. *JACM*, 47:549–595, 1993.
6. A. Blumer, A. Ehrenfeucht, D. Haussler, and M. K. Warmuth. Occam's razor. *Information Processing Letters*, 24(6):377–380, April 1987.
7. N. Bshouty, J. Jackson, and C. Tamon. More efficient PAC-learning of DNF with membership queries under the uniform distribution. In *Proceedings of COLT*, pages 286–295, 1999.
8. Y. Dodis, O. Goldreich, E. Lehman, S. Raskhodnikova, D. Ron, and A. Samorodnitsky. Improved testing algorithms for monotonocity. In *Proceedings of RANDOM*, pages 97–108, 1999.
9. O. Goldreich, S. Goldwasser, E. Lehman, D. Ron, and A. Samorodnitsky. Testing monotonicity. *Combinatorica*, 20(3):301–337, 2000.
10. O. Goldreich, S. Goldwasser, and D. Ron. Property testing and its connection to learning and approximation. *JACM*, 45(4):653–750, 1998.
11. O. Goldreich and D. Ron. Property testing in bounded degree graphs. In *Proceedings of STOC*, pages 406–415, 1997. To appear in *Algorithmica*.
12. J. Jackson. An efficient membership-query algorithm for learning DNF with respect to the uniform distribution. *JCSS*, 55:414–440, 1997.
13. A. Klivans and R. Servedio. Boosting and hard-core sets. In *Proceedings of FOCS*, pages 624–633, 1999.
14. M. Parnas, D. Ron, and A. Samorodnitsky. Testing boolean formulae. Available from: `http://www.eng.tau.ac.il/~danar`, 2001.
15. D. Ron. Property testing. To appear in the Handbook on Randomization. Currently available from: `http://www.eng.tau.ac.il/~danar`, 2000.
16. R. Rubinfeld. Robust functional equations and their applications to program testing. *SIAM Journal on Computing*, 28(6):1972–1997, 1999.

17. R. Rubinfeld and M. Sudan. Robust characterization of polynomials with applications to program testing. *SIAM Journal on Computing*, 25(2):252–271, 1996.

Equitable Coloring Extends Chernoff-Hoeffding Bounds

Sriram V. Pemmaraju*

Abstract. Chernoff-Hoeffding bounds are sharp tail probability bounds for sums of bounded independent random variables. Often we cannot avoid dependence among random variables involved in the sum. In some cases the theory of martingales has been used to obtain sharp bounds when there is a limited amount of dependence. This paper will present a simple but powerful new technique that uses the existence of small sized equitable graph colorings to prove sharp Chernoff-Hoeffding type concentration results for sums of random variables with dependence. This technique also allows us to focus on the dependency structure of the random variables and in cases where the dependency structure is a tree or an outerplanar graph, it allows us to derive bounds almost as sharp as those obtainable had the random variables been mutually independent. This technique connects seemingly unrelated topics: extremal graph theory and concentration inequalities. The technique also motivates several open questions in equitable graph colorings, positive answers for which will lead to surprisingly strong Chernoff-Hoeffding type bounds.

1 Introduction

In a seminal 1952 paper, Chernoff [3] introduced a technique that gave sharp upper bounds on the tails of the distribution of the sum of binary independent random variables. Hoeffding [5] in 1963 extended Chernoff's technique to obtain upper bounds on the tails of the distribution of the sum of bounded independent random variables. Bounds obtained by using these techniques are collectively called *Chernoff-Hoeffding bounds* (CH bounds, in short). In many situations, tail probability bounds obtained using Markov's or Chebyshev's inequality are too weak, while CH bounds are just right. CH bounds have proved extremely useful in the design and analysis of randomized algorithms, in derandomization, and in the probabilistic method proofs.

Let $\mathcal{X} = \{X_1, X_2, \ldots, X_n\}$ denote a set of random variables with $S = \sum_{i \in [n]} X_i$ and $\mu = E[S]$. We are interested in upper bounds on the upper tail probability $\text{Prob}[S \geq (1+\epsilon)\mu]$ and in the lower tail probability $\text{Prob}[S \leq (1-\epsilon)\mu]$ for $0 < \epsilon \leq 1$. When the X_i's are mutually independent binary random variables (that is, $X_i \in \{0, 1\}$ for all $i \in [n]$), Chernoff's technique leads to the following

* Department of Computer Science, The University of Iowa, Iowa City, IA 52242, sriram@cs.uiowa.edu

M. Goemans et al. (Eds.): APPROX-RANDOM 2001, LNCS 2129, pp. 285–296, 2001.

bounds on the tail probabilities (see [7] for a proof):

$$\text{Prob}[S \geq (1+\epsilon)\mu] \leq \left(\frac{e^{\epsilon}}{(1+\epsilon)^{(1+\epsilon)}}\right)^{\mu}$$
$$\text{Prob}[S \leq (1-\epsilon)\mu] \leq e^{-\mu\epsilon^2/2}$$

Following Motwani and Raghavan [7], we shall use $F^{+}(\mu, \epsilon)$ to denote the above upper tail probability bound and $F^{-}(\mu, \epsilon)$ to denote the above lower tail probability. When the X_i's are mutually independent bounded random variables (that is, $X_i \in [a_i, b_i]$ for finite positive real a_i and b_i, for each $i \in [n]$), Hoeffding's extension of Chernoff's technique leads to the following bound (see Theorem 2 in [5]):

$$\text{Prob}[|S - \mu| \geq \epsilon\mu] \leq 2e^{-2\mu^2\epsilon^2/\sum_{i\in[n]}(b_i-a_i)^2}.$$

Letting $\boldsymbol{a} = (a_1, a_2, \ldots, a_n)$ and $\boldsymbol{b} = (b_1, b_2, \ldots, b_n)$, we use $G(\mu, \epsilon, \boldsymbol{a}, \boldsymbol{b})$ for the above upper bound. We are interested in the case in which the X_i's are all identical. In this case, if $X_i \in [a, b]$ and $E[X_i] = \mu'$ for all $i \in [n]$, $G(\mu, \epsilon, \boldsymbol{a}, \boldsymbol{b})$ simplifies to the following:

$$G(\mu, \epsilon, \boldsymbol{a}, \boldsymbol{b}) = 2e^{-2\mu^2\epsilon^2/n(b-a)^2} = 2e^{-2\mu\mu'\epsilon^2/(b-a)^2}.$$

In this case, we simply use $G(\mu, \epsilon, a, b)$ to denote the above bound.

A crucial step in the derivation of CH bounds is to consider $E[e^{tS}]$ for any positive real t and rewrite this as

$$E[e^{tS}] = E[e^{t\sum X_i}] = E[\prod e^{tX_i}] = \prod E[e^{tX_i}].$$

The last of the above equalities depends on the X_i's being mutually independent. However, in a variety of applications mutual independence in not guaranteed. Hoeffding [5] himself considers situations where certain limited kind of dependence is allowed. More recently, a few researchers have extended CH bounds to allow for limited dependence in certain specific ways. Schmidt, Siegel, and Srinivasan [10] derive a bound on the tail probability of S when the X_i's are k-wise independent for k smaller than a certain function of n, ϵ, and μ. By *k-wise independence* they mean that any k-subset of $\mathcal{X}$ contains mutually independent random variables. Panconesi and Srinivasan [8] define the notion of λ-correlation and derive bounds on the tail probability of S when the X_i's are not necessarily mutually independent, but exhibit λ-correlatedness.

An extremely important technique for deriving bounds on tail probabilities of S when the X_i's are not mutually independent involves the theory of martingales (see Section 3 in [6] or Section 4.4 in [7]). Occasionally we can exploit additional structure of the X_i's and set up a sequence of random variables, $Y_0, Y_1, \ldots, Y_n$ called a *martingale sequence* with the property that $Y_0 = E[S] = \mu$ and $Y_n = S$. In addition, if the Y_i's satisfy the bounded difference property, that is, $|Y_k - Y_{k-1}| \leq c_k$, for each integer k, $1 \leq k \leq n$ and positive c_k that depends only on k, then we can use an inequality called the Azuma-Hoeffding inequality to derive sharp upper bounds on the tail probabilities of S. One of the most exciting applications of the Azuma-Hoeffding inequality is in Bollobás' 1988 result that resolved a long standing open question [1]. Let $G(n, p)$ denote a random graph with n vertices in which each pair of vertices is independently connected by an edge chosen with probability p. Bollobás showed that the chromatic number of $G(n, 1/2)$ is $n/\log_2 n$ almost surely.

In this paper, we extend CH bounds by allowing a rather natural, limited kind of dependency among the X_i's. To make this precise, we first define the notion of a dependency graph. A *dependency graph* for $\mathcal{X}$ has vertex set $[n]$ and an edge set such that for each $i \in [n]$, X_i is mutually independent of all other X_j such that $\{i, j\}$ is not an edge. For any non-negative integer d, we say that the X_i's exhibit *d-bounded dependence*, if the X_i's have a dependency graph with maximum vertex degree d. Readers might recall the notion of a dependency graph of random variables from the hypothesis of the Lovasz Local Lemma [11].

It is common in randomized algorithms to associate random variables with the vertices or the edges of a graph and as a result the dependency among the random variables is usually dictated by the structure of the underlying graph. Consider the following example.

Example: Maximum Independent Set. Here is a randomized algorithm that computes a large independent set in a given k-regular graph G (Problem 5.3 in [7]):

Step 1 Delete each vertex from the given graph independently with probability $1 - 1/k$.

Step 2 For every edge that remains, delete one of its endpoints.

The vertices that remain after Step 2 form an independent set of the original graph. Let n be the number of vertices in G. Let V_i denote the indicator random variable whose value is 1 if vertex i is *not* deleted in Step 1. Let $V = \sum V_i$ be the random variable that stands for the number of vertices that remain in the graph after Step 1. Similarly, let U_i denote the indicator random variable whose value is 1 if the edge i is *not* deleted in Step 1. Let $U = \sum U_i$ be the random variable that stands for the number of edges that remain in the graph after Step 1. It is easy to show that $E[V] = n/k$ and $E[U] = n/2k$. It is also easy to see that the size of the independent set computed by the algorithm is at least $V - U$ and hence the expected size of the independent set produced by the algorithm is at least $n/2k$. For any constant k, this is a $O(1)$-factor approximation algorithm for the maximum independent set problem.

The next step in the analysis is to show that the actual value of $V - U$ is very close to $n/2k$ with high probability. The standard approach is to use CH bounds for this purpose. It is easy to see that the V_i's are all mutually independent and therefore CH bounds can be used to derive sharp tail probability bounds on V. It is also easy to see that the U_i's are not mutually independent. In particular, U_i and U_j are mutually independent if edge i and edge j are not incident on the same vertex. This implies that the line graph $L(G)$ of G is a dependency graph of the U_i's. Since G is k-regular, $L(G)$ is $(2k - 1)$-regular and the U_i's exhibit $(2k - 1)$-bounded dependence. Using this fact one can derive sharp tail probability bounds on $\sum U_i$. The tail probability bounds on $\sum V_i$ and $\sum U_i$ are crucial in showing that the algorithm produces a large independent set with high probability.

Our main result is as follows.

Theorem 1. *For identically distributed binary random variables X_i with d-bounded dependence, for any ϵ, $0 < \epsilon \leq 1$, we have the upper tail probability bound*

$$Prob[S \geq (1+\epsilon)\mu] \quad \leq \quad \frac{4(d+1)}{e} F^+(\mu, \epsilon)^{\frac{1}{(d+1)}}$$

and the lower tail probability bound

$$Prob[S \leq (1-\epsilon)\mu] \quad \leq \quad \frac{4(d+1)}{e} F^-(\mu, \epsilon)^{\frac{1}{(d+1)}}.$$

For identically distributed bounded random variables X_i, where each $X_i \in [a, b]$ for finite positive reals a and b we have the tail probability bound

$$Prob[|S - \mu| \geq \epsilon\mu] \quad \leq \quad G(\mu, \epsilon, a, b)^{\frac{1}{(d+1)}} \cdot e^{2b^2\epsilon^2/(b-a)^2}$$

Note that we make absolutely no assumptions about the nature of correlation between random variables that are not mutually independent. Using the fact that $\epsilon \leq 1$, we can derive $F^+(\mu, \epsilon) \leq e^{-\epsilon^2\mu/3}$. This means that for $\mu = \Omega(\log^{1+\delta} n)$, for any $\delta > 0$, $F^+(\mu, \epsilon)$ is an inverse superpolynomial function. Similarly, for $\mu = \Omega(\log^{1+\delta} n)$, for any $\delta > 0$, $F^-(\mu, \epsilon)$ is also an inverse superpolynomial function. This is precisely what make the CH bounds so powerful – as n grows the probability of the upper tail approaches 0 at an inverse superpolynomial rate. Remarkably, Theorem 1 implies a similar conclusion even when the random variables in $\mathcal{X}$ exhibit d-bounded dependence for relatively large d. Specifically, for $\mu/(d+1) = \Omega(\log^{1+\delta} n)$ with $\delta > 0$ we have that $F^+(\mu, \epsilon)$ and $F^-(\mu, \epsilon)$ grow at inverse superpolynomial rate. The condition $\mu/(d+1) = \Omega(\log^{1+\delta} n)$ is equivalent to $d = O(\mu/\log^{1+\delta} n)$. For example, let us focus on the random variables U_i in the analysis of the Maximum Independent Set algorithm described earlier. We have $\mu = E[U] = n/2k$ and that the U_i's exhibit $(2k-1)$-bounded dependence. We have a inverse superpolynomial bound on the tail probabilities of U provided $k = O(n/k \log^{1+\delta} n)$ for some $\delta > 0$. This condition simplifies to $k = O(\sqrt{n}/\log^{1+\delta} n)$ for some $\delta > 0$. In other words, even when the dependency graph of the random variables is fairly dense – roughly $\sqrt{n}$ edges per vertex – we are still able to obtain a inverse superpolynomial bound on the tail probabilities. So even when the dependency graph is relatively dense, implying a large amount of correlation between random variables, Theorem 1 provides a sharp bound on the tail of the distribution.

Our proof technique is novel in its use of equitable graph colorings. A coloring of a graph is *equitable* if the sizes of any pair of color classes are within one of each other. For any positive integer t, we use *t-equitable coloring* to refer to an equitable coloring that uses t colors. Equitable colorings seem to be tucked away somewhere in the backwaters of graph coloring research. But, like with many other related areas the history of equitable colorings seems to have started with a conjecture of Erdös. The conjecture was proved by Hajnal and Szemerédi [4] via a long and complicated proof in 1970. This deep result of Hajnal and Szemerédi is at the heart of the derivation of our bounds.

Theorem 2 (Hajnal-Szemerédi). *A graph G with maximum degree Δ has an equitable coloring that uses $(\Delta + 1)$ colors.*

Our technique allows us to pay attention to the structure of the dependency graph and by doing this we reap several rewards. Specifically, we obtain sharper bounds on the tail probabilities when the dependency graph is a tree or an outerplanar graph. In fact, for these classes of graphs our bounds are essentially those that would have been obtained had the random variables been mutually independent! Not much is known about the existence of small equitable colorings for different classes of graphs. Using our technique, new results on equitable colorings of graphs immediately imply sharper tail probability bounds. We discuss this further in the conclusion of the paper.

The rest of the paper contains three sections. In the next section (Section 2), we prove Theorem 1, our main result. In Section 3, we show that certain common classes of graphs have small equitable colorings and using these results we derive sharper tail probability bounds for random variables whose dependency graphs belong to any of these classes. The results on equitable colorings in Section 3 seem to be only the tip of an iceberg and stronger results seem quite possible for a wide variety of classes of graphs. Section 4 contains a few of our speculations and conjectures in this direction.

2 Proof of the Main Theorem

In this section we derive the bounds claimed in Theorem 1. For the rest of the section we assume that $E[X_i] = \mu'$ for each i.

Lemma 1. *Suppose the X_i's are binary random variables with dependency graph G. Further suppose that G can be colored equitably with t colors. Then for any ϵ, $0 < \epsilon \leq 1$, we have*

Upper tail probability

$$\text{Prob}[S \geq (1+\epsilon)\mu] \leq \frac{4t}{e} F^+(\mu, \epsilon)^{1/t}.$$

Lower tail probability

$$\text{Prob}[S \leq (1-\epsilon)\mu] \leq \frac{4t}{e} F^-(\mu, \epsilon)^{1/t}.$$

Proof. Let $C_1, C_2, \ldots, C_t$ be the t color classes in a t-equitable-coloring of G. For each $i \in [t]$, let $\mu_i = E[\sum_{j \in C_i} X_j]$. We now rewrite the event $S \geq (1+\epsilon)\mu$

as follows.

$$\begin{aligned} S \geq (1+\epsilon)\mu &\equiv S \geq (1+\epsilon)\mu' n \\ &\equiv S \geq (1+\epsilon)\mu' \sum_{i\in[t]} |C_i| \\ &\equiv S \geq \sum_{i\in[t]} (1+\epsilon)\mu' |C_i| \\ &\equiv \sum_{i\in[t]} \sum_{j\in C_i} X_j \geq \sum_{i\in[t]} (1+\epsilon)\mu_i \end{aligned}$$

The first equivalence follows from the fact that

$$\mu = E[\sum_{i\in[n]} X_i] = \sum_{i\in[n]} E[X_i] = \sum_{i\in[n]} \mu' = n\mu'.$$

The second equivalence follows from the fact that the color classes form a partition of $[n]$. The last equivalence is the result of expressing S as the sum of the X_i's grouped into color classes. Now

$$\sum_{i\in[t]} \sum_{j\in C_i} X_j \geq \sum_{i\in[t]} (1+\epsilon)\mu_i \Longrightarrow \exists i \in [t] : \sum_{j\in C_i} X_j \geq (1+\epsilon)\mu_i.$$

Hence,

$$\begin{aligned} \text{Prob}[S \geq (1+\epsilon)\mu] &= \text{Prob}[\sum_{i\in[t]} \sum_{j\in C_i} X_j \geq \sum_{i\in[t]} (1+\epsilon)\mu_i] \\ &\leq \text{Prob}[\exists i \in [t] : \sum_{j\in C_i} X_j \geq (1+\epsilon)\mu_i]. \end{aligned}$$

The last probability in the above sequence can be simplified as follows.

$$\begin{aligned} \text{Prob}[\exists i \in [t] : \sum_{j\in C_i} X_j \geq (1+\epsilon)\mu_i] &\leq \sum_{i\in[t]} \text{Prob}[\sum_{j\in C_i} X_j \geq (1+\epsilon)\mu_i] \\ &\leq \sum_{i\in[t]} F^+(\mu_i, \epsilon). \end{aligned}$$

The first inequality above follows from the fact that the probability of the disjunction of a set of events is no greater than the sum of the probabilities of the events. The second inequality is obtained by applying the Chernoff bound on the sum of $\sum_{j\in C_i} X_j$. By the definition of a dependency graph, all X_j's belonging to a color class C_i are mutually independent and so CH bounds apply to the sum $\sum_{j\in C_i} X_j$.

The equitability of the coloring implies that $C_i \in \{\lfloor n/t \rfloor, \lceil n/t \rceil\}$ for each $i \in [t]$. This implies the following relationship between μ_i and μ.

$$\mu_i = E[\sum_{j\in C_i} X_j] = \sum_{j\in C_i} E[X_j] = |C_i|\mu' \geq \lfloor n/t \rfloor \mu' \geq (n/t - 1)\mu'.$$

Since $X_i \in \{0,1\}$, we have that $\mu' \in [0,1]$ and from this it follows that

$$(n/t - 1)\mu' \geq (n\mu'/t - 1) = \mu/t - 1.$$

So we obtain that $\mu_i \geq \mu/t - 1$ and from this we derive

$$\begin{aligned}\text{Prob}[S \geq (1+\epsilon)\mu] &\leq \sum_{i\in[t]} F^+(\mu_i, \epsilon) \\ &\leq \sum_{i\in[t]} F^+(\mu/t - 1, \epsilon) \\ &= tF^+(\mu/t - 1, \epsilon).\end{aligned}$$

The second inequality above follows from the fact that $F^+(\mu, \epsilon)$ is monotonically decreasing in μ. $F^+(\mu/t - 1, \epsilon)$ can be simplified as follows.

$$F^+(\mu/t - 1, \epsilon) = \left(\frac{e^\epsilon}{(1+\epsilon)^{1+\epsilon}}\right)^{\mu/t} \cdot \left(\frac{(1+\epsilon)^{1+\epsilon}}{e^\epsilon}\right) \leq \frac{4}{e} F^+(\mu, \epsilon)^{1/t}.$$

The last inequality is a result of the fact that $(1+\epsilon)^{1+\epsilon}/e^\epsilon$ is a monotonically increasing function of ϵ that achieves its maximum in the range $\epsilon \in (0,1]$ at $\epsilon = 1$. Substituting this simplification of $F^+(\mu/t - 1, \epsilon)$ into the above inequality we get

$$\text{Prob}[S \geq (1+\epsilon)\mu] \leq \frac{4t}{e} F^+(\mu, \epsilon)^{1/t}.$$

This completes the proof of the upper tail probability bound. The proof of the lower tail probability bound is identical and is skipped.

Theorem 3. *Suppose the X_i's are binary random variables exhibiting d-bounded dependence. Then, for any ϵ, $0 < \epsilon \leq 1$, we have*

Upper tail probability

$$\text{Prob}[S \geq (1+\epsilon)\mu] \leq \frac{4(d+1)}{e} F^+(\mu, \epsilon)^{1/d+1}.$$

Lower tail probability

$$\text{Prob}[S \leq (1-\epsilon)\mu] \leq \frac{4(d+1)}{e} F^-(\mu, \epsilon)^{1/d+1}.$$

Proof. Let G be the dependency graph of the X_i's with maximum vertex degree d. Such a dependency graph exists by definition of d-bounded dependence. By the Hajnal-Szemerédi Theorem (Theorem 2) G has a $(d+1)$-equitable coloring. Replacing t by $(d+1)$ in Lemma 1 yields the desired bounds.

We now prove our tail probability bounds for bounded random variables.

Lemma 2. *Suppose the X_i's are bounded random variables with $X_i \in [a,b]$ for each $i \in [n]$ and further suppose that the X_i's have a dependency graph G that can be colored equitably with t colors. Then, for any ϵ, $0 < \epsilon \leq 1$, we have*

$$\text{Prob}[|S - \mu| \geq \epsilon\mu] \leq tG(\frac{\mu}{t} - \mu', \epsilon, a, b) \leq tG(\mu, \epsilon, a, b)^{1/t} \cdot e^{2b^2\epsilon^2/(b-a)^2}.$$

Proof. As in Lemma 1 we can derive the inequality

$$\text{Prob}[|S - \mu| \geq \epsilon\mu] \leq \sum_{i \in [t]} G(\mu_i, \epsilon, a, b).$$

The connection between μ and μ_i is as follows:

$$\mu_i = E[\sum_{j \in C_i} X_j] = \sum_{j \in C_i} E[X_j] = |C_i|\mu' \geq \lfloor \frac{n}{t} \rfloor \mu' \geq \left(\frac{n}{t} - 1\right)\mu' = \frac{\mu}{t} - \mu'.$$

Substituting this in the inequality above we get

$$\begin{aligned}\text{Prob}[|S - \mu| \geq \epsilon\mu] &\leq \sum_{i \in [t]} G(\frac{\mu}{t} - \mu', \epsilon, a, b) \\ &= tG(\frac{\mu}{t} - \mu', \epsilon, a, b).\end{aligned}$$

This bound can be further simplified as follows.

$$\begin{aligned}G(\frac{\mu}{t} - \mu', \epsilon, a, b) &\leq exp\left[\frac{-2(\mu/t - \mu')\mu'\epsilon^2}{(b-a)^2}\right] \\ &= exp\left[\frac{-2(\mu/t - \mu/n)\mu'\epsilon^2}{(b-a)^2}\right] \\ &= exp\left[\frac{-2\mu\mu'\epsilon^2}{t(b-a)^2} + \frac{2\mu'\mu'\epsilon^2}{(b-a)^2}\right] \\ &= G(\mu, \epsilon, a, b)^{1/t} \cdot e^{2\mu'\mu'\epsilon^2/(b-a)^2}\end{aligned}$$

Since $a \leq \mu' \leq b$, we get

$$\text{Prob}[|S - \mu| \geq \mu\epsilon] \leq tG(\mu, \epsilon, a, b)^{1/t} \cdot e^{2b^2\epsilon^2/(b-a)^2}.$$

This completes the proof of the lemma.

The following theorem follows immediately from the above lemma by using the Hajnal-Szemerédi theorem (Theorem 2).

Theorem 4. *Suppose the X_i's are bounded random variables, with $X_i \in [a,b]$ for all i and for positive a and b, that exhibit d-bounded dependence. Then*

$$\text{Prob}[|S - \mu| \geq \epsilon)\mu] \leq G(\mu, \epsilon, a, b)^{1/(d+1)} \cdot e^{2b^2\epsilon^2/(b-a)^2}$$

There does not seem to be any other technique that can be used to derive bounds as sharp as these for random variables that exhibit d-bounded dependence. In some cases, a martingale can be set up and the Azuma-Hoeffding inequality can be used to derive sharp bounds. However, even in such cases, the bounds obtained from the Azuma-Hoeffding inequality are not as strong as those we have derived here. For example, in the Maximum Independent Set example discussed earlier we can derive tail probability bounds on $\sum_i U_i$ using martingales, but the bounds obtained are a weaker than the bounds we obtain by an exponential factor of $1/k$.

3 Sharper Bounds in Special Cases

As is standard, we let $\chi(G)$ denote the chromatic number of a graph. The *equitable chromatic number* of a graph G, denoted $\chi_{eq}(G)$, is the fewest colors that can be used to equitably color the graph. $\chi_{eq}(G)$ can be arbitrarily large as compared to $\chi(G)$ as can be seen by examining the star graph. For a star graph G with n vertices $\chi(G) = 2$ while $\chi_{eq}(G) = \lceil (n-1)/2 \rceil + 1$.

The strength of the equitable coloring technique in deriving tail probability bounds lies in the fact that it allows us to focus closely on the structure of the dependency graph. In particular, a small equitable chromatic number for a dependency graph leads to sharp tail probability bounds. Not much seems to be known about the equitable chromatic number of different graph classes. The connection between equitable colorings and tail probability bounds presented in this paper provides strong motivation for deriving stronger upper bounds on the equitable chromatic number of different graph classes.

The star graph example seems to imply that $\chi_{eq}(G)$ is bound to be large for any interesting graph class. However, we present two examples of widely used graph classes in which "most" graphs have equitable chromatic numbers bounded above by a constant. Using these bounds we essentially get tail probability bounds on S that are roughly as good as those we would have obtained had the X_i's been mutually independent!

The first result comes ready-made from Bollobás and Guy [2], who show that almost all trees can be equitably colored with 3 colors. As is standard, we let $\Delta(G)$ stands for the maximum degree of any vertex in a graph G.

Theorem 5 (Bollobás-Guy). *A tree T with n vertices is equitably 3-colorable if $n \geq 3\Delta(T) - 8$ or if $n = 3\Delta(T) - 10$.*

The theorem essentially implies that if $\Delta(T) \leq n/3$ then T can be equitably 3-colored. This immediately translates, via Lemma 1 and Lemma 2 to the following tail probability bounds on tree structured dependency graphs.

Theorem 6. *Suppose the X_i's are identical random variables that have a tree-structured dependency graph with maximum degree no greater than $n/3$. Then we have the following bounds of the tail probabilities of $S = \sum_{i=1}^{n} X_i$.*

(i) If the X_i's are binary then

$$\text{Prob}[S \geq (1+\epsilon)\mu] \leq 3F^+(\mu,\epsilon)^{1/3}$$
$$\text{Prob}[S \leq (1-\epsilon)\mu] \leq 3F^-(\mu,\epsilon)^{1/3}$$

(ii) If the X_i's are bounded random variables with $X_i \in [a,b]$ for all i then

$$\text{Prob}[|S-\mu| \geq \epsilon\mu] \leq 3G(\mu,\epsilon,a,b)^{1/3} e^{2b^2\epsilon^2/(b-a)^2}$$

Here ϵ satisfies $0 < \epsilon \leq 1$.

In our second example we consider outerplanar graphs Elsewhere [9], we have shown that most outerplanar can be equitably colored with 6 colors. The main result obtained there is as follows.

Theorem 7. *A connected outerplanar graph with n vertices and maximum vertex degree at most $n/6$ has a 6-equitable coloring*

As in the case of trees, this result leads to strong tail probability bounds on S if the X_i's have a dependency graph that is outerplanar. This result also hints at the fact that there may be many other commonly used classes of graphs with constant equitable chromatic number.

The crucial stepping stone to our result on the equitable coloring of outerplanar graphs is a result on equitable forest partitions of outerplanar graphs. A partition $\{V_1, V_2, \ldots, V_k\}$ of the vertex set V of a graph $G = (V,E)$ is called an *equitable forest partition* if (i) $V_i \in \{\lfloor |V|/k \rfloor, \lceil |V|/k \rceil\}$ and (ii) each induced subgraph $G[V_i]$ is a forest. In [9] we show the following.

Theorem 8. *Any outerplanar graph has an equitable 2-forest partition.*

This theorem along with the Bollobás-Guy Theorem leads to Theorem 7. This in turn leads to the following concentration result for random variables with outerplanar structured dependency graphs.

Theorem 9. *Suppose the X_i's are identical random variables whose dependency graph is outerplanar with maximum degree no greater than $n/6$. Then we have the following bounds of the tail probabilities of $S = \sum_{i=1}^n X_i$.*

(i) If the X_i's are binary then

$$\text{Prob}[S \geq (1+\epsilon)\mu] \leq 6F^+(\mu,\epsilon)^{1/6}$$
$$\text{Prob}[S \leq (1-\epsilon)\mu] \leq 6F^-(\mu,\epsilon)^{1/6}$$

(ii) If the X_i's are bounded random variables with $X_i \in [a,b]$ for all i and for some positive constants a and b then

$$\text{Prob}[|S-\mu| \geq \epsilon\mu] \leq 6G(\mu,\epsilon,a,b)^{1/6} e^{2b^2\epsilon^2/(b-a)^2}$$

Here ϵ satisfies $0 < \epsilon \leq 1$.

This result gives us probability tail bounds on outerplanar structured dependency graphs that are essentially as strong as those obtainable had the X_i's been mutually independent.

In the results on equitable coloring of n-vertex trees and n-vertex outerplanar graphs we required upper bounds of $n/3$ and $n/6$ respectively on the maximum vertex degree. As the star example shows, some such bounds are required in order to obtain constant size equitable colorings. The question we are interested in is whether these vertex-degree bounds are required to obtain sharp concentration results? That the answer is in the negative is seen by relaxing the requirement of an equitable coloring in two specific ways. First, we allow a constant number of vertices to be without color. Second, we allow the sizes of the color classes to be proportional, that is, for some constant $\alpha \geq 1$, the size of a color class is within α times the size of any other color class. To be precise, for non-negative c and $\alpha \geq 1$ we define a (c, α)-coloring as a vertex coloring of a graph G in which all except at most c vertices are colored and the coloring is such that for any pair of color classes C and C', $|C| \leq \alpha|C'|$. Given this definition, it is possible to extend the Bollobás-Guy Theorem to show the following result.

Theorem 10. *Every tree has a $(1, 5)$-coloring with two colors.*

In other words, we can properly color all but at most one vertex in a tree with two colors such that the larger color class is at most 5 times the smaller color class. Using this result and the fact that every tree has an equitable 2-forest partition (Theorem 8) yields the following coloring result for outerplanar graphs.

Theorem 11. *Every outerplanar graph has a $(2, 5)$-coloring with four colors.*

Bounds similar to those in Lemma 1 (with different multiplicative and exponential constants) can be obtained if the X_i's have a dependency graph that can be (c, α)-colored for some constants $c \geq 0$ and $\alpha \geq 1$. Details appear in [9].

4 Conclusions

In this paper we have presented a novel technique involving equitable graph coloring for deriving sharp tail probability bounds on the sums of bounded random variables that exhibit a certain kind of dependence. For the cases we consider, these bounds are sharper than those obtained using any other known technique. More importantly, this technique allows us to focus more carefully on the structure of the dependency graph of random variables and allows us to prove bounds that are as strong as those obtainable had the random variables been mutually independent.

The results in this paper seem, in some sense, to be the tip of the iceberg. It is our conjecture that $\chi_{eq}(G)$ depends not on the maximum vertex degree (as in the Hajnal-Szemerédi Theorem), but on the average vertex degree. In particular, we conjecture the following.

Conjecture 1. There is a positive constant c such that if an n-vertex graph G has maximum vertex degree at most n/c and average degree d then

$$\chi_{eq}(G) = O(\chi(G) + d).$$

The truth of this conjecture will have many ramifications. For example, it will immediately imply an $O(1)$ equitable chromatic number for most planar graphs and that will translate into extremely sharp tail probability bounds for the sum of random variables that have a planar dependency graph.

References

1. Béla Bollobás. The chromatic number of random graphs. *Combinatorica*, 8:49–55, 1988.
2. Bela Bollobás and Richard K. Guy. Equitable and proportional coloring of trees. *Journal of Combinatorial Theory, Series B*, 34:177–186, 1983.
3. Herman Chernoff. A measure of asymptotic efficiency for tests of a hypothesis based on the sum of observations. *Annals of Mathematical Statistics*, 23:493–507, 1952.
4. Hajnal and Szemérredi. Proof of a conjecture of erdös. In P. Erdoös, A. Rényi, and V.T. Sós, editors, *Combinatorial Theory and Its Applications, Vol II, Volume 4 of Colloquia Mathematica Societatis János Bolyai*, pages 601–623. North-Holland, 1970.
5. Wassily Hoeffding. Probability inequalities for sums of bounded random variables. *American Statistical Association Journal*, 58:13–30, 1963.
6. Colin McDiarmid. Concentration. In M. Habib, C. McDiarmid, J. Ramirez-Alfonsin, and B. Reed, editors, *Probabilistic Methods for Algorithmic Discrete Mathematics*, pages 195–248. Springer, 1998.
7. Rajeev Motwani and Prabhakar Raghavan. *Randomized Algorithms.* Cambridge University Press, 1995.
8. Alessandro Panconesi and Aravind Srinivasan. Randomized distributed edge colouring via an extension of the chernoff-hoeffding bounds. *SIAM Journal on Computing*, 26(2):350–368, 1997.
9. Sriram V. Pemmaraju. Equitable colorings, proportional colorings, and chernoff-hoeffding bounds. Technical report, TR 01-05 Department of Computer Science, The University of Iowa, 2001.
10. Jeanette Schmidt, Alan Siegel, and Aravind Srinivasan. Chernoff-hoeffding bounds for applications with limited independence. *SIAM Journal on Discrete Mathematics*, 8(2):223–250, 1995.
11. Joel Spencer. *Ten Lectures on the Probabilistic Method.* SIAM, Philadelphia, 1987.

Author Index